47.60100K

Nuclear Physics with Stored, Cooled Beams
(McCormick's Creek State Park, Indiana, 1984)

AIP Conference Proceedings
Series Editor: Hugh C. Wolfe
Number 128

Nuclear Physics with Stored, Cooled Beams
(McCormick's Creek State Park, Indiana, 1984)

Edited by
P. Schwandt and H. O. Meyer
Indiana University

American Institute of Physics
New York 1985

Copying fees: The code at the bottom of the first page of each article in this volume gives the fee for each copy of the article made beyond the free copying permitted under the 1978 US Copyright Law. (See also the statement following "Copyright" below.) This fee can be paid to the American Institute of Physics through the Copyright Clearance Center, Inc., Box 765, Schenectady, N.Y. 12301.

Copyright © 1985 American Institute of Physics

Individual readers of this volume and non-profit libraries, acting for them, are permitted to make fair use of the material in it, such as copying an article for use in teaching or research. Permission is granted to quote from this volume in scientific work with the customary acknowledgment of the source. To reprint a figure, table or other excerpt requires the consent of one of the original authors and notification to AIP. Republication or systematic or multiple reproduction of any material in this volume is permitted only under license from AIP. Address inquiries to Series Editor, AIP Conference Proceedings, AIP, 335 E. 45th St., New York, N. Y. 10017.

L.C. Catalog Card No. 85-71167
ISBN 0-88318-327-7
DOE CONF-8410202

Proceedings of the
Workshop on Nuclear Physics
with Stored Cooled Beams
McCormick's Creek State Park
Spencer, Indiana
October 15-17, 1984

PREFACE

By now, one may call it a tradition: The fourth annual fall Workshop organized by the Indiana University Cyclotron Facility was once again held in the pastoral surroundings of the Canyon Inn at McCormick's Creek State Park at the peak of southern Indiana's beautiful fall colors. The topic of this year's workshop was chosen appropriately in connection with the construction of a cooler-storage ring currently underway at IUCF.

Phase-space cooling of stored beams was first suggested almost 20 years ago. In recent years, however, there has been a sharp increase in attention devoted to the subject. Not only do we now know of several dissipative mechanisms that can be used for cooling, such as electron cooling, stochastic cooling and radiation cooling, but also the principle of cooling has been successfully applied in physics research as a tool. Possibly the best known example is the discovery of the Z- and W-bosons, made possible by stored, cooled antiprotons at CERN. Serendipitously, two of the proponents of this research effort, C. Rubbia and S. van der Meer, were awarded the Nobel Prize while our workshop was in progress.

Storage rings for confining and cooling ion beams are currently under construction or are considered as additions to accelerators at several laboratories in Europe and Japan. Beam parameters in such an environment are adjustable and in many respects much superior to conventional ion beams used for nuclear research. While these new facilities are expected to offer a range of novel and unique experimental possiblilites, it is at the present time difficult to reliably predict the capability of ion coolers and their possible uses and impact in nuclear physics research. Hence one purpose of this workshop was to assess the current understanding of electron cooling and competitive mechanisms, to exchange ideas on predictions of the parameters of stored, cooled beams and to discuss the status of various Cooler projects in Sweden, Switzerland, Germany, Japan and the U.S.

Another important facet of the workshop was the simultaneous consideration of research-oriented issues and technical aspects associated with the physics uses of stored, cooled beams, with the expectation that such discussions might lead to the formulation of an initial experimental program of research with the IUCF Cooler ring. Although the anticipated date for completion of the new IUCF facility was still about two years away, the fall of 1984 was considered an appropriate time to hold a workshop in order to stimulate and precipitate concrete ideas for experiments since it would allow sufficient lead time for the design, funding and development of unconventional experimental equipment which is likely to be required.

The workshop was well attended. A total of 112 participants enjoyed the informal atmosphere and the exchange of ideas on both physics and technology. Because of the strong current international interest in the application of beam cooling techniques for storage rings, the workshop attracted strong international participation: 44 of the participants were from outside the US, from 8 foreign countries.

The topics of the workshop were arranged into six sessions, each devoted to a relevant physics idea. In each session specific technical features of cooled beams were discussed and their implications with regard to nuclear physics considered. The motivation underlying this format was to get scientists concerned with current problems in nuclear physics and researchers developing the technology of cooled beams in storage rings to talk with each other. In particular, invited nuclear theorists and experimentalists were asked to speculate on new physics which may be revealed or made amenable to study once the novel techniques become available. In the spirit of a workshop, ample time for informal discussion was set aside to stimulate a free exchange of ideas.

The workshop was clearly successful in establishing a productive and mutually beneficial dialogue between the participants concerned with physics-oriented issues on the one hand and those concerned with technical aspects of stored, cooled ion beams on the other hand. Much less successful was our attempt to use the workshop as a forum for formulating specific, practical ideas for new and exciting physics experiments which would exploit the unique features of stored cooled beams; in retrospect it it clear that more time must pass for the participants to digest the ideas exchanged at this workshop, to realize fully the implications and promises of the novel and unusual features of cooler-storage ring technology and translate them into concrete proposals for experiments.

The major credit for the success of the workshop goes to the speakers for their stimulating talks and to the participants for their contributions to the animated discussions.

While many people contributed their time and effort to the organization of the workshop, we single out Phil Thompson for special mention and heartfelt thanks for his central role in setting up and supervising the workshop arrangements and for his invaluable assistance beyond the call of duty. From the early planning stages of the workshop to the final editing of the proceedings, a time period of well over a year, our IUCF secretaries, Diana McGovern and Becky Westerfield, dedicated a sizable fraction of their time to this workshop. We greatly appreciate their efforts and commend them for their enthusiasm and competence.

We also acknowledge here the contributions of time and skills by R. Woodley, D. Duncan, B. Starks, J. Self, G. Bing, and W. Thomas from the IU Cyclotron. We were also very pleased with the active role played by the physics graduate students, S. Aziz, V. Cupps, J. Gering,

A. Opper, K. Pitts, J. Templon, and T. Throwe; their help is greatly appreciated. Photographs of the workshop reproduced here were taken by K. Berglund who deserves all the credit and our thanks for his efforts.

Finally, our thanks are also due to EG&G/Ortec and Harshaw/Filtrol for their financial contributions toward the expenses of the social functions.

>   Peter Schwandt
>   Hans-Otto Meyer
>   Bloomington, IN
>   February 1985

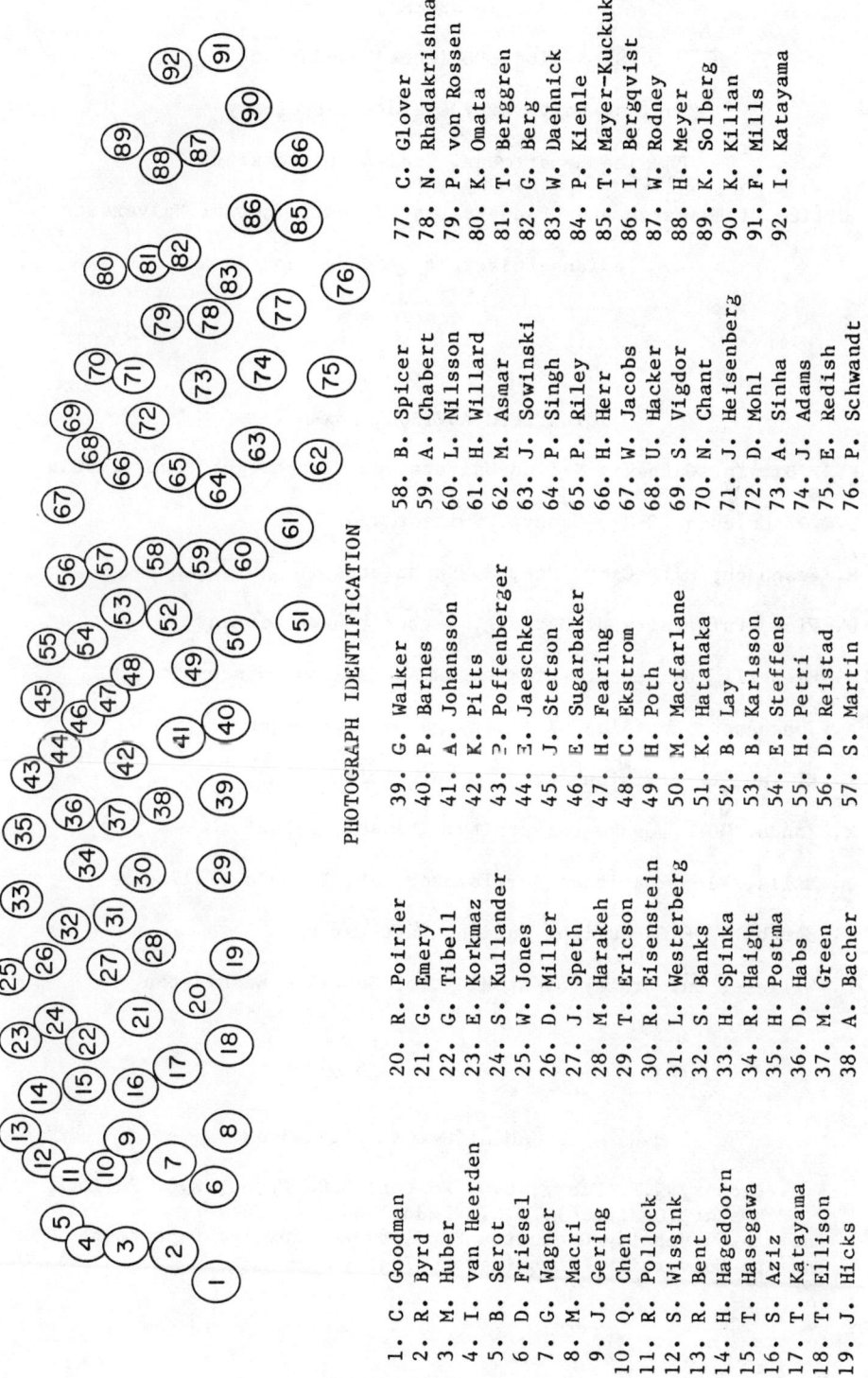

PHOTOGRAPH IDENTIFICATION

1. C. Goodman
2. R. Byrd
3. M. Huber
4. I. van Heerden
5. B. Serot
6. D. Friesel
7. G. Wagner
8. M. Macri
9. J. Gering
10. Q. Chen
11. R. Pollock
12. S. Wissink
13. R. Bent
14. H. Hagedoorn
15. T. Hasegawa
16. S. Aziz
17. T. Katayama
18. T. Ellison
19. J. Hicks
20. R. Poirier
21. G. Emery
22. G. Tibell
23. E. Korkmaz
24. S. Kullander
25. W. Jones
26. D. Miller
27. J. Speth
28. M. Harakeh
29. T. Ericson
30. R. Eisenstein
31. L. Westerberg
32. S. Banks
33. H. Spinka
34. R. Haight
35. H. Postma
36. D. Habs
37. M. Green
38. A. Bacher
39. G. Walker
40. P. Barnes
41. A. Johansson
42. K. Pitts
43. P. Poffenberger
44. E. Jaeschke
45. J. Stetson
46. E. Sugarbaker
47. H. Fearing
48. C. Ekstrom
49. H. Poth
50. M. Macfarlane
51. K. Hatanaka
52. B. Lay
53. B. Karlsson
54. E. Steffens
55. H. Petri
56. D. Reistad
57. S. Martin
58. B. Spicer
59. A. Chabert
60. L. Nilsson
61. H. Willard
62. M. Asmar
63. J. Sowinski
64. P. Singh
65. P. Riley
66. H. Herr
67. W. Jacobs
68. U. Hacker
69. S. Vigdor
70. N. Chant
71. J. Heisenberg
72. D. Mohl
73. A. Sinha
74. J. Adams
75. E. Kedish
76. P. Schwandt
77. C. Glover
78. N. Rhadakrishna
79. P. von Rossen
80. K. Omata
81. T. Berggren
82. G. Berg
83. W. Daehnick
84. P. Kienle
85. T. Mayer-Kuckuk
86. I. Bergqvist
87. W. Rodney
88. H. Meyer
89. K. Solberg
90. K. Kilian
91. F. Mills
92. I. Katayama

SPONSORS

U.S. National Science Foundation

Indiana University Cyclotron Facility

Physics Department, Indiana University

Office of Research and Graduate Development, Indiana University

Indiana University Foundation

SCIENTIFIC ADVISORY BOARD

P.D. Barnes, Carnegie-Mellon University, Pittsburgh, Pennsylvania

T.E.O. Ericson, CERN, Geneva, Switzerland

H. Feshbach, MIT, Cambridge, Massachusetts

D. Fick, University of Marburg, Marburg, West Germany

W. Haeberli, University of Wisconsin, Madison, Wisconsin

A. Johansson, Uppsala University, Uppsala, Sweden

K. Kilian, EP-Division, CERN, Geneva Switzerland

M. Kondo, RCNP, Osaka University, Ibaraki, Japan

F. Mills, Fermi National Accelerator Lab, Batavia, Illinois

J. Speth, IKP-KFA Julich, Julich, West Germany

L. Willets, University of Washington, Seattle, Washington

ORGANIZING COMMITTEE

A.D. Bacher, G.T. Emery, C.C. Foster, H.O. Meyer, R.E. Pollock
P. Schwandt (Chairman), S.E. Vigdor and G.E. Walker,
Indiana University Cyclotron Facility and Physics Department,
Indiana University, Bloomington, Indiana

TABLE OF CONTENTS

SESSION A

INTRODUCTION

COSY: A cooler-synchrotron for the KFA Jülich: G. Berg 2

Status of CELSIUS: D. Reistad and A. Johansson 14

The TARN II project: T. Katayama 24

Status report on the IUCF Cooler: R.E. Pollock 34

SESSION B

HIGH-RESOLUTION STUDIES

Nuclear structure physics at high resolution: J. Heisenberg ... 47

Resolution limits with cooled beams: H. Herr 68

Cooled beams with internal targets: H.O. Meyer 76

Gas jet internal targets: M. Macri 89

    Contribution:

Target developments for CELSIUS: C. Ekström, B. Holmquist
and H. Sterner 106

SESSION C

SHORT-RANGE AND OFF-SHELL STUDIES

Short-range effects in nucleon-nucleus scattering:
E. Redish 112

Bremsstrahlung and radiative processes at medium energies:
H. Fearing 138

Direct reaction studies of the nuclear continuum: G. Wagner ... 157

SESSION D

ENERGY DEPENDENCE STUDIES

Energy-dependent nuclear phenomena: T.E. Ericson 172

Techniques of energy changing and energy monitoring of cooled
  beams: D. Möhl 180

  Contribution:

On the possible existence of narrow (N^*N) resonances:
M. Huber, H.G. Hopf, and M. Dillig 188

SESSION E

PHYSICS WITH POLARIZED BEAMS AND TARGETS

The nucleon-nucleon interaction at medium energies:
  H. Spinka 198

Few-body reactions at medium energies: M. Bleszynski 220

Storing and cooling of polarized ions: E. Steffens 241

Free and stored atomic beams as internal polarized targets:
  W. Haeberli 251

Prospects for a deuterium internal target tensor polarized by
  optical pumping spin exchange: M. Green 268

SESSION F

ATOMIC AND PARTICLE PHYSICS WITH THE COOLER

Physics with energetic H° beams: I. Katayama 278

Nuclear and atomic physics experiments with cooled
  heavy-ion beams: P. Kienle 291

Laser application in electron cooling: H. Poth 305

New possibilities with recoilless kinematics using high-
  quality proton beams: K. Kilian 319

  Contribution:

Some particle-physics problems for cooler rings:
  E. Hagberg and S. Kullander 327

WORKSHOP SUMMARY

R. Eisenstein 333

LIST OF CONFERENCE PARTICIPANTS 346

AUTHOR INDEX 349

SESSION A

INTRODUCTION

COSY: A COOLER-SYNCHROTRON FOR THE KFA JOLICH

G.P.A. Berg*
Institut für Kernphysik, Kernforschungsanlage Jülich
D-5170 Jülich, Fed. Rep. of Germany

ABSTRACT

The concept and the status of the cooler-synchrotron COSY under study at the Institute for Nuclear Physics of the KFA Jülich are presented. This storage ring with phase space cooling and RF acceleration is designed to accept light ions injected from the existing cyclotron JULIC or protons from the LINAC of the proposed neutron spallation source (SNQ). The lay-out of COSY was developed in cooperation with the Universities in Nordrhein-Westfalen and meets the experimental requirements of variable and high quality beams which are necessary for the future nuclear research program under discussion.

INTRODUCTION

The motivation for the cooler-synchrotron COSY emerged from discussions of nuclear physicists in Nordrhein-Westfalen on the future nuclear research program. The "Workshop on Electron Cooling" in 1982 initiated by Sig Martin came to the conclusion[1] that a cooler-synchrotron is an excellent tool for advanced nuclear research experiments. It should be built at the KFA in cooperation with the surrounding Universities. In the beginning of 1983 the minister of research and technology (BMFT) in Bonn made the decision on the construction site of the SNQ in favour of the KFA in Jülich. The decision on the funding of the SNQ is expected in 1985. This event influenced also the plans for COSY, because the SNQ consisting of a 1.1 GeV high current proton LINAC will be an ideal injector for the COSY ring. Before the SNQ-LINAC comes into operation, the existing isochronous cyclotron JULIC will serve as injector for light and medium mass ions. The three essential properties of the storage ring will be:

---

* For the COSY-Group: G.P.A. Berg, J. Meissburger, F. Osterfeld, D. Prasuhn, G. Riepe, M. Rogge, P. von Rossen, O.W.B. Schult, J. Speth, P. Turek (IKP-KFA Jülich); U. Hacker, A. Hardt, S.A. Martin (ABT/SNQ-KFA Jülich); M. Köhler (ZEL-KFA Jülich); F. Hinterberger, Max Huber, R. Jahn, T. Mayer-Kuckuk (ISKP, Univ. Bonn); N.L. Hagedoorn, J.A. van der Heide (Univ. Eindhoven); H. Paetz gen. Schieck (IKP, Univ. Köln); G. Gaul (IKP, Univ. Münster)

- high luminosities and very efficient use of the beam in the storage ring by thin internal targets
- energy variability in the range of 20 MeV to 1.5 GeV by RF acceleration
- extremely high beam quality through phase space cooling.

Two classes of experiments will be feasible in the first step of the construction. At proton energies up to about 200 MeV it will be possible to excite selectively fundamental nuclear spin isospin modes and to resolve the corresponding nuclear levels with the existing high resolution spectrometer BIG KARL[2]. At higher energies single nucleons in the nucleus can be excited internally and coherent production of various types of mesons can be studied, processes which reflect details of the subnuclear structure.

In common with other cooler storage rings the COSY project is a technological challenge and has scientifically promising experimental possibilities. These projects became possible because of the inventions of electron cooling by Budker[3], the stochastic cooling by van der Meer[4] and the pioneering work of the groups[5] in Novosibirsk, CERN and FERMI-Lab. The first cooler storage ring in operation for nuclear physics applications is the LEAR built at CERN for experiments with low energy antiprotons.

THE LATTICE

The basic structure of COSY is a hexagon as shown in fig. 1. It consists of 6 unit cells which are connected by two long insertions built as telescopes to provide beam characteristics appropriate for the electron cooler on one side and the high resolution target spot for the BIG KARL spectrometer[2] on the other side. Each unit cell contains four 15° bending magnets (B), two quadrupoles (F) and drift sections (O) in the following arrangement OFOBBOBBOFO. The betatron phase advance in one unit cell is relatively small $0.06 \leq \Delta\nu \leq 0.3$ rad. This gives horizontally and vertically a narrow and parallel beam in the bending magnets. Depending on the tune the maximum vertical acceptance of 30 π mm mrad in the unit cell is limited by the 4 cm vertical opening of the vacuum chamber. The cost optimization is essentially done by minimizing the gap height of the dipoles and by keeping the quadrupole strength relatively small ($\leq$ 7.5 T/m).

Table I summarizes the main parameters of the lattice. The characteristic lattice functions (β_x, β_z, D) for the working point with a tune of $Q_x = 3.75$ and $Q_z = 3.30$ are displayed in fig. 2. For this working point and a dispersion of $D_T = -10$ m at the target $\gamma_{tr} = 1.0026$ calculated with the programme MAD.

Table I Parameters of the preliminary COSY lattice

---

6 unit cells of the structure: OFOBBOBBOFO
Two telescopes of 35 m length with magnification +1

Circumference		163.6 m
Bending radius		7.0 m
Dipole field for 1.5 GeV protons (p = 2.25 GeV/c)		1.1 T
Free length for cooling region		6.5 m
Free length for BIG KARL		2.6 m
Dispersion at target	variable	$1m \leq D_T \leq 30m$
Maximum acceptance ($\Delta p/p=0$)	$\varepsilon_x = 280$ π mm mrad	$\varepsilon_z = 30$ π mm mrad
β-function at the target	variable	$0.27m \leq \beta \leq 10.25m$

The following values correspond to the target dispersion $D_T = -10$ m

Q-values $\qquad Q_x = 3.75 \qquad\qquad Q_z = 3.30$

acceptance: $\Delta p/p = 0$: $\qquad E_x = 270$ π mm mrad $\qquad E_z = 30$ π mm mrad

$\qquad\qquad\quad \Delta p/p = 0.25$ % for $E_x = 47$ π mm mrad

β-values: at the target: $\qquad \beta_x = 2.79$ m $\qquad \beta_z = 6.23$ m

$\qquad\qquad$ maxima in the unit-cells: $\beta_x = 20.8$ m $\qquad \beta_z = 13.3$ m

transition energy: $\qquad\qquad\qquad\qquad \gamma_{tr} = 1.0026$

Fig. 3 shows the dispersions along the ring for different working points and the corresponding γ_{tr}.

The target telescope provides a variable dispersion at the target position as can be seen from fig. 4. This telescope allows four different target dispersions without changing the working point of the ring because the tune shift of the insertions is always 2π. This allows "dispersion matching" between the ring and the spectrometer necessary for high resolution experiments. At the same time the Twiss parameter α can be varied in a wide range ($-1.25 \leq \alpha \leq 0.5$) so that so-called "kinematic matching" between the ring and the spectrometer is possible up to very high K-values of $K \lesssim 0.5$. Again this correction can be done with the telescope without changing the tune.

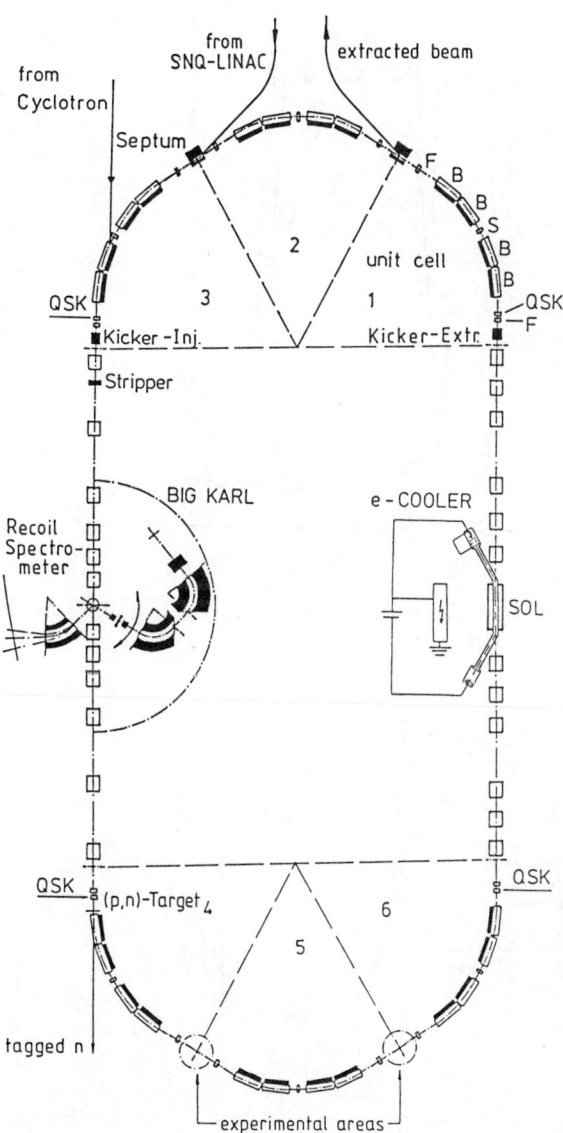

Fig. 1. COSY layout with 6 unit cells and two insertions (35 m) and the experimental areas.

The polarization of particles with axial spin directions can be preserved by using a solenoid to counteract the spin precession in the cooler solenoid (SOL). The effects of spin resonances[6,7] have to be studied and can be taken into account to maintain the spin direction. The spin of longitudinally polarized beams is precessed by the field of the dipole magnets. If longitudinal spin has to be maintained special dipole arrangements[6] (Siberian snake) have to be included in the lattice.

The purpose of the skewed quadrupoles (QSK) is to correct deviations of the dispersive plane from the horizontal plane. The sextupoles (S) between the dipole pairs allow chromaticity corrections and they are used for slow and ultra slow resonance extractions[8]. Octupoles may be necessary for the generation of a tune spread if Landau damping[9] is required.

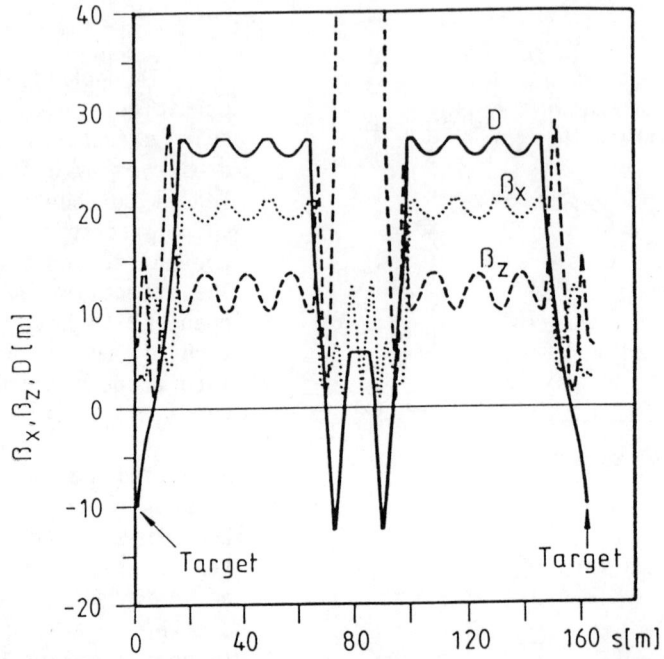

Fig. 2. Lattice functions of the COSY ring for a working point with the dispersion D_T = -10 m at the target (s = 0 m).

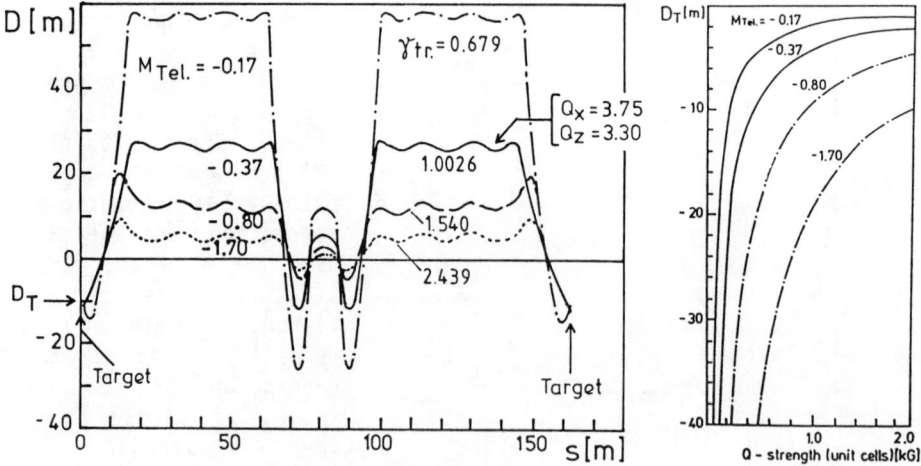

Fig. 3. Dispersion D along the ring for four different target telescope settings. The target dispersion is D_T = -10 m.

Fig. 4. Target dispersion D_T as function of the quadrupole strength in the unit cell for the telescope settings shown in fig. 3.

INJECTION AND EXTRACTION

The injection into the COSY storage ring will be accomplished at first from the existing isochronous cyclotron JULIC[10] which delivers at the moment unpolarized p,d,α, and ^{3}He-particles from an internal source with variable energies from 22.5 to 45 MeV/N. Light ions up to about ^{20}Ne will be available from the cyclotron at the beginning of 1986 from the ISIS injection system[11] (external injection and ECR source) which is under construction. The cyclotron delivers about 10^6 particles every 30-40 ns with an emittance of $E_x \sim E_y \sim 7\pi$ mm mrad and a momentum spread of $\Delta p/p = \pm 1.5 \cdot 10^{-3}$. This beam is limited by the heat dissipation in the extraction septum and can be increased for injection by a factor of about 10 by pulsing the ion source. Injection by stripping of e.g. H_2^+ will allow higher intensities of typically 10^9 particles in the storage ring. For other particles such as α's stacking by kicker and septum magnets has to be applied.

For high luminosity proton experiments the SNQ-LINAC will be an ideal injector which enables the ring to be filled to the space charge limit ($\gtrsim 10^{11}$ particles). This LINAC provides a 250 μs long bunch of $6 \cdot 10^9$ protons every 5 ns so that 20 bunches in 100 ns will fill the ring. This is smaller than the revolution time of about 600 ns allowing stripping injection of H$^-$ or kicker-septum injection of protons up to the space charge limit within less than a revolution.

With the injection scheme as shown in fig. 1 counter-clockwise circulation allows measurement of the forward angles ($\Theta_{lab} \lesssim 90°$) with the high resolution spectrometer BIG KARL. Special modifications have to be done for the smallest forward angles. Clockwise injection would allow measurements of backward angles ($\Theta_{lab} \gtrsim 90°$). If cooled beams are needed also in this direction the electron cooling direction has to be reversed.

For high luminosity experiments internal targets have to be used. However, it is also possible to install a slow resonance extraction system if an external average beam of a few nA for cyclotron injection can be used in external experiments. If the COSY ring is used as stretcher for the LINAC beam (spill time $\lesssim 1$ msec) currents of 0.1 to 1 μA can be extracted with a duty factor of about 90 % and good energy resolution of $\Delta p/p \sim 10^{-4}$ without cooling.

PHASE SPACE COOLING

Phase space cooling will be an important part of the COSY ring. Without cooling the increase of the brilliance (= beam current/emittance) in the recirculator mode[12] is limited to a factor of $10-10^5$ depending on energy and particle. In this mode the beam passes many times through an internal target of typically 50-100 µg/cm^2. While the average energy loss can be corrected by a small RF acceleration multiple scattering causes an emittance growth until the resolution becomes so poor that the beam has to be dumped in a beam stop. Electron cooling will be able to keep the emittance of the beam small if the target thickness is typically smaller than 0.1 µg/cm^2. This implies the use of very thin targets e.g. gas jet targets.

In the first construction step a 100 keV electron gun will be installed which allows 10^9 particles stored from the cyclotron to be cooled within about one second. This gun will be designed and constructed in cooperation with the group of Helmut Poth at LEAR (CERN and Karlsruhe). Protons up to 180 MeV can be cooled with this gun.

Stochastic cooling[13] with amplifier bandwidths of 200-700 MHz is about a factor of 10 slower than electron cooling for the same number of particles. With the possible development of 2-10 GHz bandwidths the cooling time will be decreased by a factor of 10. Because for large emittances stochastic cooling works faster than electron cooling a combination of both cooling methods will be installed in the COSY ring.

ACCELERATION

The basic requirements for acceleration in COSY are given in table II.

Table II Acceleration parameters

	injection		maximum energy
E [MeV]	40	350	1400
Bρ [T·m]	0.924	2.945	7.144
accel. time [sec]	-	0.1	0.3
β	0.283	0.685	0.916
revolution time [µs]	1.755	0.725	0.542
Δp/Δt	6058.97 (MeV/c)/sec		
field ramp [T/sec]	2.9		
ΔE per turn	3.0 keV		

A 3 keV energy gain per turn accelerates the beam within 0.1 sec from 40 up to 350 MeV. The acceleration time up to 1400 MeV is about 0.3 sec. Two ferrite loaded tuned cavities with a peak voltage of 4 kV and a frequency change of a factor 3 perform the acceleration. The ramp rate of the magnetic field is less than 3 T/sec.

CONSTRUCTION STEPS AND MODES OF OPERATION

Construction and operation of the storage ring should be realized in the following steps:

1) Construction of the storage ring with RF-structures for the correction of the average energy loss in the target, matching and installation of the BIG KARL spectrometer and a recoil spectrometer at the ring, injection from the cyclotron. Operation of the system in the recirculation mode.
2) Installation of ferrite loaded tunable RF-structures. Operation in the extended energy range.
3) Installation and operation of an electron cooler ($E_e \lesssim 100$ keV) for $E_p \lesssim 180$ MeV. Use of gas jet and atomic beam targets. Development of new target techniques. Installation of stochastic cooling for higher energies.
4) Injection from the SNQ-LINAC. RF-cavities for the energy range up to 1.5 GeV.

For the future requirements, electron cooling for higher energies can be developed and installed in the ring. Depending on the construction step and the experimental requirements the ring can be operated in different modes:

- as recirculator
- as beam cooler
- with energy variation
- with external beam
- as beam stretcher for proton beams of the SNQ-LINAC.

The recirculator mode has been described previously[12] in detail. It will allow the increase of the brilliance $B=I/\epsilon$ at the target by a factor of $10-10^5$ depending on energy and particle. Conventional targets (50-100 µg/cm^2) can be used. Special requirements in this mode are: energy loss correction, dispersion matching to the spectrometer and scrapers for controlled removal of particles which fall out of the ring acceptance. Beam cooling which is able to compensate the heating of the beam by target and residual gas effects will allow operation with higher luminosities (possibly up to $L = 10^{33}$ cm^{-2}s^{-1}) and very high resolutions ($\Delta E/E \approx 10^{-4}-10^{-5}$) depending on particle numbers in the ring. Thin targets ($\lesssim 10^{17}$ atoms/cm^2) are needed because of the limited cooling speed. The extension of the physics experiments to higher energies with good resolution ΔE (e.g. $\Delta E \approx 10$ keV at $E_p \approx 200$ MeV) will be possible after the installation of the RF cavities using our existing high resolution spectrometer BIG KARL.

The properties in the cooler ring available for nuclear and atomic physics are summarized in the following:

- Very high luminosity for experiments with extremely thin targets ($\leq 10^{17}$ atoms/cm^2).
- Variable duty factor up to 100 % (DC beam).
- Fine tuning of the primary energy (1 MeV/min) with the electron cooler.
- Very accurate absolute energy definition by the measurement of electron energy.
- A stripping injection efficiency of about 50 %, i.e. measurement during injection.
- No beam stop in the recirculation mode (internal target) hence reduced background.

As internal targets polarized atomic beams[15] including existing polarized heavy ion beams can also be used. For polarized protons and deuterons densities of about 10^{11} cm^{-2} are available[15]. Higher densities up to about 10^{14} cm^{-2} might be feasible. For target materials with high evaporation temperature other techniques have to be developed.

EXPERIMENTS

Except for the spectrometer target area including a recoil spectrometer with specially designed variable beam properties, there are several places in the ring which will be used as experimental areas (see fig. 1).

With the beam properties of COSY a variety of nuclear and atomic physics experiments becomes possible. With the proton energy range up to 1.5 GeV the COSY ring will be an ideal instrument to study subnuclear degrees of freedom in nuclei.

While atomic physics experiments with heavy ions will be the working domain of designed cooler storage rings like ESR[16] at GSI and HSR[17] at Heidelberg our laboratory first plans experiments in high resolution nuclear physics up to a few hundred MeV. In the following some of the experimental possibilities are listed:

- Fine energy tuning for sharp resonances and threshold energies.
- Measurement of heavy recoil nuclei and reaction products.
- Tagging of secondary beams by recoil measurements (e.g. tagged neutrons).
- Nuclear reactions with extremely high resolution.
- Coincidence experiments with high luminosity and good energy resolution in three body reactions.

- Experiments with polarized atomic beams as target.
- Storage of exotic reaction products (e.g. tritons [18]).
- Nuclear reactions and γ-spectroscopy with low background and very small cross sections.
- Spin excitations in the energy window of 150-300 MeV.
- Production of neutrons with good energy resolution by ^{7}Li(p,n) or ^{12}C(p,n) reactions at 0° for neutron induced reactions.
- Low lying hole states at E_p = 400-500 MeV.
- Polarization transfer experiments, spin-spin interactions with high resolution.

With protons in the energy range up to 1.5 GeV the COSY ring will be an ideal instrument to study the meson production mechanism in complex nuclei. COSY is particularly suited to investigate the threshold behaviour of these reactions. In fig. 5 we show various meson production thresholds as a function of the target mass M_A, which will be covered by COSY. Particularly interesting is the pro-

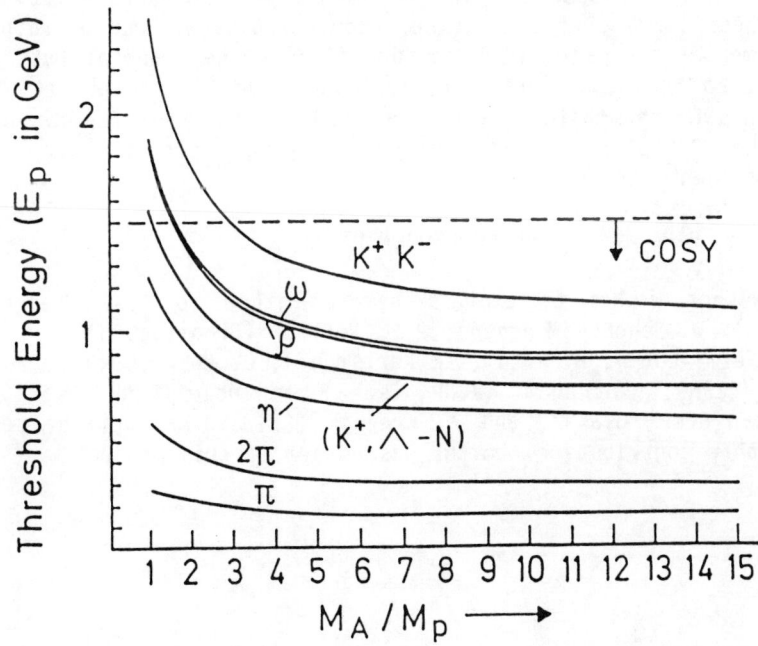

Fig. 5. Meson production thresholds as function of target mass M_A covered by COSY. M_p is the proton mass.

duction of the isoscalar η-meson (m_η = 550 MeV) and of the K^+-meson (m_K = 495 MeV) which, when produced in the $A(p,K^+)$ reaction, might give information on hypernuclear physics. These reactions at energies typically above 500 MeV do not require very high energy resolution, but very high luminosities since the cross sections, e.g. for the (p,K^+) reaction will be very small.

STATUS

After intensive discussions and studies in collaboration with the Universities of Nordrhein-Westfalen the COSY group will present a proposal for the COSY project by the end of this year after a final meeting with the potential users. The KFA recognized the high priority of the cooler storage ring for our institute and supports the collaboration with the universities. We expect a budget decision soon. The man power will come to a large fraction from our institute, the universities and from KFA infra-structure. The latter resource is limited, however, due to the large neutron spallation source project (SNQ) of the KFA. The hardware costs without building and experimental equipment will be specified in the proposal but is anticipated to be of the order of 30 million DM. If the budget is assigned in 1985 the building could be ready by the end of 1987. In parallel the COSY planning and construction group could start to order the components of the ring.

ACKNOWLEDGEMENT

The author wishes to express his gratitude to C.A. Wiedner (Heidelberg), W. Schott (München), R.E. Pollock (Bloomington), F. Mills (Fermi-Lab.), H. Poth (KfK Karlsruhe), D. Möhl (CERN), K. Kilian (CERN), H. Ikegami (RCNP, Osaka), M. Inoue (RCNP, Osaka), I. Katayama (RCNP, Osaka), and P. Krejcik (ABT/SNQ-KFA Jülich) for their valuable contributions to the discussion of this project.

REFERENCES

1. Workshop on Electron Cooling, Bad Honnef, May 1982, Jül-Spez-159 Juli 1982, Kernforschungsanlage Jülich, ed. G.P.A. Berg, W. Hürlimann, J.G.M. Römer, J. Debrus, F. Hinterberger, and S.A. Martin.
2. S.A. Martin, A. Hardt, J. Meissburger, G.P.A. Berg, U. Hacker, W. Hürlimann, J.G.M. Römer, T. Sagefka, A. Retz, O.W.B. Schult, K.L. Brown, and K. Halbach, Nucl. Instr. Meth. $\underline{214}$, 281 (1983).
3. G. Budker and A. Skrinsky, Sov. Phys. Usp. $\underline{21}$, 277 (1978) and report of Novosibirsk group: CERN 77-08, April 1977.
4. S. van der Meer, Stochastic Damping of Betatron Oscillations, CERN(ISR-PO/72-31.
5. F.T. Cole and F.E. Mills, Ann. Rev. Nucl. Part. Sci. $\underline{31}$, 295 (1981).
6. D. Husmann, Acceleration and Storage of Polarized Particle Beams, in ref. 1.
7. E. Steffens, Storage and Cooling of Polarized Particles, contribution to ECOOL84, Karlsruhe, Sept. 1984.
8. W. Hardt, Ultraslow Extraction out of LEAR, pp LEAR Note 98, and P. Lefèvre, Lecture given at the CERN Accelerator School 1983, Geneva 1983.
9. N.L. Hagedoorn and N.F. Verster, Nucl. Instr. Meth. $\underline{18}$, 201 (1962).
10. L. Aldea, W. Bräutigam, R. Brings, C. Mayer-Böricke, J. Reich and P. Wucherer, Proc. 9th Int. Conf. on Cyclotrons and their Applications, Caen (France) (1981) p. 103.
11. H.-G. Mathews, H. Beuscher, W. Krauss-Vogt, ECR-Ion-Source Development at the Jülich Cyclotron, Proc. Int. Ion Engineering Congress, Kyoto 1983.
12. S.A. Martin, D. Prasuhn, W. Schott, and C.A. Wiedner, A Storage Ring for the JULIC Cyclotron, contribution to the IX All Union National Conference on Particle Accelerators, Dubna, USSR, Oct. 1984.
13. D. Möhl, Comparison between Electron Cooling and Stochastic Cooling, contribution to ECOOL84, Karlsruhe, Sept. 1984.
14. D. Möhl, Stochastic Cooling Methods, Lecture given at the CERN Accelerator School, Geneva 1983.
15. W. Haeberli, contribution to this conference
16. B. Franzke, H. Eickhoff, B. Franczak, B. Langenbeck, Zwischenbericht zur Planung des Experimentier-Speicherringes (ESR) der GSI, GSI-SIS-INT/84-5, August 1984.
17. E. Jaeschke, Heavy Ion Cooling Test Facility at Heidelberg, contribution to ECOOL84, Karlsruhe, Sept. 1984.
18. R. Jahn, Tritonenstrahl in COSY, COSY-Arbeitstreffen, 26.-27. Okt. 1983, ed. G.P.A. Berg, D. Prasuhn, J.G.M. Römer, Jül-Spez-253, KFA Jülich 1984

STATUS OF CELSIUS

D. Reistad and A. Johansson

Tandem Accelerator Laboratory, Uppsala University, Sweden

INTRODUCTION

A Swedish national accelerator center for research in particle, nuclear, and atomic physics as well as biology and medicine is presently being established in Uppsala. The accelerator center will be based on the Uppsala Synchrocyclotron which is presently undergoing a reconstruction[1]. The cyclotron will be used for fixed target experiments and also as injector to the storage ring CELSIUS.

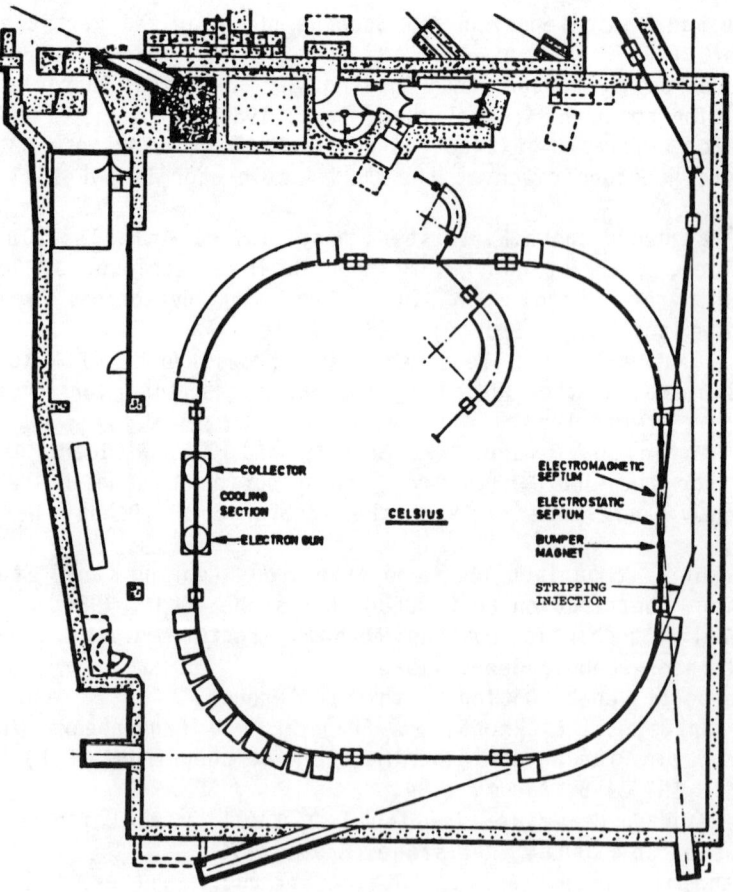

Fig. 1. Floor plan of the CELSIUS hall.

CELSIUS will be a storage ring for protons and ions with internal targets and electron cooling. It will be used for experiments in particle, nuclear, atomic and molecular physics[2,3].

Partners in the CELSIUS project are the Tandem Accelerator Laboratory and the Gustaf Werner Institute of Uppsala University, the Department of Accelerator Technology of the Royal Institute of Technology in Stockholm (electron gun and collector), and the Studsvik Science Research Laboratory (targets[4]). CELSIUS will be constructed using the former ICE[5] bending magnets from CERN, and there is a formal agreement about collaboration on CELSIUS between CERN and Uppsala University.

The floor plan of the CELSIUS hall is shown in fig. 1. Some of the the basic parameters of CELSIUS are shown in table 1.

TABLE 1
SOME OF THE BASIC PARAMETERS OF CELSIUS

Magnet ring	
Circumference	81.8 m
Max. rigidity	6.25 Tm
Bending radius	7.0 m
Max. momentum / Z	1.875 GeV/c
Max. kinetic energy (protons)	1.158 GeV
Working point (approx.)	
Q_x 1.59	
Q_z 1.95	
γ_{tr} 2.52	

	Targets	Cooler and injection
β_x	1.1 m	8.3 m
β_z	1.1 m	8.3 m
D_x	6.6 m	-0.9 m

Electron beam	
Voltage	20-300 kV
Current	Up to 2.5 A
Diameter	2 cm

The maximum energy per mass unit for other ions than protons is shown in fig. 2. The maximum electron energy of 300 keV limits the maximum proton energy that can be cooled to 550 MeV.

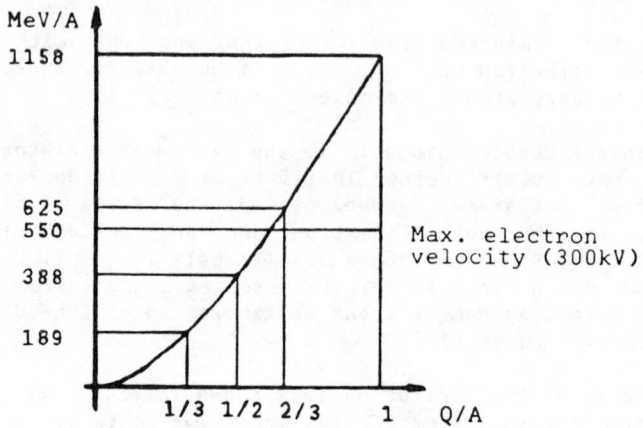

Fig. 2. Maximum energy per mass unit as a function of Q/A. The maximum energy of the ions that is compatible with cooling with 300 keV electrons is shown.

THE CYCLOTRON AS INJECTOR TO CELSIUS

The K-value of the cyclotron is going to be 200. The cyclotron will be able to operate either with a fixed frequency or with frequency modulation[1]. The FM mode will be used for protons in the energy range from 110 to 200 MeV. When the cyclotron will be used as injector of protons to CELSIUS its pulse length will be compressed to about 3 µs and the number of protons per pulse may be up to $6 \cdot 10^{10}$.

When the synchrocyclotron gets into operation in the summer of 1985 it will be equipped with an internal PIG ion source. With this ion source the charge states will be limited to 5 or 6, and the heaviest ions available will be Ar and Ca (A=40). Examples of estimated intensities are given in table 2.

TABLE 2

Ion	Energy (MeV)	Acc mode	Energy res. (%)	Hor. emitt. mm mr	Estimated intensity (eµA)
p	185	FM	0.2	8π	10
$^3He^{2+}$	267	FM	0.2	8π	2
$^3He^{2+}$	250	CW	0.5	5π	20
D	100	CW	0.2	20π	40
$^{12}C^{4+}$	267	CW	0.2	20π	5
$^{16}O^{5+}$	312	CW	0.2	20π	10
$^{20}Ne^{7+}$	490	CW	0.2	20π	0.1

MAGNETS

As mentioned above, the CELSIUS ring will be built of the former ICE magnets from CERN. These are 40 magnets, to be arranged as four quadrants with 10 magnets each. The coils are common for all magnets in each quadrant.

These are combined function magnets. There are six D-magnets and four F-magnets in each quadrant. The k-values of the D and F-magnets are nominally 0.1242 m^{-2} and -0.133 m^{-2} respectively. The magnets are 1.1 m long, and have a bending radius of 7 m. The magnets are made of solid non-laminated steel. The cross section of a D-magnet is shown in fig. 3.

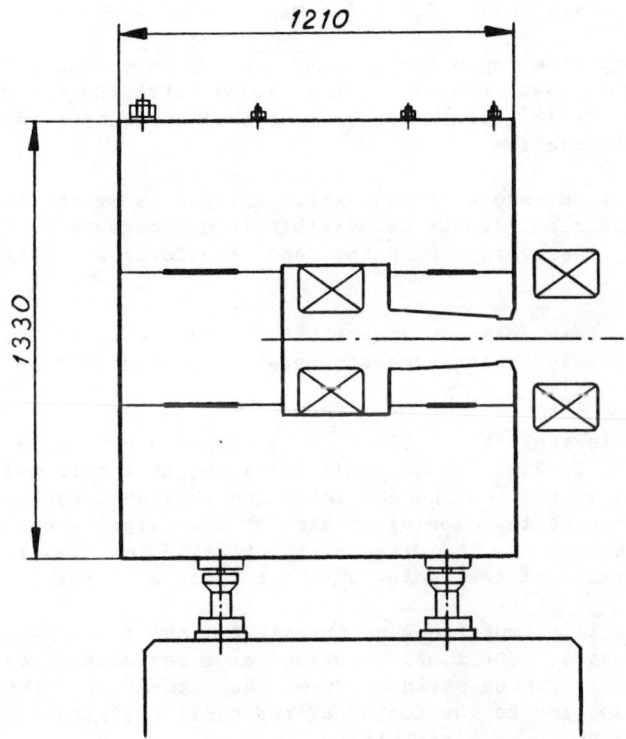

Fig. 3. Cross section of D-magnet.

LATTICE

Like in ICE the bending magnets will be arranged as four quadrants, and there will be four straight sections (see fig. 1). The straight sections will be slightly longer than at ICE; the injection and cooling straight sections will be 9.6 m each and the two target straight sections will be 9.3 m each.

The requirements on the lattice are:

1) The value of the Twiss β-function at the target must be small in order to have a small equilibrium emittance, good acceptance of Coulomb scattering in the target, and good resolution in the experiment. We have chosen to aim for $\beta = 1$ m.

2) For optimum initial cooling conditions the β-function in the cooling region should be just large enough that the initial proton beam divergence is not much larger than the electron beam divergence$_6$. We have optimised for an assumed emittance of 10π mm mr, and have chosen to aim for $\beta = 8.6$ m.

3) An acceptance angle for Coulomb scatter on the target θ of 10 mr has been judged useful. This means that the the value of the expression $\theta \cdot \sqrt{\beta_T \cdot \beta(s)}$ must be made smaller than the useful aperture of the machine.

4) In order to make a machine which accepts as much imperfections in the$_7$ bending magnets as possible (e.g. because of Eddy current effects), the average β in the bends should be as small as possible.

5) For the same reason the horizontal and vertical tune of the machine should be in a as good location as possible in the working diagram.

6) We should avoid transition. Therefore γ_{tr} should be chosen above our working range. It turns out that this means that the dispersion at the cooling and injection straight sections becomes about -1 m, and that the dispersion at the target straight sections becomes about 7 m. This is consistent with the requirement that the dispersion at the cooler should not be too large.

Many possible permutations of the six D and four F-magnets have been evaluated. The final solution has a periodicity of two, where the two cells can be defined from the center of the injection straight section to the center of the cooling straight section, and back. The cells have reflection symmetry.

There are quadrupole doublets at each end of the target straight sections, and quadrupole singlets at each end of the other straight sections. The permutation is QD-FDFDDFDFDD-QF-QD. For good flexibility of the lattice functions it is necessary to provide pole face windings in several bending magnets. A typical example of β-functions and dispersion are plotted in fig. 4, and the corresponding aperture requirements for a scattering angle of 10 mr in the target are shown in fig. 5.

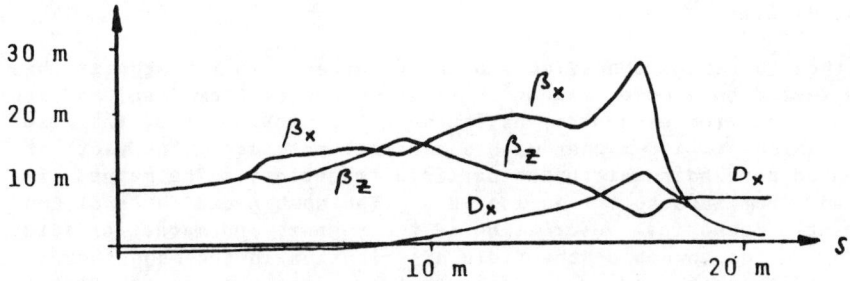

Fig. 4. β-functions and dispersion.

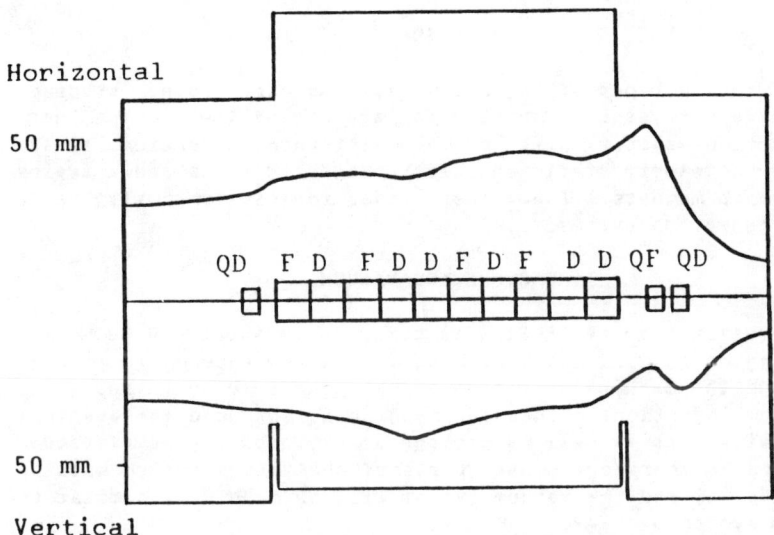

Fig. 5. Aperture requirements for 10 mr scatter at the target.

EDDY CURRENT EFFECTS

The Eddy current effects in the magnets were measured when the magnets were still at CERN[1]. Analysis of the measurements show that during ramping of the magnets from 0.1 T to 0.3 T during 7 s the Eddy currents contribute about -0.0015 m^{-2} to the quadrupole components of the fields in the magnets and about -0.015 m^{-3} to the sextupole components of the fields in the magnets. We expect this to be true also for ramps from e.g. 0.3 T to 0.9 T (corresponding to acceleration of protons from 185 MeV to 1158 MeV) since the magnets do not get very saturated.

If uncorrected, the change in the quadrupole components of the magnetic fields would cause Q-shifts of the order of 0.1, and the sextupole components would make the dynamic aperture of the machine very small.

A method to reduce the effects of Eddy currents in the magnets has been tested on a model magnet[8]. Slots which are 5 cm deep, and are spaced 5 cm from each other have been cut in the poles of the magnet, which is a C-magnet with a parallel pole gap. The slots are oriented perpendicular to the particle trajectory. The magnet was ramped from zero to 1 T during 10 s. The combined effects of Eddy currents, mechanical deformation of the magnet, and magnetic saturation on the change of the field distribution in the magnet during the ramping decreased by a factor of two. It is not yet decided wether to make such cuts in the CELSIUS magnets. Further compensation of the Eddy current effects through pole-face windings and through active quadrupoles is fore-seen.

INJECTION

Two complementary types of injection systems are being studied. Details have not yet been settled for any one of these. The "normal" injection system will be a multi-turn injection system consisting of electrostatic and electromagnetic septa and a system of four bumper magnets. The other injection system will be a stripping injection system.

TIME CYCLE AND VACUUM

The acceleration time of CELSIUS is going to be about 10 s. The data taking phase will last anything from a few seconds to several minutes, and the machine will need a few seconds to get ready to accept a new injection. Since the machine is intended for eventual use with heavy ions as well as protons a very good degree of vacuum is required. Therefore most parts of the vacuum system will be baked to 350 °C, and the vacuum system will be made very similar to the vacuum system of LEAR.

The possibility to use Non-Evaporable Getter (NEG) Pumps is being studied. In particular this is studied for the cooling straight section, where it is not possible to interrupt the solenoid with a pumping port, but where it may be possible to integrate a NEG pump in the tubular vacuum chamber.

ELECTRON GUN

The design of the electron gun is made by Sedlaček of the Royal Institute of Technology in Stockholm. The design is similar to Oleksiuk's proposal 1981[9]. A cross section of the gun is shown in fig. 6. The voltage range for the gun is going to be from 20 kV to 300 kV. The potential distribution along the axis of the gun is illustrated for a total voltage of 100 kV and 300 kV in fig. 7. The anode voltage is kept at a constant value of about 30 kV above the cathode for all accelerating voltages above this voltage and 300 kV, and the design current is 2.5 A in this voltage range.

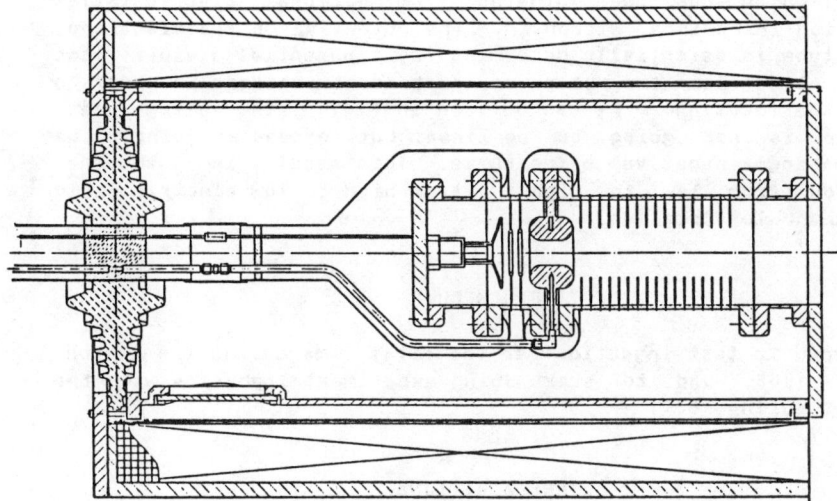

Fig. 6. Cross section of the electron gun.

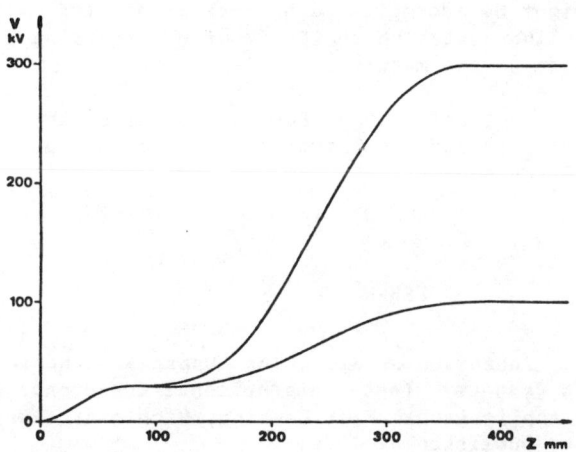

Fig. 7. Potential distribution along
axis of gun for 100 kV and 300 kV.

Around the cathode there is a Pierce electrode shaped so that the
computer simulation gives almost negligible scallop near the
cathode. There are two small ring-shaped electrodes near the
cathode in order to equalize the electric field distribution.

The anode is shaped so that the electric field distribution at the
output side compensates the disturbance on the electron orbits
which is produced by the electric field distribution at the input

side. After the anode follows a so-called high-gradient accelerator tube, which has been ordered from National Electrostatics Corporation in Madison, Wisconsin. The potential on the electrodes in this tube is essentially determined by a potential divider (two of the electrodes can be given independent voltages in order to control the focusing). As can be seen in fig. 7 the voltage distribution is not going to be linear but instead is going to be approximating a negative cosine curve. This results in a smoother transition from low to high and from high to low electric field strength and less scallop.

TIME SCHEDULE

It is hoped to test injection for the first time during the second half of 1986, and to start doing experimental physics with the equipment during 1987.

FUTURE AMBITIONS

Hopes for the future beyond the present design of CELSIUS are of course to extend the cooling capability up to the full maximum energy, which might be accomplished by increasing the voltage of the electron cooling system up to 600 kV or by providing a stochastic cooling device on the machine.

Another hope is to put an external ECR ion source on the cyclotron, and thereby reach heavier ions, than those which are possible with the internal PIG ion source.

REFERENCES

1. S. Holm, A. Johansson et al. The Uppsala Synchrocyclotron and Storage Ring Project. Tenth International Conference on Cyclotrons and their Applications, East Lansing, Michigan, April 1984. Proceedings to be published by IEEE.

2. B. R. Karlsson, G. Tibell (editors). Contributions to the Workshop on the Physics Program at CELSIUS, Nov. 7-9, 1983, Uppsala.

3. B. Badelek, L. Bergström, C. Ekström, S. Fredriksson, E. Hagberg, M. Jändel, S. Kullander, P. A. Tove, L. Westerberg, P. Zielinsky. Letter of Intent for a Research Program on Elementary Particle Physics at CELSIUS.

4. C. Ekström, B. Holmqvist and H. Sterner. Target Developments for CELSIUS. 1984 IUCF Workshop: Nuclear Physics with Stored Cooled Beams, Bloomington, Indiana, October 1984.

5. M. Bell, J. Chaney, H. Herr, F. Krienen, P. Møller-Petersen, G. Petrucci. Electron Cooling in ICE at CERN. Nuclear Instruments and Methods 190 (1981) 237.

6. H. Herr. A First Approach to find Parameters for the Uppsala Storage Ring. CELSIUS-Note 83-2.

7. W. Hardt. Field shape during acceleration in CELSIUS. CELSIUS-Note 84-27. CERN PS/DL/Note 83-4.

8. P.-U. Renberg. Private communication.

9. L. Oleksiuk. Design Study of a 750 kV Electron Gun for Electron Cooling. Fermilab p̄ NOTE 124

THE TARN II PROJECT

T. Katayama

Institute for Nuclear Study, University of Tokyo
Tanashi, Tokyo 188, Japan

ABSTRACT

On the basis of the achievement of the accelerator studies at present TARN, it is decided to construct the new ring TARN II which will be operated as an accumulator, accelerator, cooler and stretcher. It has the maximum magnetic rigidity of 7 T·m corresponding to the proton energy 1.3 GeV and the ring diameter is around 23 m. Light and heavy ions from the SF cyclotron will be injected and accelerated to the working energy where the ring will be operated as a desired mode, for example a cooler ring mode. At the cooler ring operation, the strong cooling devices such as stochastic and electron beam coolings will work together with the internal gas jet target for the precise nuclear experiments. TARN II is currently under the construction with the schedule of completion in 1986. In this paper general features of the project are presented.

INTRODUCTION

TARN (Test Accumulation Ring for Numatron) is a storage ring, built for the development of accelerator technique of beam accumulation and stochastic cooling with a view of future application in a scheme of high energy heavy ion accelerator complex, Numatron. The idea[1] of beam accumulation both in the horizontal betatron and in the longitudinal phase spaces as many ions as possible has been experimentally verified.[2] The momentum cooling with the stochastic system has been also successfully tested.[3,4] For the accumulation experiments twenty turn beam were injected in the betatron phase space of 87 πmm·mrad with a multiturn injection method and 15 pulses are RF stacked in the momentum space $\Delta p/p$ of 2.2 %. Overall stacking number was then attained at around 300 turns. The stochastic cooling experiment was performed on the RF stacked beam to decrease its momentum spread. With the feedback system of active gain 107 dB and band width 100 MHz, the initial momentum spread 1.4 % was cooled down to 0.2 % within 20 seconds for the 7 MeV proton beam of 7×10^7.

On the basis of these successful achievements of accelerator studies, we have decided to construct the improved TARN, TARN II which will accumulate, accelerate and stretch the ion beam from the cyclotron and/or the planned injector linac. The maximum $B\rho$ of TARN II is 7 T·m and the corresponding proton beam energy is 1.3 GeV and the energy of full stripped ions such as C^{6+} is 450 MeV/u. Cooling devices of both the stochastic and electron beam will cool the horizontal and vertical emittances and momentum spread of the stored beam. With these cooling devices it is expected to obtain a quasi-monochromatic ion beams of the momentum spread better than 1×10^{-4}.

0094-243X/85/1280024-10 $3.00 Copyright 1985 American Institute of Physics

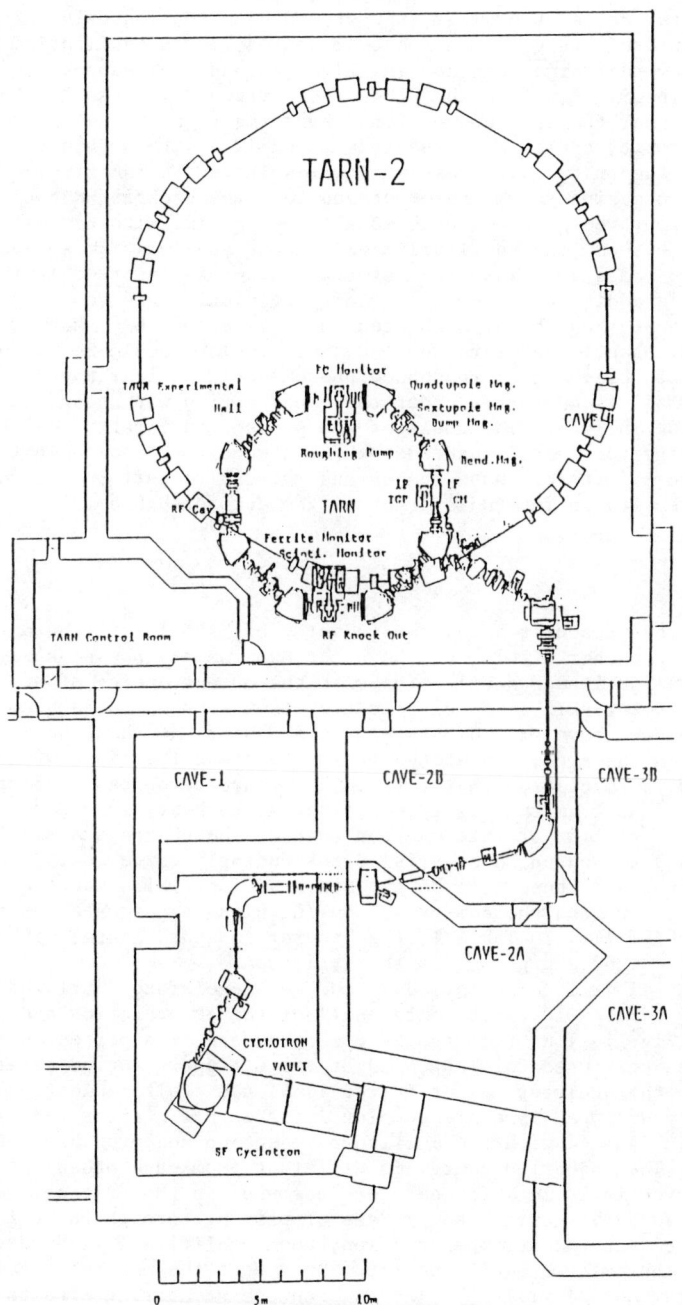

Fig. 1 Layout of TARN II

In the cooler ring mode, the stored beam continuously circulate through a thin internal target, and the beam approaches an equilibrium between heating by the target and cooling with stochastic feedback and/or cold electron beam. Although the product of target thickness and beam current is not higher than the conventional experiments with the extracted beam, the luminosity can be increased to around 10^{31} with a thin target of 10^{14} atms/cm² due to the advantage of the revolution in the ring. The beam is not dumped during the measurement and then the background conditions will be much improved over a conventional experiment. With use of the fast cooling devices, the equilibrium emittances and momentum spread can be maintained as small even when the internal target is inserted in the ring orbit. In addition to the well energy resolution and good angular resolution, the ion energy can be also changed in small steps less than 10 KeV with use of the stochastic acceleration technique, which should be useful for the studies of threshold or resonance phenomena in nuclear physics.

TARN II is currently under the construction with schedule of the completion in 1986. The goals of the project is firstly to boost up the beam energy to several hundreds MeV/u, secondly to cool down the beam temperature in three phase spaces and thirdly to perform the nuclear experiments with an internal target or the extracted beam.

INJECTOR AND AVAILABLE IONS

In the present scheme, the injector of TARN II will be an SF cyclotron with the K number of 67. On the other hand we are now planning to construct the heavy ion linear accelerator with the output energy of around 5 MeV/u, probably being able to accelerate the heavy ions up to Xe beams. This linac is now, however, the stage of discussion and then the injector of TARN II is the cyclotron at the first stage and the linac will take the place of the cyclotron in future. The SF cyclotron can accelerate various kinds of ions from the light ions of p, α, to heavy ions Fe^{8+}. However due to the restriction of internal ion source, the charge state of heavy ions is low and the output energy is correspondingly quite low. Among these heavy ions, Ne^{4+} beam will be the heaviest ions which has the reasonable current $\sim$ 1 μA and the energy 2.6 MeV/u which is adequate for the acceleration in TARN II. In Table 1, the heavier ions are listed with the available intensities and energies from the cyclotron.

TARN II will be installed in the new accelerator hall which is made by clearing up the Old Experimental Hall of the FM cyclotron and present TARN room. (Fig.1) Ions from the SF cyclotron are transported through the beam line presently used for TARN, and at the stripper section located just prior to the analyzer magnet in the line, the orbital electrons of partially stripped ions are completely taken off. After that ions are injected in TARN II with the multiturn injection and probably the RF stacking method. The injection energy is different from each other for various ions as is given in Table 1 for heavier ions whereas the proton injection energy will be 20 MeV. During the process of passing through the thin carbon foil $\sim$ 50 μg/cm² at the stripper section, beam qualities will be degraded, for example the emittance will be grown up due to the multiple scattering and the energy spread will be enlarged with the straggling effects. As a most

Table 1 Heavy Ion Beam from SF Cyclotron

Charge Stage	2	3	4	5	6	7
^{6}Li	7.6	17.0				
	4.5	1.0				
^{7}Li	5.6	12.5				
	1.0	0.3				
^{11}B	2.2	5.1	9.0			
		3.0				
^{12}C		4.3	7.6	11.8		
		5.1		0.025		
^{14}N		3.1	5.6	8.7	12.5	
		10.0	5.5	3.0	0.006	
^{16}O			4.3	6.6	9.6	13.0
				1.5	0.07	
^{20}Ne			2.7	4.3	6.1	8.3
			1.7		0.1	0.01

Notation: Kinetic energy (Mev/u)   ...upper figure
Extracted beam intensity (eμA)...lower figure

serious case, Ne^{4+} beam of 2.6 MeV/u is examined which shows the emittance 20 πmm·mrad of the beam from the cyclotron will be increased up to 35 πmm·mrad and the energy straggling will be around 1×10^{-3}. The fraction of full stripped ions is estimated at one third.

The cyclotron is essentially the CW machine, not having the high peak current and does not fit to the injection of the pulsed operating accelerator such as synchrotron. For the present TARN, the peak current at the injection point is around 1 pμA (p, α) and 0.1 pμA (heavier ions) after passing through the beam transport line with magnetic analyzer system. The momentum spread of the injected beam is around ± 0.1 % which comes from the narrow phase spread ± 2 degree in the RF acceleration field of the AVF cyclotron. The horizontal and vertical emittances are 10 πmm·mrad for p, d, α and 35 πmm·mrad for heavier ions. As the acceptance of TARN II is designed at 400 πmm·mrad and the dilution factor during the process of multiturn injection is assumed at 2.5, being experimentally verified at the present TARN, and the expected beam intensities are around 1×10^8 for p, α, d and 1×10^6 for heavy ions. However if one uses both the horizontal betatron and longitudinal phase spaces, the expected intensity will be increased up by the order of two. This is mainly due to the fact that the AVF cyclotron beam has a quite small longitudinal phase spaces $\varepsilon_\ell = \Delta\phi \cdot \Delta T/T$, around 5×10^{-4} (rad).

Beam life time at the injection energy is mainly determined, for heavy ions by the charge capturing process of full stripped ions through the collisions with residual gases and for the light ions such as proton and α by the Rutherford scattering. Assuming the vacuum pressure in the ring as 1×10^{-10} Torr, the beam life is estimated as follows: p (20 MeV);

3300 sec, C^{6+} (7.6 MeV/u); 760 sec and Ne^{10+} (2.7 MeV/u); 12 sec. From the arguments on this beam life at the injection energy, it is expected that there is enough time for the beam manipulation such as RF stacking or fast stochastic cooling even at the flat base, injection period.

GENERAL DESCRIPTION OF THE RING

Modes of Operation

The ring will be used in three modes;
1) normal synchrotron operation,
2) long spill operation, strethcer mode, and
3) cooler ring mode.

In the synchrotron mode, the repetition cycle is 0.5 Hz with the acceleration period of 0.75 sec, the flat-top of 0.5 sec and the falling period of 0.75 sec. This repetition rate is determined mainly due to the available electric power at the present INS electric station. In the stretcher mode, the acceleration period will be around 10 sec, whereas the flat-top will be long enough, say 1 hour for 500 MeV protons, which sould be a good compromise between the beam life and the ultraslow ejection method such as stochastic extraction. At the cooler ring mode, the operation scheme will be the same for stretcher mode while the strong beam cooling devices will be operated as well as the internal gas jet target.

Lattice

A set of preliminary lattice parameters for stretcher and cooler modes is given in Table 2, and the lattice functions are shown in Fig.2. The

Table 2   Specification of TARN II

Maximum Beam Energy	proton	1300 MeV
	ions with $\varepsilon = 1/2$	450 MeV/u
Circumference		69.908 m
Average Radius		11.650 m
Radius of Curvature		3.820 m
Focusing Structure		FBDBFO
Superperiodicity		6
" for Cooler Ring Mode		3
Betatron Tune Value		around 1.75
" for Cooler Ring Mode		ν_H around 2.25
Transition γ		1.87
Repetition Rate for Synchrotron Mode		1/2 Hz
Maximum Field of Dipole Magnets		18 kG
Deflection Angle of Dipole Magnets		15°
Maximum Gradient of Quadrupole Magnets		70 kG/m
Revolution Frequency		0.38 - 3.75 MHz
Acceleration Frequency		0.76 - 7.50 MHz
Harmonic Number		2
Maximum RF Voltage		6 kV
Vacuum Pressure		better than 10^{-10} Torr

circumference 69.908 m is the maximum size of a ring that fits into the new

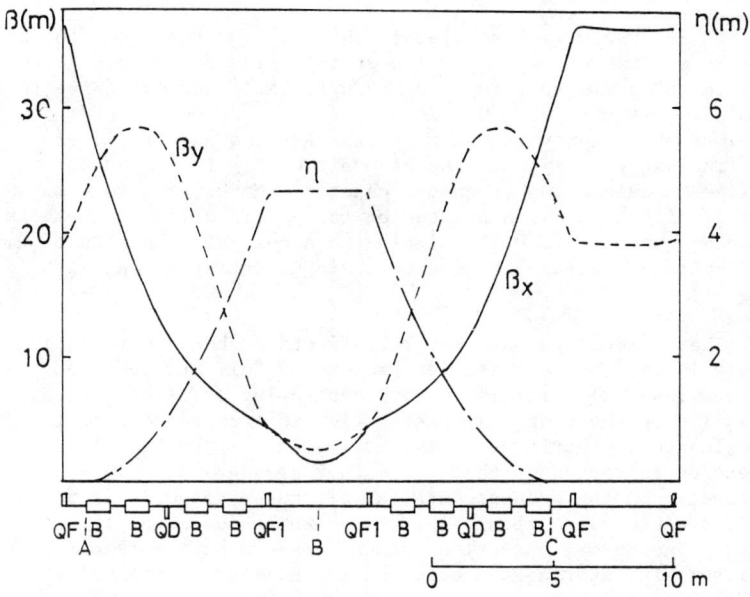

Fig. 2   Lattice function of cooler ring mode.

accelerator hall. A symmetric three-period lattice with the six long straight sections of 4 m long each, was adopted. Hence there are three dispersion-free straight sections and three large dispersion sections, of which the feature is adequate for the beam cooling and the internal target experiments. In the dispersion-free straight sections, an electron cooling device, stochastic cooling kickers and an RF accelerating cavity will be installed, while the large dispersion sections are prepared for stochastic cooling pickups, internal target system and the electric inflector for the beam injection.

Magnets

In the renovation process of TARN, the whole dipole magnets are rebuilt, while the quadrupole magnets being used in the present TARN will be again used for TARN II. The magnetic structure of the new ring is made up of 24 dipole magnets and 18 quadrupole magnets. Each dipole magnet is an H type structure with the straight core length of 1 m. The edge shape at the end of yoke is approximately Rogowsky curve, cutting the yoke with four steps. The designed field region is 200 mm in width and 60 mm in vertical directions. So as to realize the large good field region for the wide excitation range up to 18 kG, the side pole edges are shaped with B constant curve. Also the small shims are attached to suppress the falling down of the magnetic field at pole ends at the high field of saturated condition.[5]

Acceleration

An RF system will accelerate the ion from the injection energy to the desired working energy. The lowest injection energy among the various ions from the SF cyclotron, is 2.58 MeV/u for Ne^{4+} corresponding to the revolution frequency of 0.307 MHz. At the top energy of $\sim$ 500 MeV/u, the revolution frequency is 3 MHz and the RF frequency ratio of the initial and final stages is ten. The harmonic number is chosen as two and the designed acceleration frequency changes from 0.6 MHz to 6 MHz. An RF voltage of 6 KV seems adequate for the acceleration of the beam with the momentum spread of $\pm$ 0.5 % within the acceleration period of 0.75 sec. This RF voltage is produced using a cavity loaded with ferrite 2.5 m long.[6,7]

Vacuum

The residual gas has several effects on beams circulating in TARN II, namely 1) the charge capturing process of full stripped heavy ions leading to beam loss, 2) multiple Coulomb scattering of light ions determines the beam life in the ring, 3) contribution to background when the jet target experiments are performed. The estimation of the beam life at the injection energy shows that the vacuum pressure of $\sim 10^{-11}$ Torr should be achieved. In the present TARN, an all metal vacuum system bakable at 200°C are used with eight sputter ion pumps and eight titanium getter pumps. The normal operating vacuum pressure better than 1×10^{-10} Torr is achieved with storing the beam and the similar system will be used also for TARN II.

Slow Ejection

In the synchrotron mode operation, the beam will be extracted with use of one-third resonance at the flat-top period of $\sim$ 0.5 second. The extraction system consists of following elements. 1) four bumps magnets for the closed orbit distortion, 2) four sextupole magnets as a chromaticity

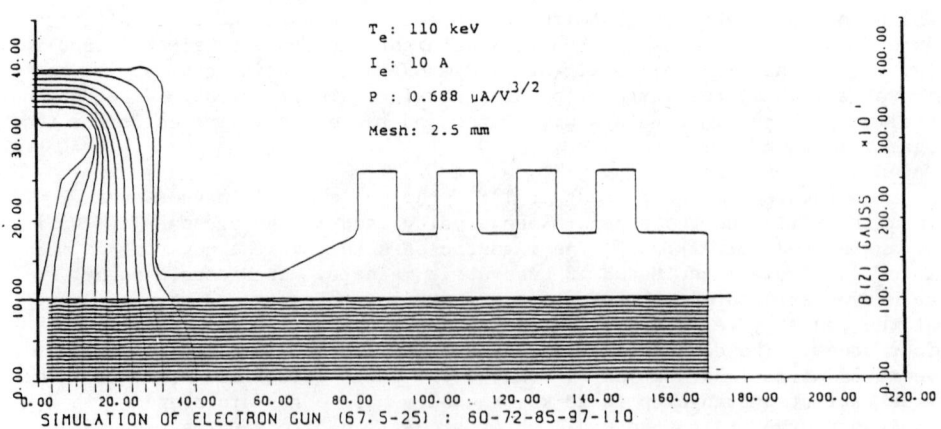

Fig. 3   Calculated electron trajectories in gun.

adjustment and a resonance excitor, 3) an electrostatic and three magnetic septum in the extraction channel. Comparing with the several extraction resonances, one-third resonance is chosen as it can give the extraction efficiency around 90 % and the small beam emittance. When the ring is operated as stretcher mode, the spill time would be order of 100 seconds which requires the beam being far off the linear resonances to avoid the sudden beam loss. To perform this ultra-slow ejection, the stochastic extraction system practised at CERN, LEAR, will be used with the combination of stochastic cooling system and normal extraction equipments.

BEAM COOLING AND INTERNAL TARGET

Both the stochastic and electron beam coolings will be used to obtain the high quality beam. For the stochastic cooling, there will be used two systems, one being the precooling and the other the high energy cooling. With a precooling system, the momentum spread of the injected beam from the cyclotron and/or the linac will be improved. Especially when the RF stacking is employed as an injection method, this precooling is indispensable to keep the acceleration RF voltage as reasonably small as possible. The RF stacked beam will have the momentum spread of ± 1 % and it should be decreased to ± 0.2 % to accelerate the beam within the designed RF voltage 6 kV. The stochastic cooling system currently used at present TARN, the band width of $\sim$ 100 MHz, the system gain of $\sim$ 150 dB, the pickup and the kicker being helical type, will be used for this pre-cooling purpose. After the acceleration, the high energy cooling system with the band width $\sim$ 1 GHz can be reasonably used to attain the momentum spread of $\sim 10^{-4}$. In the high energy cooling system, pickups and kickers are loop coupler type with 16 pairs of coupler of $\lambda/4$ length. The calculated coupling impedance is around 100 Ω in the concerned frequency region 1 $\sim$ 2 GHz. Pre-amplifier is composed of Ga.As. Field Effect Transistor, being cooled down to the temperature of liquid nitrogen. The expected noise figure (NF) is around 0.5 dB. With these parameters the optimum or fastest cooling time is 33 msec when the system gain should be 197 dB with the unrealistic power of 2.5 GWatt. With the reduction of gain to 137 dB, the cooling time increases up to 33 second with the power of 2.5 kW which would be a good compromise between the cooling time and needed RF power. In this case, the rms momentum spread is expected to be 5×10^{-5}.

Electron cooling is most effective at lower energies, say $\sim$ 100 MeV/u and for beams which are already relatively cool. It can thus complement the stochastic system, especially at the experiments with the internal circulating beam of the momentum spread $\sim 10^{-5}$ and of the beam size smaller than a cm resulting from an equilibrium between the cooling and the heating through the internal target.

The main parameters of the electron cooling system are listed in Table 3. The system is designed to cool down the ions from H^+ to Ne^{10+}, up to 200 MeV/u limited by the maximum electron energy of 120 keV. The electron energy is variable from 12 to 120 keV and the maximum current density 0.5 A/cm^2 is available at the voltage higher than 60 keV whereas the current at the lower collector voltage is determined with the perveance. The length of the interaction region is 1.8 m which is limited by the length

Table 3   Electron cooling parameters

Maximum working energy	ions	200 MeV/A
	electrons	120 keV
Cooled ions		$H^+ - {}^{20}Ne^{10+}$
Gun optics	Pierce type + resonance focussing electrodes	
Length of interaction region		1.8 m
Maximum electron current density		0.5 A/cm^2
Cathode diameter		50 mm
Maximum current		10 A
Maximum solenoid field		1000 G

of straight section ∿ 4 m. As the beam size at the cooling section after the acceleration is less than 50 mm, the cathode diameter is designed as 50 mm, which is a type of flat cathode rather than a spherical one. The cathode is immersed in the uniform solenoidal field of the maximum field strength of 1 kG. An example of electron trajectories in the region of the electron gun is given in Fig.3, where the collector voltage is 110 keV, electron current is 10 A and the perveance is 0.688 µA/V$^{3/2}$. The transversal electron temperature is designed in this case as less than 1 eV. A layout of the electron cooling device is so-called U scheme (Fig.4), where the electron beam is injected and ejected over the beam line of the ring.[9]

An internal gas jet target or polarized beam target will be used to perform the precise nuclear experiments taking advantage of the good beam energy resolution and 100 % duty of the circulating beam in the ring. With the wide band stochastic cooling system and the electron cooling system, both having the cooling time constants around 10 seconds, it is possible to use the thick target ∿ 10^{15} atoms/cm^2 and the luminosity up to 10^{31} (cm^{-2}·s^{-1}) may be attained. The target should be placed in a low beta section to suppress the beam loss scattered by the target.

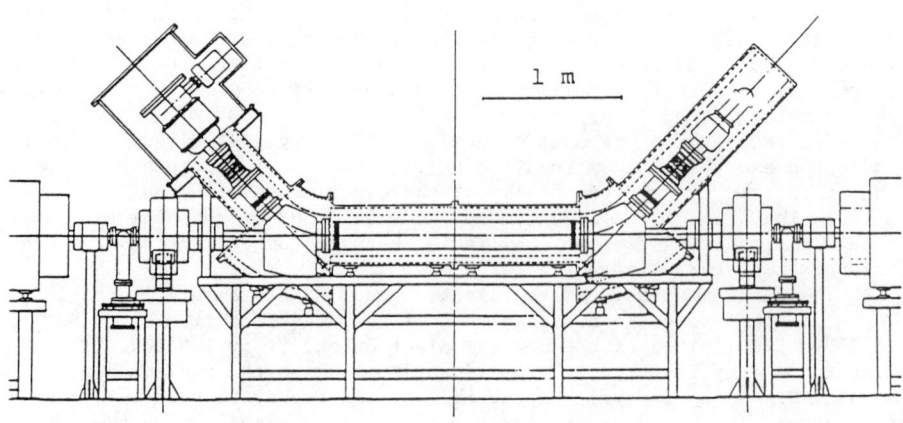

Fig. 4   Layout of the electron cooling device.

ACKNOWLEDGEMENT

TARN II is the collaborative work of the member of accelerator research division at INS. It is the pleasure for the author to thank all of them for their collaboration.

REFERENCES

1) T. Katayama, T. Nakanishi and S. Yamada, IEEE Trans. NS-28 (1981) 2608.
2) T. Katayama et al., IEEE Trans. NS-30 (1983) 2080.
3) T. Katayama, Proc. of the Xth Int. Conf. on Cyclotrons and Their Applications (1984).
4) N. Tokuda et al., Proc. of the 5th Symp. on Acc. Science and Technology, KEK (1984).
5) A. Noda et al., ibid. in ref.4).
6) A. Itano et al., ibid. in ref.4).
7) K. Sato et al., ibid. in ref. 4).
8) M. Takanaka et al., ibid. in ref.4).
9) T. Tanabe et al., Proc. of ECOOL 84, Kahlsruhe (1984).

STATUS REPORT ON THE IUCF COOLER*

Robert E. Pollock
Physics Department, Indiana University, Bloomington, Indiana, 47405.

ABSTRACT

The "Cooler", a storage ring with electron cooling, is under construction at the Indiana University Cyclotron Facility (IUCF). Stored beams in the Cooler will be used with internal targets for research in intermediate energy nuclear physics. The improved properties of the cooled beam should make possible an interesting extension of our experimental capabilities within the existing IUCF research program. The present status of the Cooler design and of the construction project is reviewed.

INTRODUCTION

IUCF is a national user facility for nuclear research which has been in operation for about 8 years. Light ions beams from a 2.2 Tesla-meter separated-sector isochronous cyclotron have been used in a wide variety of experiments. The polarized hydrogen beams, and especially the proton beams in the energy range from 80 to 200 MeV have been shown to be particularly useful. The excellent beam quality, which makes possible the resolving of individual nuclear states, and the beam energy, which covers the lower half of the 150-350 MeV energy range in which the nucleon most easily penetrates to the nuclear interior, have led to a productive early life of the facility.

Although the IUCF beams are arguably among the best that have ever been available in the intermediate energy range from the standpoint of the experimenter, the pursuit of better data quality encounters the fundamental limit set by ion source brightness. A new technology to overcome this limit would open up new experiments which are at present most difficult or impractical.

The IUCF Cooler can be viewed either as a facility upgrade intended to make available beams of even better quality over a wider energy range, or as a large experiment attached to the existing facility, with which it will be possible to explore the limits of cooled storage techniques in nuclear research applications. When Cooler experiments begin in 1987, we shall try to establish the extent to which the electron-cooled storage concept can be used to improve the quality of nuclear measurements in the intermediate energy range.

Beam cooling systems and storage rings both originate in the high energy physics domain. Their application in the discovery of the W boson is a recent shining example of the value of this technology

---

* This work was supported by the U.S. National Science Foundation under grants NSF PHY 81-14339 and 82-11347.

in making practical a facility for antiproton-proton collisions. The application to physics at lower energies has more recently begun to attract attention. Numerous recent workshops and facility proposals make clear that the nuclear and atomic physics communities are actively exploring the application of cooling and beam storage techniques in their own fields of interest.

In this contribution we will summarize the current status of the IUCF Cooler project. An earlier paper[1] reported the Cooler design status as of eighteen months ago. A recent companion publication[2] discusses some of the limitations of the electron cooling technique and the implications for the design of Cooler experiments.

LATTICE

A storage ring must contain bending elements, such as magnetic dipoles, to bend the beam through 360° to form a closed path, and focussing elements, such as magnetic quadrupole lenses, adjusted for stable confinement of the stored beam to a region of limited volume about the closed orbit. Other elements, for steering adjustments and for control of higher-order aberrations, are also necessary.

The maximum beam momentum per unit particle charge is directly related to the magnetic rigidity (0.3 GeV/c per Tesla-meter). The Cooler is designed for a maximum beam rigidity of 3.6 T-m with a dipole field of 1.52 T, corresponding, for example, to protons up to kinetic energies of 490 MeV and particles of Q/A = 1/2 to 144 MeV/amu (at which v/c = 1/2). To exploit the higher stored energies which are above the cyclotron rigidity limit, the ring is designed with laminated magnets and with a ferrite-tuned rf acceleration cavity so that operation as a synchrotron is possible. The option to operate the Cooler at an energy different from that of the cyclotron also opens the prospect of two users sharing the cyclotron beam with independently chosen energies. The efficiency of beam use within the Cooler and the cyclic nature of its operation permits the filling of the Cooler to take place as an intermittent diversion of beam from the cyclotron experiment.

The parts of the Cooler ring where injection, acceleration, and electron cooling hardware are placed, and other parts where experiments are carried out, have quite different requirements for stored beam properties and for clear space. Electron cooling, for example, requires a long empty space with a fairly parallel beam and no dispersion (η = p·dx/dp = 0). Longitudinal stacking, and spectrograph operation with dispersion matching, both require fairly high momentum resolving power (ratio of dispersion to beam size). Injection has an additional special constraint associated with the operation of kickers, that a space at ±1/4 betatron wavelength with respect to the injection point be open to kicker installation.

The IUCF Cooler lattice design philosophy is to construct a ring of given maximum rigidity with as much open space as possible for internal target experiments, and with properties of the beam at the waist in the center of each open space tailored to the differing requirements arising from the assigned use of that location. The bending and focussing elements are concentrated in six corner regions separated by six long "straight" sections, three of which are set

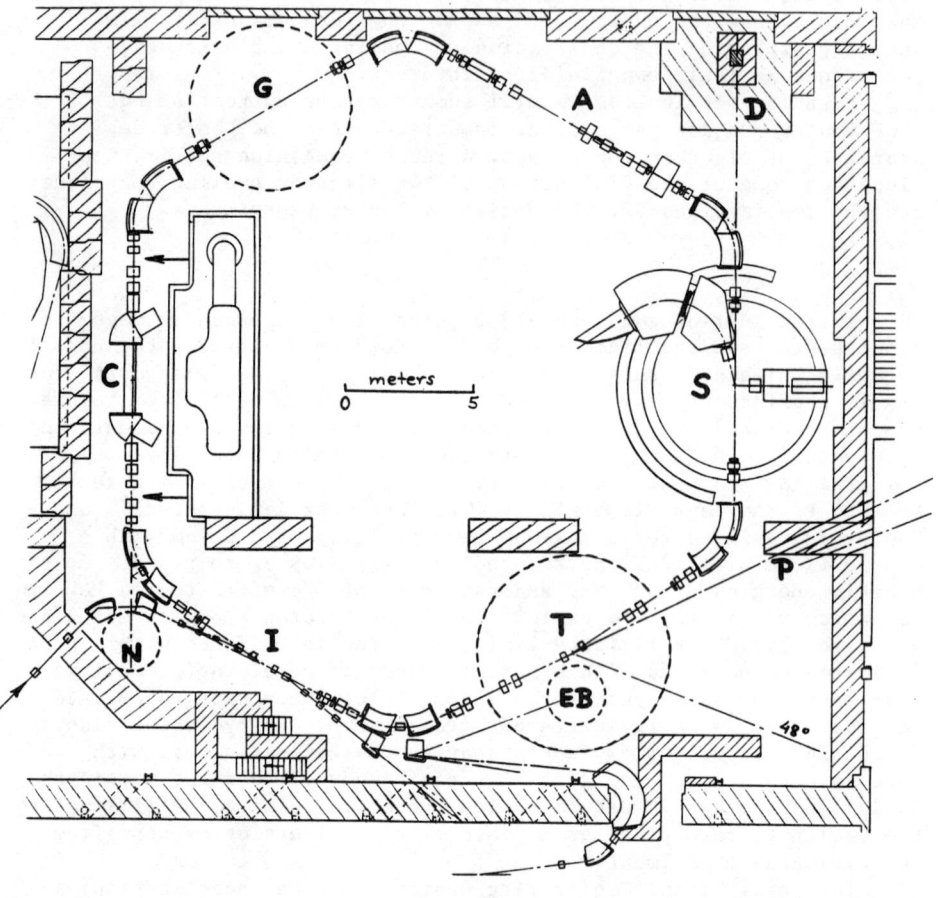

Fig. 1. Layout of the Indiana Cooler Ring. The labels denote the following: I - beam injection and extraction; T - target station (tagged secondary beams); S - target station (double-arm spectrometer); A - acceleration and diagnostic equipment; G - target station (general purpose); C - electron cooling; EB - extracted beam area; D - dump for one-pass experiments; N - neutral atom area; P - exit ports for long flight paths.

aside for experiments, and one each for cooling, for injection and extraction, and for acceleration, diagnostics, and tune adjustment. Each corner has six quadrupole singlets, the minimum needed to fix a given dispersion and zero angular dispersion at each waist, to obtain a double waist with reflection symmetry at the midpoint of the straight section and to vary the beam size (more correctly the aperture functions β_x, β_y) at the waist. The dispersion function alternates between zero and about 4 meters at successive waists. Target waists lie in the range of $0.1 \leqslant \beta \leqslant 1$ m with some range of adjustment possible. The cooling waist is adjustable over the range $3 \leqslant \beta \leqslant 13$ m. A satisfactory betatron tune is obtained by changing the beam size at the sixth (adjustment) waist.

This design is quite unconventional in contrast to the high symmetry of large machine lattices, but is well suited to the specialized needs of a ring of modest size designed for optimal performance in internal target experiments. Some considerable effort has been expended over the past two years in exploring the behavior of an extended family of variations on this basic structure. We have learned for example that while it is possible to tune the ring for a minimum aperture function $\beta_x = 0.1$ m at three dispersed waists[1], the resulting chromaticity (change of tune with momentum) is large enough to require correction by a family of hexapoles whose strength drives unacceptable particle losses via higher order effects which become apparent in orbit tracking. It seems possible to have this small β_x at one waist (eg. for a spectrograph experiment) with some loss in acceptance, but not possible to do so while maintaining three-fold symmetry.

The lattice has another unusual feature in that two of the waists are bracketed by 3° dipoles, accompanied by a reduction in the bend angle from 30° to 27° in the adjacent large dipoles at the ends of the "straight" section. The small bends are of great utility in the injection straight in providing for charge-changing injection and for improving clearance of the injection path to the side of the ring quadrupoles. The small bends in the experimental "straight" provide access at 0° and 180° for detection of reaction products of rigidity widely different from that of the beam.

Hexapoles pairs are located symmetrically at opposite ends of dispersed straights for chromaticity control while three other hexapole pairs at non-dispersed locations (selected, for example, for large β_x and small β_y) control the nearest third integer resonance terms and the coupling term. The design momentum acceptance is rather small, being related to the delta electron knockon momentum, so that it will be possible to operate the ring in some modes without hexapole excitation.

The kickers move the stack in the hexapoles in the injection straight, leading to emittance doubling after a few hundred kicks.

Steerer families must be employed to control the closed orbit position at injection and at the location of experiment targets. In addition, the closed orbit distortion at each hexapole must reduced to about 1 mm or less before the hexapoles can be excited. Trim windings on the ring dipoles provide both global and local[3] steering control in the bend plane while discrete vertical steerers are employed for the second transverse dimension. The hexapoles, steerers, and position sensors are designed to be interchangeable in location witha common insertion length of about 0.1 m.

INJECTION AND STACKING MODES

A cyclotron which produces a continous beam of small phase space volume is not the ideal injector for a storage ring. The low peak current means that multiturn injection is necessary to build up a stored beam intensity suitable for nuclear research. Pioneering experiments on filling a ring from a cyclotron have been carried out at INS Tokyo[4], while theoretical studies of alternative stacking modes have been carried out at two of the meson factories[5].

The injection scheme for the IUCF Cooler allows the flexibility to explore as many stacking methods as possible. The beam is sent to the new building down a beam line branch which splits off from the main switchyard in a manner which allows time sharing with other experiments on the basis of the cyclotron rf microstructure. A small accumulator ring[6] has been built which can be used to increase the number of particles in an individual rf burst prior to injection into the two cyclotrons.

The normal stacking method uses the large ratio of longitudinal phase space of the Cooler to the small longitudinal emittance of the cyclotron beam. Several beam bunches are injected onto the Cooler equilibrium orbit by firing two full-aperture kickers to perturb the equilibrium orbit as each bunch clears the injection septum. A debunching synchrotron rotation narrows the momentum spread and a slow deceleration moves the group of new bunches onto the top of the stack. The operation can be repeated about 40 times before the momentum aperture is filled.

The cyclotron can deliver ion beams with strippable electrons. The ring injection path allows for three different changes in magnetic rigidity: 2/3 (for $Li^{++} \to Li^{+++}$), 1/2 (for $He^+ \to He^{++}$, and for $H_2^+ \to 2H$), and -1 (for $H^- \to H$). The stacking procedure in this stripping mode is simply to add beam continuously until multiple passes through the stripping foil have filled both longitudinal and transverse acceptances. For protons, the limit is reached after about $2-3 \cdot 10^4$ turns. Rf buckets help reduce the rate of momentum growth. The stacked beam is moved away from the stripping foil before acceleration or cooling begins.

For polarized beams, which constitute a large fraction of normal cyclotron operation, the stripping methods would require a polarized H^- beam, which is not at present available. It appears possible, for some experiments which have slow beam consumption, to use the electron cooling in conjunction with longitudinal stacking to circumvent the Liouivillian limitations of the longitudinal stacking scheme. Cooling at the injection energy is used to contract the width of the stack at the same rate as the newly decelerated turns are added. A rate limit is set by the phase displacement acceleration during deceleration onto the stack which must be overcome by the cooling force. A filling rate of about 100 turns/second appears possible. This cooled stacking method requires minutes rather than

seconds to accumulate to the maximum beam intensity. When
acceleration is also used, the electron beam has to alternate between
initial and final beam velocities.

During startup, the ring can be operated without kickers in a
one-turn-and-out non-accumulating mode parasitic to ongoing
experiments. In this mode, simple destructive diagnostic devices such
as phosphor screens with TV viewers can be used for surveying,
steering and focussing studies. This mode should also be quite useful
for localizing hardware faults, such as a dropped bit in a DAC
control, because it restores the upsteam/downstream cause-and-effect
relationship which is obscured when the beam path is closed.

ACCELERATION

The rf waveform used for debunching during the stacking process
is provided by a ferrite-tuned cavity which can be ramped over a
frequency range to accelerate protons from 49 to 490 MeV. The magnets
in the ring can be ramped at a rate up to about 30% of full-scale
rigidity per second, limited by the power supply, which has an
overvoltage mode to provide the additional energy stored in the
dipole inductance during the up-ramp. The ramp time scale is similar
to the cooling time scale so that an experiment operating at an
energy different from the injection energy does not suffer much
reduction in macroscopic duty factor.

The rf energy gain per turn is modest because of the slow ramp
cycle. The rf amplitude is largely determined by the large phase
space volume of the stacked beam after injection.

The Cooler circumference, and the rf frequency, have to be
selected to accommodate the pulsed nature of the cyclotron beam, and
in our case the large number of cyclotron harmonic numbers ($3 \leq h \leq 8$
and $12 \leq h \leq 16$) which are used to cover the wide cyclotron beam
velocity range with a cyclotron rf system that tunes over a
relatively narrow range of $25.2 \leq f \leq 35.4$ MHz. We have chosen to fix
the Cooler circumference at 4.5 times the cyclotron extraction
circumference, and the normal Cooler rf harmonic to be 9. This means
that when the cyclotron is operated on $h = 4$ for protons from 120 to
200 MeV, with every other beam burst suppressed by use of an $f/2$
buncher, the beam bursts fall in adjacent Cooler rf buckets during
injection. The small accumulator ring circumference and the
splitting hardware permit two intense bunches in succession to reach
the Cooler, so the kickers fire four times to fill eight of the nine
buckets. The empty bucket is useful for the diagnostic pickups.

For other harmonics a variety of matching conditions of somewhat
less simplicity can be employed. For example, for the cyclotron $h = 5$
mode, the Cooler rf is locked to 2/5 of the cyclotron rf frequency
during injection and the bunching/chopping/splitting system delivers
beam from every fifth cyclotron rf burst into every second Cooler rf
bucket. Because the ring harmonic is odd, the ring buckets can all
be filled after 9 repetitions.

The refilling time of the low energy accumulator ring is of
order 10 µs so the narrow time structure of the incident beam would
be lost before all buckets were filled. A high frequency cavity
synchronized to the cyclotron rf, and with bucket shape matched to

the cyclotron burst, is excited to hold the longitudinal phase space shape of the beam bursts constant until the 8 or 9 buckets (at the acceleration frequency) are full. The debunching rotation takes about 100 to 150 μs with the rf amplitude changing times extended to satisfy the adiabatic condition that stack ripples lead to little dilution of the longitudinal phase space.

COOLING

The straight section for electron cooling has a clear length of 7.1 m, bounded by quadrupole quartets that give some degree of beam size control. Within this long straight section are the magnetic elements which create the guide field for the intense cold electron beam that performs the cooling of the stored ion beam.

Cooling is required for electron kinetic energies from 7.5 to 275 keV. Optimal performance is desired over the more limited range from 75 to 150 keV. The design goal for the electron transverse temperature is 0.2 eV. The regulation specification for the high voltage supply that largely determines the electron energy is about 10 ppm. The electron beam will be formed by an indirectly heated, dispenser-type, flat cathode of 25 mm diameter of commercial manufacture, with a nominal maximum current capability of 4 Amperes. Beam from the cathode is accelerated to 30-50 kV in a Pierce geometry in the cathode region into a hole in the gun anode. The entire gun region is immersed in a uniform magnetic field of strength up to 0.15 T. The cathode-to-anode space contains two tuning electrodes to obtain a low electron temperature over a range of cathode-to-anode voltages which covers the low velocity end of the Cooler operating range. The lenses also allow some variation of the perveance if desired.

The beam leaves the anode and enters an accelerating tube where the energy can be increased by 0-250 keV. The electric field gradient is non-uniform and is selected to give a low final temperature at all tube voltages. Resonant focussing at the foot of the tube may be employed to minimize the output temperature for each operating energy.

The electron beam next enters the first of two toroidal field elements that merges the electron trajectory with that of the ion beam at the upstream end of the cooling region and then separate the paths at the downstream end. The electron beam can be collected and most of its power recovered. The main part of the cooling region consists of a 3 m long solenoid, of maximum field up to 0.15 T., which will be carefully constructed and shimmed to give an accurately straight and uniform magnetic field so that the electrons can travel parallel to the ion beam over as long a distance as possible. The field must also be free of spatial Fourier components at the electron gyrowavelength to avoid resonant heating effects.

The electron guide fields have a non-negligible effect on the stored ion beam. The solenoid focussing can be compensated to a large extent by small changes in the adjacent lattice quadrupoles. In our Cooler, the strong coupling of the two transverse dimensions arising from rotation in the solenoid interior and in its end fields is compensated by strong short solenoids of opposite polarity located on

either end of the long straight, just beyond the region occupied by
the long solenoid and the toroids. The toroidal field which bends the
electrons through a 60° angle in the horizontal plane also bends the
ions by about 0.6° in the vertical plane. The magnetic return
structure of the compensating solenoids contains a set of steering
coils to give an opposite bend, while the solenoid axis is tilted
upward slightly to avoid the need for yet more steerers. Computer
simulations show that to obtain the minimum electron temperature over
a wide range of electron energies will require confinement magnetic
field strengths which do not scale directly with beam rigidity. To
cover a range of ion energies with a fixed tilt to the long solenoid,
there are small dipole steering components in the solenoids and
toroids to tilt the electron trajectory.

EXPERIMENTAL STATIONS

Two of the target stations are identical in their properties.
Each has a clear length of 6.5 m, and a dispersion of 4 m. The
nominal aperture functions are 0.3 to 0.5 m in the horizontal plane
and 0.7 to 1.0 m in the vertical plane. With the design acceptance of
25π mm-mr in both transverse planes and momentum acceptance of
±0.20%, these parameters translate to a beam spot "keepaway"
dimension (kept clear of material for long beam life and low beam
background) which is 2 cm wide and 1 cm high. The bright part of the
beam distribution, from which the bulk of the reaction events will
originate, is very much smaller. For an assumed emittance and energy
distribution as outlined below, the beam spot is about 0.5 mm in
diameter.

The third "T" target station has zero dispersion and a fixed
aperture function of 0.85 m in both planes. This target region has 3°
bends upstream and downstream of the target for access to reaction
products at 0° and 180°. The keepaway area for the tails of the beam
distribution is in this case a circle of 1 cm diameter. Beyond the
small dipole magnets, the "straight" section is clear of lattice
elements for a distance of 2 m in each direction. This station has
provision for long flight paths at small angles for neutron
time-of-flight. The paths extend into a field to the north of the
Cooler building to a distance of about 100 m. for a production angle
range $0° \leq \theta \leq 45°$.

The area around a target waist is kept clear of ring services
such as water pipes and cable trays on a circle of 2.5 m radius. For
one of the dispersed target straights, the clear area has been
enlarged (half circles of 7.5 m and 4.5 m respectively) , and the
floor specially reinforced to accommodate a pair of magnetic
spectrometers of total weight 200 tons.

TARGETRY

Despite the rather large electron beam current and the long
immersion length within the guide solenoid, the electron cooling
force is quite weak, and if it is used to add back the energy
transferred by the beam to electrons in the target, the force is only
strong enough to permit a target thickness of about 10^{16} or a few

times 10^{16} target electrons/cm^2. If an auxiliary conservative force is employed to add back the mean energy lost to electrons during each passage through the target, with the cooling force taking care of the fluctuations, the target thickness can be increased by a factor of about three, and the beam loss rate caused by successive delta ray events is reduced[7]. The auxiliary force could be, for example, the phase displacement acceleration[8] used to increase the energy of DC beams in the ISR.

Since the maximum allowable target thickness is less than that of self-supporting foils, a variety of other target technologies must be considered. It is possible to show that stationary targets intercepting a small portion of the beam cross section will not work[9], but by sweeping the beam across the target (or vice versa) the average thickness seen by a beam particle in one cooling time can be made arbitrarily low. The duty factor is reduced in proportion to the ratio of average thickness to maximum thickness, so this method is not appropriate to all experiments.

Gas and vapor targets either in the form of effusion through slits, or as jets, or as windowless bottles cooled to just above the condensation temperature, may be the most commonly used target. These techniques have in common the need for a variety of pumping arrangements and baffles which will be specific to the target material and the detector geometry. Polarized atomic beam targets may be thought of as special cases of the gas target, although present-day intensity limitations would seem to make these lower density targets easier to deal with from the viewpoint of maintaining an adequately low ring vacuum, than would be the case for denser jets.

The possibility of using dust targets is being considered. Such a target may be viewed as intermediate between the wiggling fibre and the cluster jet. A cloud or stream of sub-micrometer diameter particles of the target material can be adjusted to satisfy the average thickness condition while still giving a good duty factor. Unlike the clusters in a cluster jet, these larger particles can be of materials which do not add to the pumping load. There are a number of interesting electrostatic problems to be dealt with in the practical design of a dust particle target. Some of these have been encountered in attempts to make dust strippers and micrometeorite simulators. A stream of small droplets formed by the technology used in ink jet printers may also be usable if the droplets can be made small enough.

MONITORING

The experimenter must know the luminosity, which is the product of target thickness and beam current, in order to extract a cross section from a measured event rate. Small angle elastic scattering can be used to normalize to Rutherford scattering, but is very sensitive to beam alignment and to detector geometry.

Another luminosity monitoring method which has been considered is the detection of forward neutral production through electron pickup. This allows a normalization to a fairly well-known and smoothly varying atomic cross section[10]. This method is not possible for the lightest targets and highest proton energies because of the low event rate. To detect a neutral beam at zero degrees downstream of each target station requires clear access via a port in the vacuum chamber of the corner dipole.

The electron pickup process sets an effective lower bound to the energy range and upper bound to the beam atomic charge range in which cooling rings can be used for nuclear physics. The slowness of the cooling process requires high efficiency of beam use. Atomic loss cross sections larger than nuclear total reaction cross sections by more than about three orders of magnitude will lead to unacceptably low nuclear event rates.

It may be possible to use X-ray emission following electron knockout from the target atoms as a luminosity monitor. The detector geometry would have to be calibrated against known cross sections.

A neutral detector downstream of the cooling straight is a valuable diagnostic for the cooling process. Capture by beam particles of cooling electrons is maximized under the same conditions of low relative velocity that give rise to the best cooling rate.

POLARIZED BEAM OPERATION

The Cooler ring is designed to store a hydrogen beam with a polarization axis normal to the bend plane. The compensating solenoids described above serve the second purpose of returning the polarization direction of a stored polarized beam to the vertical after the beam passes through the electron guide field. The variable ring tune allows depolarizing resonances to be avoided at any fixed energy. For the acceleration mode, a proton beam has to be taken through one depolarizing resonance between 250 and 300 MeV. The methods of passing through such resonances have been tested in the weak-focussing ZGS[11] and in the strong-focussing Saturne ring[12].

It may be possible to obtain longitudinal polarization in the target straight section opposite to the cooling region by a "Siberian snake" which makes use of the cooling and compensation solenoids operated so as to aid in spin rotation rather than cancel. For the installed hardware the solenoid strength limits this mode to a feasibility demonstration with low energy protons. The compensating solenoids could in future be replaced by stronger superconducting elements to cover the full proton energy range. The strong coupling between transverse optical planes which is caused by such a strong solenoidal field may not be amenable to compensation, however, so this future option remains speculative before further study.

BEAM PROPERTIES

The beam properties at each target location are determined by the tune of the lattice, the operation of the electron cooling system, by the interaction of the beam with the target material, and by the interactions within the beam itself. The scattering of beam particles from one another while confined in a strong-focussing lattice, known in the literature as intrabeam scattering, can lead to a heating of all beam dimensions. The heating rate depends on the beam phase space brightness. The electron cooling can counteract the heating by intrabeam scattering, leading to an equilibrium emittance and energy spread. The presence of an internal target increases the equilibrium values.

Estimation of the target-out intrabeam scattering-cooling equilibrium for the IUCF Cooler indicate[2] that an emittance of order 0.1π millimeter-milliradians can be achieved at intensities of order 10^{16} s^{-1} for proton energies above 100 MeV. It is not clear whether the conventional intrabeam scattering theory, which is based on two-particle collisions, is reliable at the high beam brightness associated with this prediction.

The equilibrium in the third phase space dimension is more difficult to describe. The longitudinal momentum distribution for a beam in the presence of a target but in the absence of rf is expected to exhibit two distinct components[9]. Delta electron knock-on events and intrabeam scattering involving the tails of the transverse distribution may be expected to generate a low momentum tail on the stored beam distribution. This low momentum tail may help stabilize the beam against rapid growth of the microwave instability at the higher stored currents.

The second component is a narrow spike whose width for low beam emittance is set by the longitudinal effective temperature, which may be limited by the stability of the electron beam high voltage power supply. The width of this narrow spike may be as small as a few keV. For larger emittance, the spike is broadened by averaging over different electron energies arising from the space charge depression in the intense electron beam. The spike width is predicted to be independent of target thickness although the fraction of the beam contained in the spike is proportional to the target thickness.

As the target thickness is increased, a larger proportion of the stored beam moves from the narrow spike to the broad tail. Particles well out into the tail can be lost by a delta electron event moving them beyond the momentum acceptance limit of the ring lattice. The reduced lifetime as more particles move into the tail sets a practical limit to the internal target thickness of order 10^{16} electrons/cm^{-2}.

BUILDING ADDITION

The Cooler will be housed in an addition to the present IUCF building. The addition consists of a main floor of dimension 31 by 51 meters and a second floor of 10 by 50 meters covering the eastern third of the main floor. The western two-thirds has access from above for a 30 ton crane. The crane rails extend through a large door into a storage yard to the south so the crane can be used outdoors for unloading and storage of heavy equipment and movable shielding. The second floor will be used for power supplies and controls and also for data acquisition and the experimenter's computers.

The building design was fixed in the summer of 1983, with the formal ground-breaking ceremony in October of that year. The construction work was nearing completion at the time of this workshop one year later, with occupancy anticipated by early 1985.

The building is equipped with an oversized evaporative cooling tower, which is designed for close regulation of the supply water temperature, and which will take over cooling of all accelerators and research equipment as well as the Cooler ring. The older tower will continue to serve the building air conditioning equipment in the summer months.

STATUS SUMMARY

The IUCF Cooler is approaching the midpoint of its construction phase. Installation of ring components within the building addition will begin in the spring of 1985 and is expected to continue for about 20 months. The electron cooling system will be assembled close to its final position in the ring and will be tested in parallel with the uncooled operational tests of the ring in stacking, storage and acceleration modes. During 1987 we hope to be ready to begin an experimental program with stored, cooled beams and internal targets. Our Program Advisory Committee has already begun to review proposals for Cooler beam time. This process will guide priority decisions during construction, for example in the choice of the order of implementation of components specific to particular experiments.

Meanwhile the work to increase our understanding of factors, both practical and fundamental, which will limit the ultimate performance of the Cooler as a research tool are continuing. Even as construction goes forward, we find new physics arguments that the design can be further improved with yet more modifications. It is clear that a balance has to be struck between flexibility and complexity, and between our aspirations for the best possible device, tempered with the realization that each step in construction restricts our choices markedly.

REFERENCES

1. R.E. Pollock, IEEE Trans. Nucl. Sci. NS-30, 2089 (1983).
2. R.E. Pollock, to be published, Proceedings of 1984 Karlsruhe "E-COOL 84" Workshop.
3. S. Ohnuma, Workshop on Intersections Between Particle and Nuclear Physics (Steamboat Springs, 1984), A.I.P. Conf Proc. 123, 415 (1984).
4. Y. Hirao, Proc. Ninth International Conf. on Cyclotrons and their Applications, Caen, France 1981, G. Gendreau, ed., p. 477.
5. W. Yoho, Proc. Tenth International Conf. on Cyclotrons and their Applications, East Lansing, Michigan 1984, M. Mallory, ed., p. 611; R. Laxdal et al., IEEE Trans. Nucl. Sci. NS-30, 2013 (1983).
6. D.L. Friesel, op. cit., p. 207.
7. H.-O. Meyer, to be published, Nucl. Instr. & Methods, (1985).
8. E. Ciapala, S. Myers, & C. Wyss, IEEE Trans. Nucl. Sci. NS-24, 1431 (1977).
9. H.-O. Meyer, to be published in Workshop on Nuclear Physics with Stored Cooled Beams, A.I.P. Conf. Proc. (1985).
10. A.S. Schlachter et al., Phys. Rev. A 27, 3372 (1983).
11. T. Khoe et al., Particle Accelerators 6, 213 (1975).
12. Nouvelles de Saturne (newsletter), N° 7, 6 (1982).

SESSION B

HIGH-RESOLUTION STUDIES

NUCLEAR STRUCTURE PHYSICS AT HIGH RESOLUTION[*]

J. Heisenberg
University of New Hampshire, Durham, NH 03824

The topic of this talk has two elements (a) what is the nuclear structure we are after, and (b) how does high resolution help in this effort. The second question can be answered easily in a general sense. There is a lot of nuclear structure physics that we can learn even with the modest resolutions presently available; and in the discussion of the nuclear structure, I will give some examples from experiments that were done with modest resolution. The improvement in resolution will do two things (a) It will allow us to extend experiments into regions with much higher level density which means higher excitation energy, and (b) it will allow us to measure cross sections of much weaker states. There is inevitably some background in the measured spectra and the signal/background ratio is just inversely proportional to the resolution. Fig. 1 shows an ee' spectrum taken from ^{142}Ce with 15 keV resolution at 370 MeV incident energy. Here the major background is the radiation tail from mostly the elastic scattering.

If this high resolution would lead only to an increased number of cases that could be measured, it might be questionable whether such significant effort is warranted to achieve this resolution.

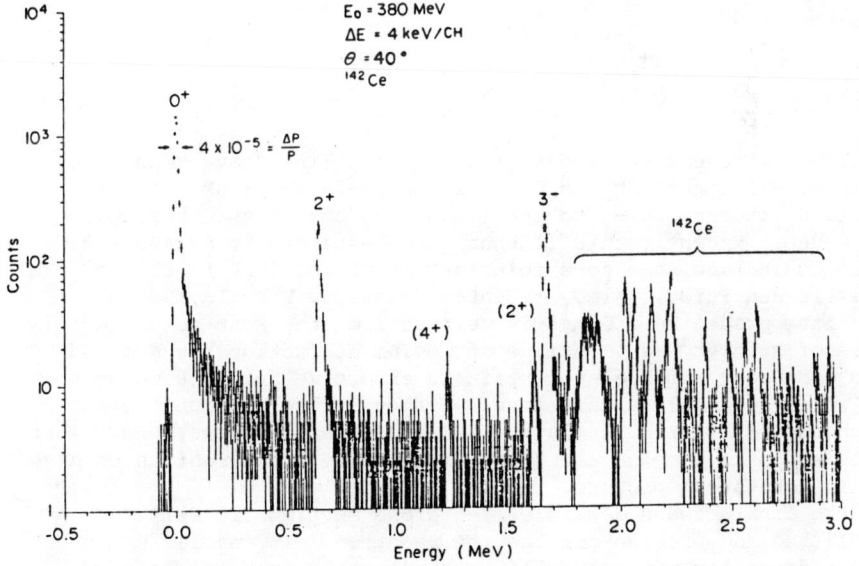

Fig.1 Spectrum of scattered electrons from a target of ^{142}Ce.

[*] Work supported by the US-DOE Contract DE-AC02-79ER 10338

Thus I will try to argue that there will be not only an increased number of cases but that it will give us access to levels that supply significantly different information than can be obtained from the low lying collective levels.

Turning now to nuclear structure, Fig. 2 gives a simplified diagram of what I would consider an understanding of nuclear structure:

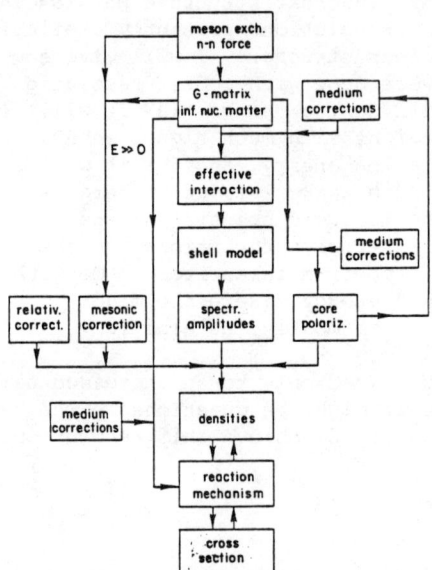

Fig.2: Schematic diagram for nuclear structure calculations

In this scheme we find that almost all links have been worked on: e.g. Wildenthal[1] has studied extensively the link from an effective interaction -- to shell model -- and to spectroscopic amplitudes. Recently this link has been expanded by Sagawa and Brown[2] to include some core polarization corrections to obtain more realistic densities. Also, recently Nakayama, Krewald, Speth and Love[3] have presented a G-matrix derived from the Bonn-potential. A series of studies has been made on medium corrections by Kuo and Brown[4], McGrory and Kuo[5] to obtain an effective interaction to be used in shell model calculations. I should also mention the work of Dubach and Friar[6] on relativistic and mesonic corrections. What is needed is to combine all these aspects under one roof as to give a fully consistent picture.

The link between densities and cross sections is well established and precise for the ee' reaction. It enables us to present densities with uncertainties extracted from ee' data. Considerable progress has been made in the study of the reaction mechanism for pp' so that a fairly quantitative description is now possible.

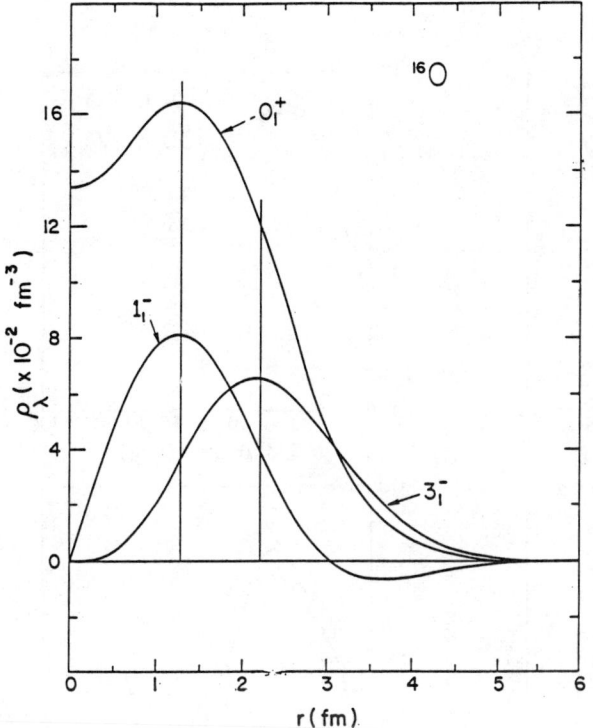

Fig.3: Charge densities extracted from (e,e') experiments.

Fig. 3 shows densities extracted from electron scattering on ^{16}O.[7] Since these are dominently isoscalar transitions, the assumption that neutron densities equal the proton densities is a very accurate theoretical prediction of neutron densities. Knowing the densities and measuring the cross sections lets us study the reaction mechanism. Fig. 4 shows the cross sections calculated with a phenomenological interaction. We cannot expect that the G-matrix interaction will be an accurate representation of the actual interaction. Since there are always medium corrections to an effective interaction, we can use the theoretical G-matrix interaction only as a guide to the important features of the effective interaction. The needed adjustments to the interaction are some measure of the needed medium corrections. To establish the effective interaction not only in the natural parity isoscalar channel we have to be able to resolve states that are non-collective and usually found at higher excitation energy.

Let me now turn to densities measured via (ee'). Fig. 5 shows the cross sections measured in forward direction from the first two 2^+ levels in ^{88}Sr.[8] We see that these cross sections have quite

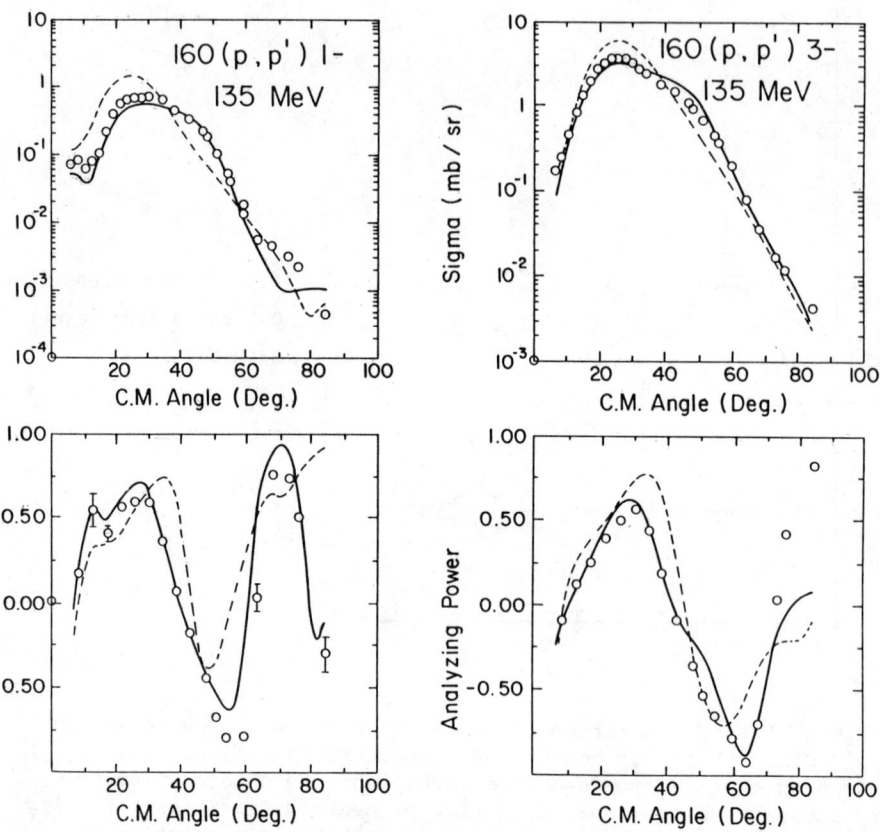

Fig.4: Cross sections and analyzing power for the pp' experiment on ^{16}O.

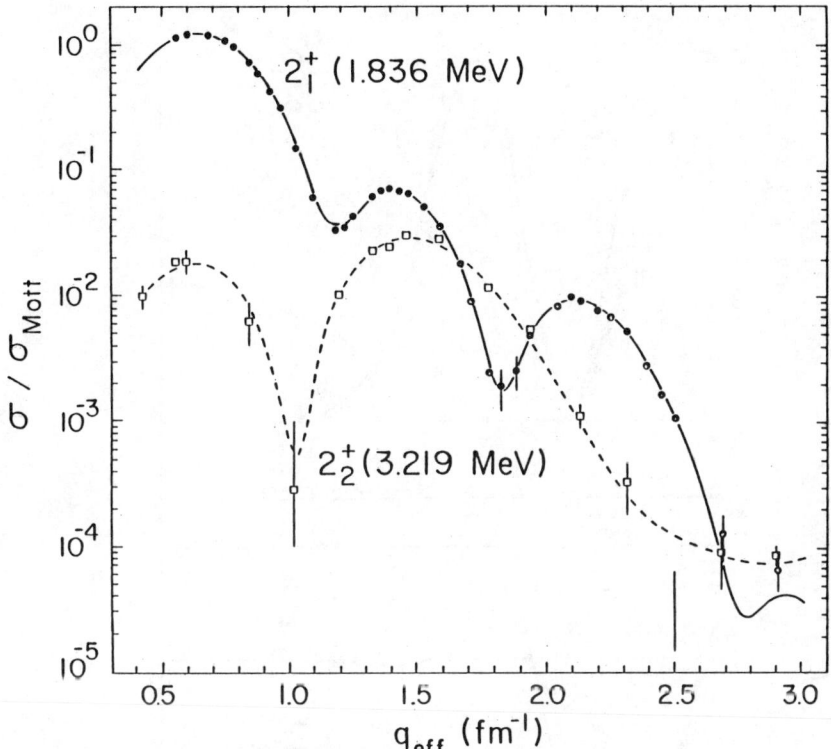

Fig.5: Cross sections for inelastic electron scattering from the first two 2^+ levels in ^{88}Sr.

different q-dependence, indicating the sensitivity of such data to the details of the nuclear structure. Fig. 6 contains the densities extracted from these cross sections. They indicate the high precision now available from ee' data, and we have to ask what physics we learn from the measurement of such densities.

The most simple case appears to be when we look at single particle densities first. The neighboring nucleus ^{89}Y displays such a single particle structure. Fig. 7 shows the simplified shell structure of ^{89}Y. The neutrons with N = 50 form an inert core. The single hole in the $2p_{1/2}$ orbit allows for the proton particle transitions from $2p_{3/2} \to 2p_{1/2}$ and $1f_{5/2} \to 2p_{1/2}$ giving rise to low lying $3/2^-$ and $5/2^-$ states. Since this might be an oversimplification, we have carried out shell model calculations in a space of 4 proton orbits of $1f_{5/2}$, $2p_{3/2}$, $2p_{1/2}$ and $1g_{9/2}$. Table 1 shows the results from such calculations using the adjusted interaction from Gazzaly et al.[9] and our own adjustments.[10]

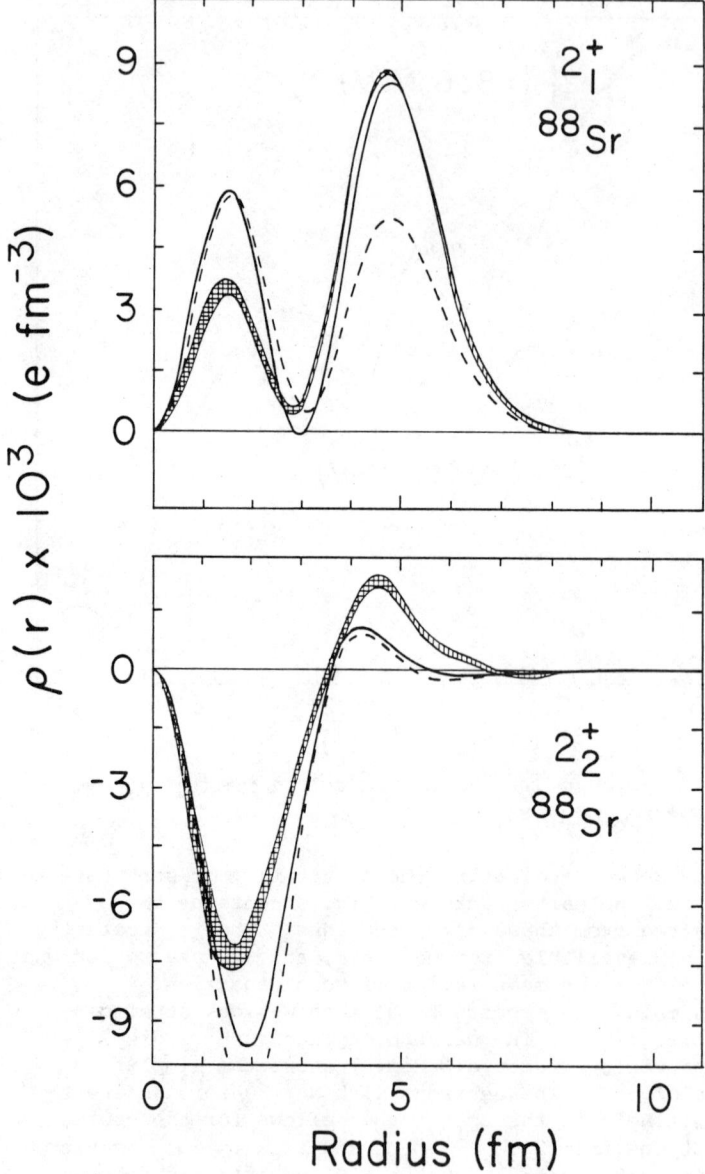

Fig.6: Transition charge densities for the first two 2^+ levels in ^{88}Sr.

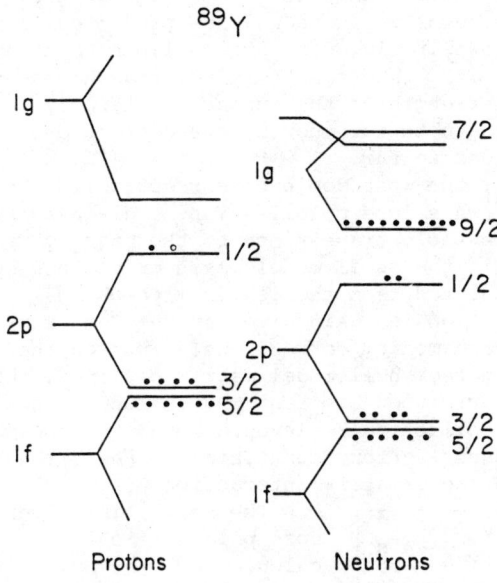

Fig.7: Shell structure for ^{89}Y.

Table 1: Spectroscopic amplitudes for the two lowest E2 transitions in ^{89}Y.

transition	$3/2^-$		$5/2^-$	
	(a)	(b)	(a)	(b)
$2p_{1/2} \to 2p_{3/2}$	0.08161	0.09841	0.00439	0.00359
$2p_{1/2} \to 1f_{5/2}$	0.00351	0.00399	-0.04956	-0.09868
$2p_{3/2} \to 2p_{1/2}$	0.76846	0.57032	-0.00170	-0.00146
$2p_{3/2} \to 2p_{3/2}$	0.00131	-0.01480	0.00103	-0.01334
$2p_{3/2} \to 1f_{5/2}$	0.00402	-0.02176	0.00552	-0.00801
$1f_{5/2} \to 2p_{1/2}$	0.00120	-0.00183	0.75136	0.74228
$1f_{5/2} \to 2p_{3/2}$	0.00508	0.00494	-0.00119	-0.00168
$1f_{5/2} \to 1f_{5/2}$	-0.00113	-0.01471	0.00111	-0.00523
$1g_{9/2} \to 1g_{9/2}$	-0.02672	0.03695	0.03472	0.06168

As can be seen, both calculations verify the dominance of the single particle aspect of these transitions, even though the absolute numbers differ somewhat. The densities predicted from these calculations are shown in Fig. 8 as broken lines together with the experimental results. Whereas the predicted shapes are quite similar, the total strength at the surface is largely underpredicted. Thus we have to turn now to the corrections to these densities not included so far.

For charge densities, the most dominant correction is the core polarization correction. This core polarization is given as the coupling of the valence particle transitions to the high lying ph transitions not included in the shell model basis or valence space. This coupling can be calculated from the residual ph-ph interaction. This interaction is again given by the G-matrix interaction modified by some medium corrections. Through the core polarization correction in the shell model matrix elements, the strength of this medium correction has a strong influence on the calculated energy of the collective 3^- level. Thus, the known energy of the 3^- level poses a strong constraint on the strength of the medium corrections in the G-matrix interaction.

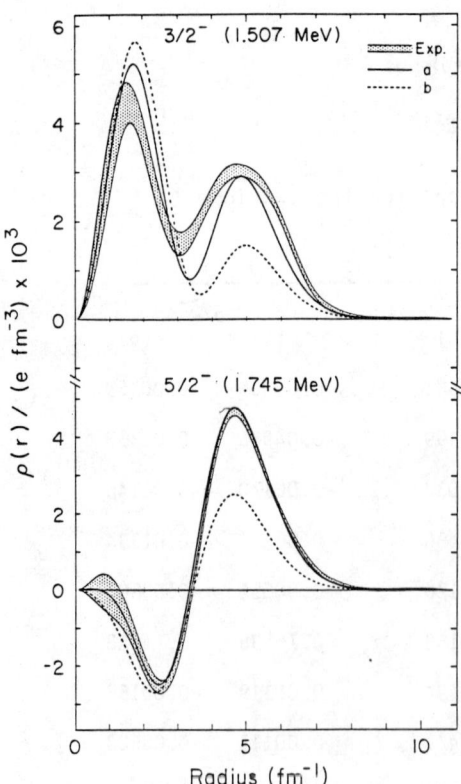

Fig.8: Transition charge densities for the first two E2 transitions in ^{89}Y.

The solid line gives the core polarization calculation. We have not yet carried out the calculation with the G-matrix interaction. Instead, we have used a phenomenological interaction from the Juelich group[11] given by a π- and ρ- exchange interaction in addition to a density dependent zero range interaction. This density dependent interaction was adjusted to reproduce the collective 3^- energy, thereby avoiding the ambiguity in strength of phenomenological interactions.

In cases where the dominant configurations are neutron configurations, the measured transition density is purely core polarization. Fig. 9 shows the measured transition charge densities for the lowest 2^+ excitation in the nuclei ^{208}Pb, ^{206}Pb, and ^{204}Pb.[12] ^{208}Pb is a doubly closed shell nucleus, and the lowest 2^+ state is

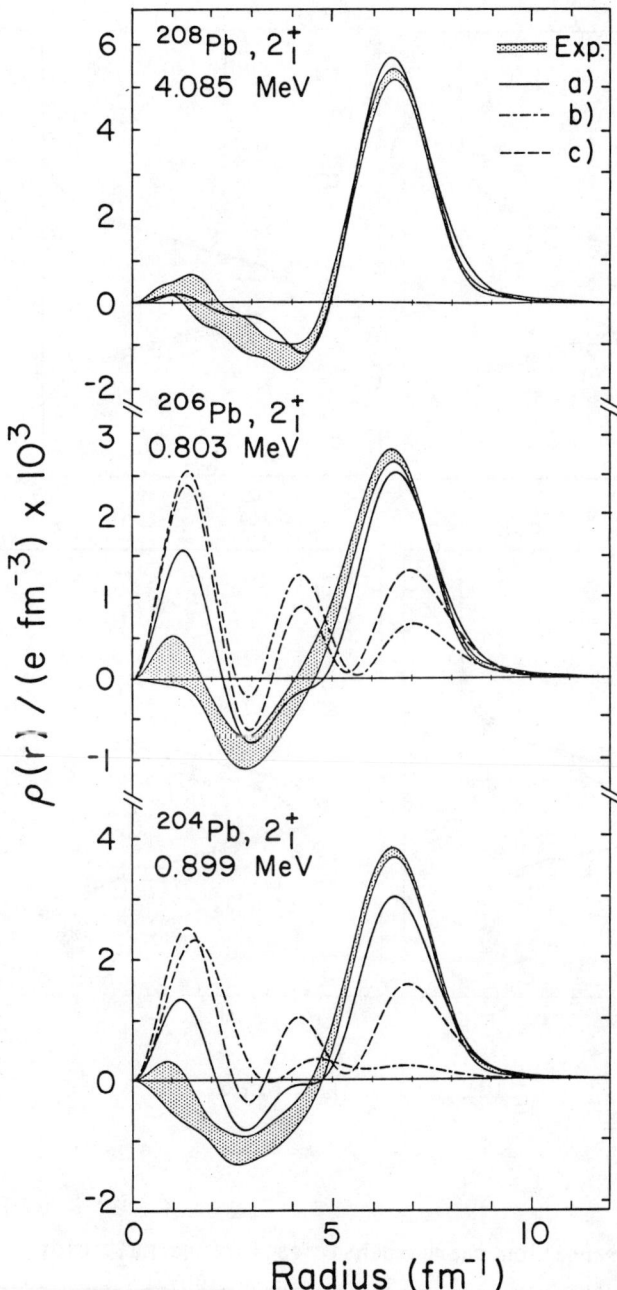

Fig.9: Transition charge densities from the first 2^+ levels in the even Pb-isotopes.

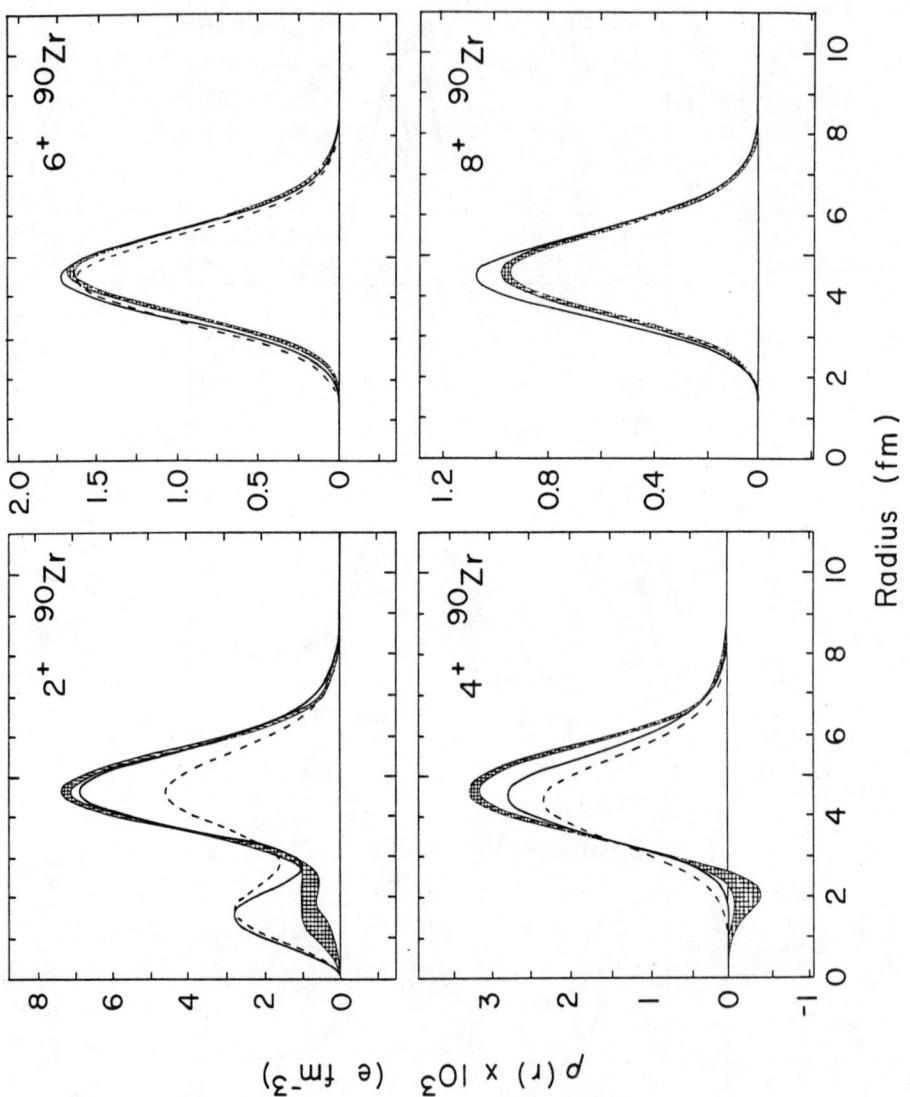

Fig.10: Transition charge densities for the multiplet formed by the $(1g_{9/2})^2$ configuration in ^{90}Zr.

described as a quadrupole phonon, a quadrupole vibration which is well described in RPA. For ^{206}Pb or ^{204}Pb the lowest states are well understood in terms of pairing effects between the neutron hole pairs. Thus the dominant configurations are neutrons. The neutron densities are shown in Fig. 9 as the broken line. The measured charge density shows no resemblance with the calculated neutron densities, instead it is, in shape, very similar to the transition density for the 2^+ phonon in ^{208}Pb. This is also reproduced by the core polarization calculations. It shows that one cannot account for core polarization by giving neutrons an effective charge, but that we can do reliable calculations which produce usually surface peaked transition densities.

There are strong multipolarity dependent effects in the core polarization corrections. Fig. 10 shows the densities extracted for the sequence of levels 2^+, 4^+, 6^+, and 8^+ in ^{90}Zr[13], which all originate from the recoupling of a proton pair in the $g_{9/2}$ orbit coupled to 0^+ in the ground state. Shown are also the calculations with (solid line) and without (broken line) core polarization. Whereas the effect of core polarization is such as to double the transition density at the surface and thereby to increase the strength by a factor of four, with increasing multipolarity the core polarization contributions become smaller and smaller and are negligible for the 8^+ transition. Fig. 11 shows the experimental results from the equivalent 8^+ transition in ^{92}Mo.[14] It shows that high resolution makes these measurements possible. Also shown are the data as a function of the momentum transfer and the best fit using a $g_{9/2}$ Woods Saxon wave function. These data do not prove that core polarization contributions are negligible; this statement comes from the theoretical calculations. The data supply a radial size for the Woods Saxon well and a spectroscopic amplitude. For ^{90}Zr this spectroscopic amplitude is consistent with calculated results for a shell model that also reproduces many other experimental quantities that are strongly correlated to the spectroscopic amplitude such as the $2p_{1/2} \rightarrow 1g_{9/2}$ E5 transition etc. Thus these data seem to be quite consistent with the absense of core polarization.

In spite of this negligible contribution to the charge density, there is significant contribution to the neutron density. The proton scattering experiment of Gazzaly et al.[15] has been analyzed using the transition charge density as established from electron scattering. These experiments allow to determine the ratio of neutron/proton transition matrix elements. The results are given in Table II and are compared to the predictions of the core polarization calculations. It should be noted that there is excellent agreement between the prediction and the experiment

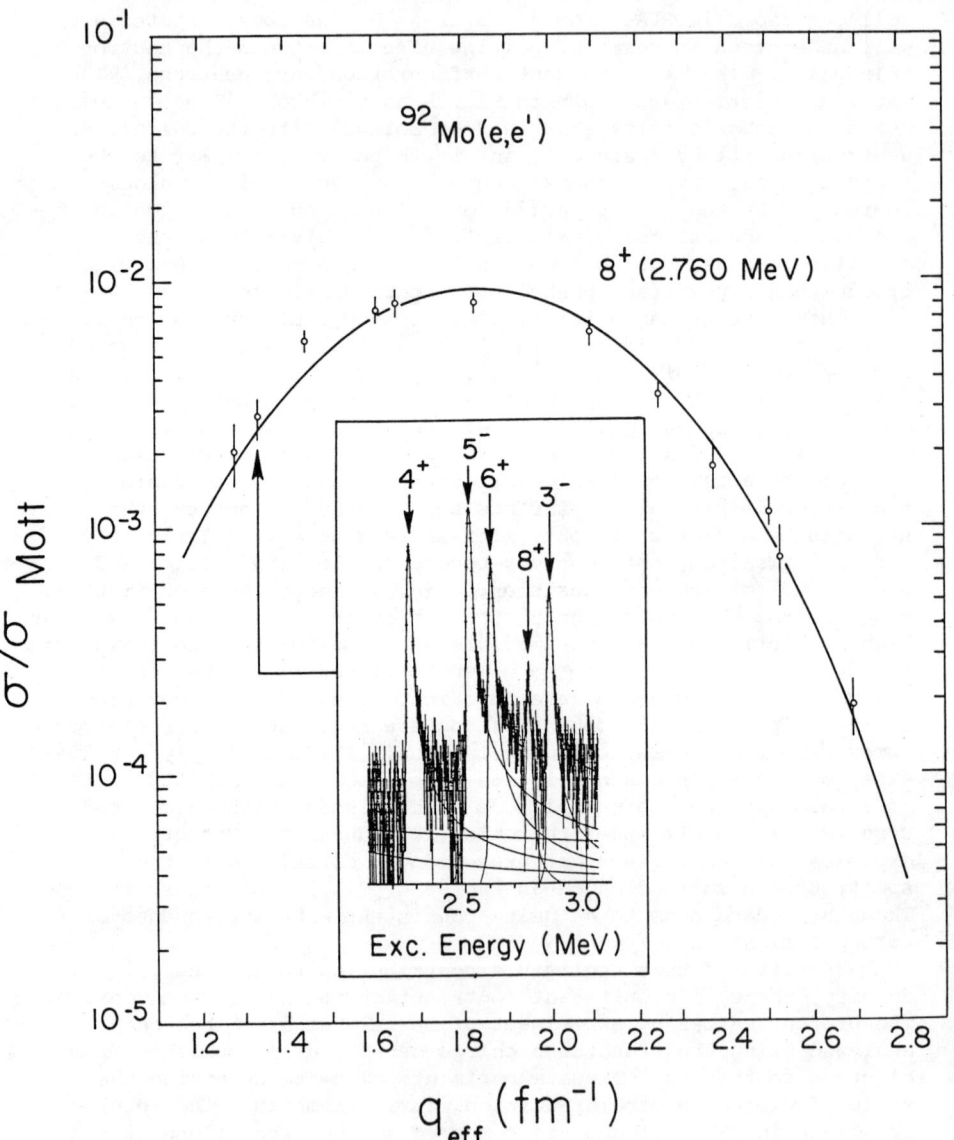

Fig.11: Cross sections for the 8+ level in ^{92}Mo. The insert shows a spectrum from the (e,e') experiment from ^{92}Mo.

Table II. Neutron/proton transition matrix element

level	M_n/M_p (exp)	calculation
2^+ (2.186 MeV)	1.12 +/- 0.04	1.04
3^- (2.748 MeV)	1.06 +/- 0.05	0.95
4^+ (3.076 MeV)	0.78 +/- 0.12	0.93
2^+ (3.307 MeV)	1.93 +/- 0.16	1.14
6^+ (3.448 MeV)	0.63	0.82
8^+ (3.590 MeV)	0.56	0.55

indicating the quality of the calculation. The discrepancy in the second 2^+ level is mostly due to the incorrect shape used in the analysis of the proton scattering data. A more recent preliminary analysis of proton scattering data from Indiana by J. Kelly[16] gives quite a different ratio.

The 8^+ level in ^{86}Sr is similar in nature. It arises from recoupling of a neutron $g_{9/2}$ hole pair. The cross section measured in electron scattering arises only from the core polarization charge contribution induced in this dominantly neutron transition. Fig. 12 shows a spectrum of scattered electrons from ^{86}Sr at a momentum transfer where the 8^+ form factor has a maximum.[17] There is no peak visible at the energy of the 8^+ level, and preliminary fits show that the cross section for exciting this 8^+ level is about a factor 50 smaller than the corresponding cross section in ^{92}Mo. Thus we find again that there is no charge core polarization in this 8^+ transition. Nothing is known yet about the neutron density and the neutron core polarization in this dominant neutron transition. Table III summarizes these results on core polarization. It shows that the dominant factor in core polarization is the nuclear ground state, and one may not assume that contributions from neutrons equal those of protons.

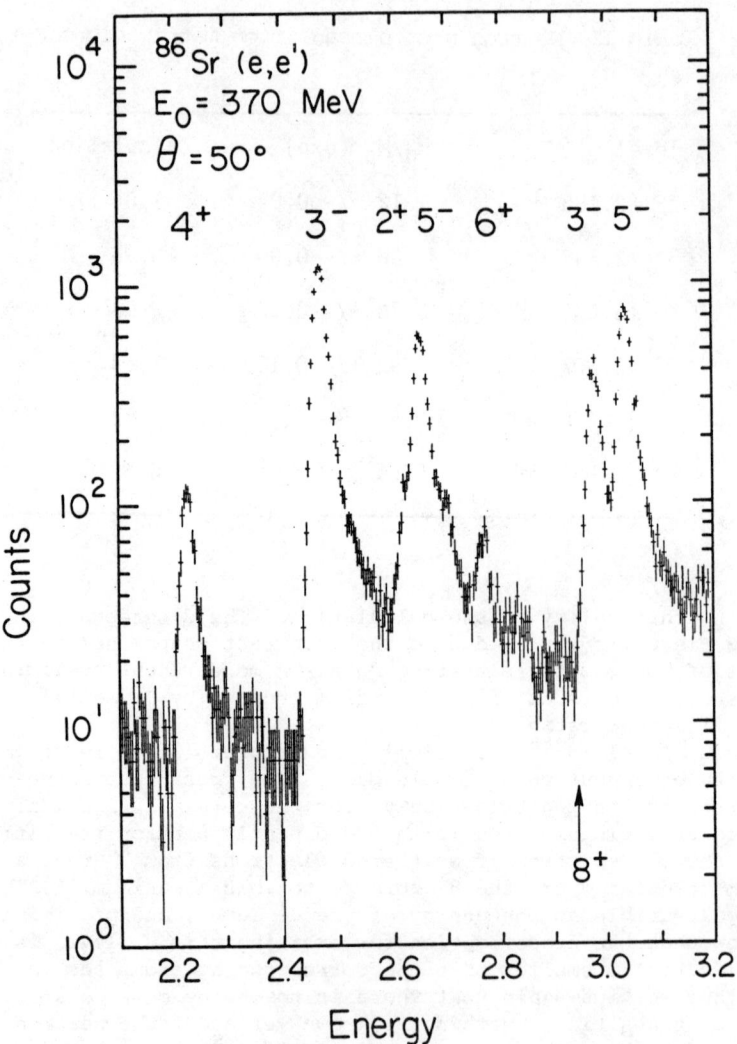

Fig.12: Spectrum of scattered electrons from ^{86}Sr.

Table III. Core polarization

dominant configuration	core polarization proton	neutron
proton $(g_{9/2})^2{}_{8^+}$	0	0.6
neutron $(g_{9/2})^2{}_{8^+}$	0	?
proton $(g_{9/2})^2{}_{2^+}$	1	2.2

An interesting question is whether we can learn more from proton scattering than just the transition matrix element. The comparisons on ^{16}O proton scattering results shown above demonstrate that the reaction mechanism is reasonably well understood. Thus we have to ask whether proton scattering data are sensitive to the nuclear interior. Fig. 13 shows the proton scattering cross sections from Hersman et al.[18] obtained on the same two 2^+ excitations in ^{88}Sr for which the electron scattering data were shown in Fig. 5. The asymmetries also differ, but are not shown here. Again, we find that the cross sections for these

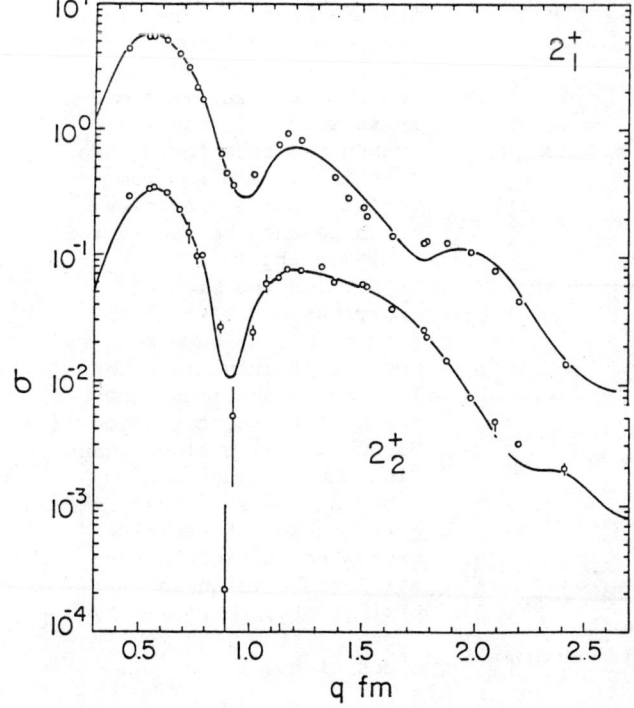

Fig.13: (p,p') Cross sections for ^{88}Sr.

two levels do not scale and thus are sensitive to differences in the nuclear structure. A preliminary fit to the neutron density[19] is shown in Fig. 14. Even though the fit starts with the assumption that neutron densities equal the proton densities, the fit requires a neutron density that differs from the proton density and is much more like the theoretically expected density. The apparent discrepancy between the fit and the calculation cannot yet be taken seriously. It seems that the strength of the empirical proton nucleus interaction is not yet right. Thus we can only conclude that there is considerable sensitivity to the interior of the nucleus, but the reaction mechanism is not yet well enough established to allow reliable extractions of neutron densities. I believe, though, this is only a matter of time.

In the last examples I have already shown the usefulness of knowing more than one density for the same transition. In the following, I will discuss some examples where two separate densities, the transition charge density and the transition current density have been measured via electron scattering. In electron scattering we have two form factors in natural parity transitions: the longitudinal form factor which dominates forward scattering and the transverse form factor that dominates backward scattering. Correspondingly one can extract two densities, the transition charge density and the transition current density.[20]

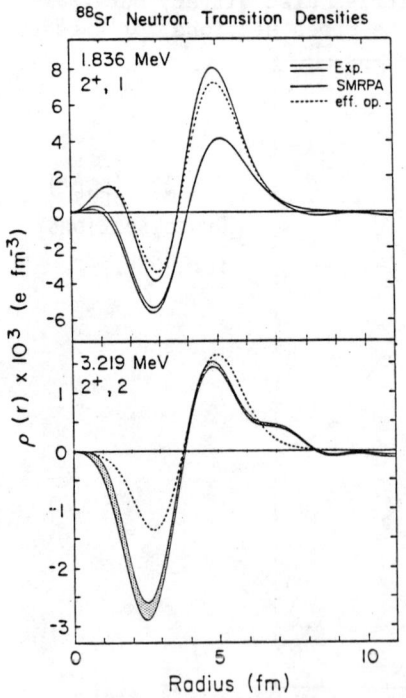

Fig.14: Neutron transition densities

We have analyzed data on ^{89}Y to extract current densities on those transitions that are single particle transition shown in Fig. 8.[21] Since ^{89}Y is an odd-even nucleus with ground state spin 1/2, the transverse form factor has contributions also from magnetic multipolarities which have to be subtracted out. In the 1.51 MeV transition the M1 contribution is small and does not contribute much to the uncertainty. For the 1.75 MeV transition, however, the magnetic contribution is M3 and is quite significant. Since this background subtraction is not yet fully established, the results presented in Fig. 15 have still to be taken as preliminary, particularly those for the 1.75 MeV transition.

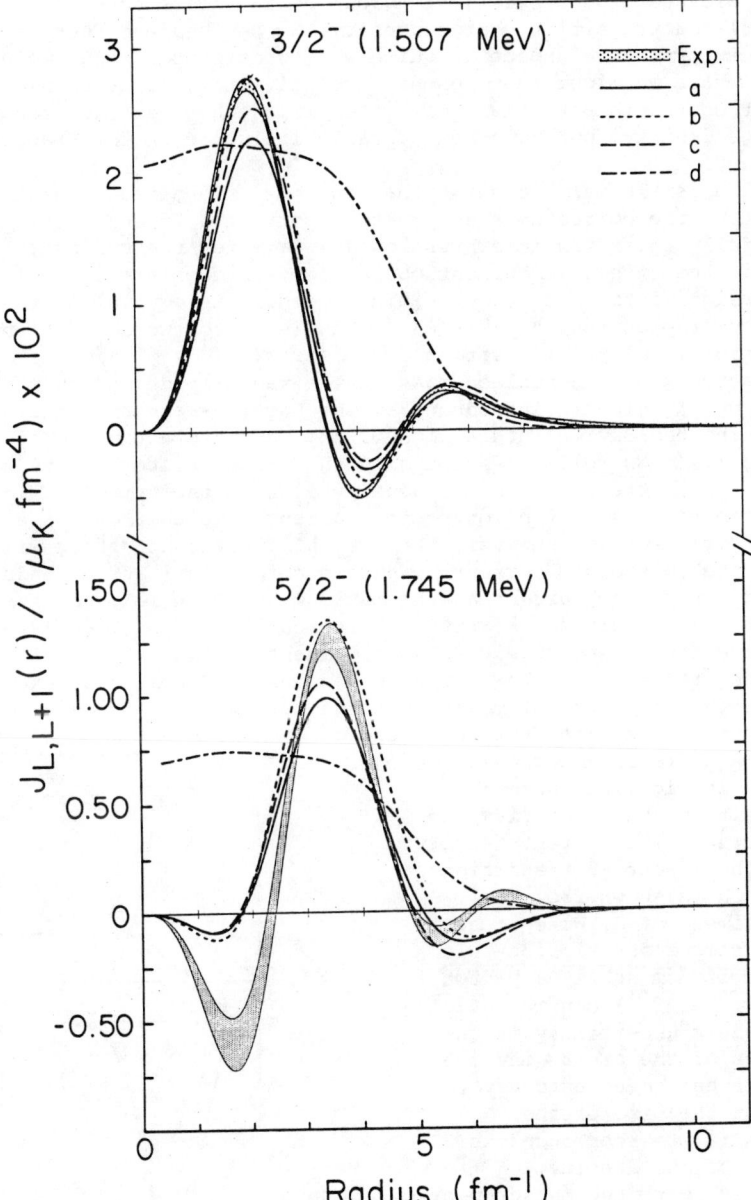

Fig.15: Transition current densities for the two lowest E2 transitions in ^{89}Y.

Because of the excellent agreement in the measured and calculated transition charge density, we can believe that the nuclear structure aspect is quite well understood. Thus we have a case where we might test mesonic, relativistic, or other higher order corrections that are of little importance for the transition charge density, but have considerable influence on the transition current.

In particular, it seems that in these transitions some relativistic corrections are quite important. If one starts from the fully relativistic expression $J = \bar{\psi}\gamma_\mu\psi$ for the current, one can obtain the effective current operator via a complete non-relativistic reduction. This procedure is equivalent to minimum replacement $\vec{\nabla} \rightarrow \vec{\nabla} - e\vec{A}$. This leads to two corrections: an additional spin orbit current $\vec{J}_{SO} = e \cdot r \cdot W(r)(\vec{\sigma} \times \hat{r})$ and the replacement of the nucleon mass by the radially dependent effective mass $m^*(r)$ which is introduced in the Hartree-Fock treatment such as using the Skyrme interaction or the DME interaction. This effective mass follows essentially the ground state density. For that reason the ground state density is also shown in Fig. 15. The contributions from the convection current in either of the cases are fairly small. However, there is a significant effect arising from the spin-orbit current. Whereas core polarization leads to a reduction of the current at the maximum by about 10%, the spin orbit current gives a significant increase of about 20-30% to improve the agreement with the experiment in the $3/2^-$ level. It may be premature to draw definite conclusions from these examples since the influence of meson exchange corrections or center of mass corrections have not yet been studied.

A different example is shown in Fig. 16. Here the transition charge density and the transition current density for the lowest E5 transition in ^{48}Ca is shown from the experiment of J. Wise et al.[22] In a simple description this state is dominantly a proton $(1f_{7/2}, 1d_{3/2}^{-1})$ configuration. One basic uncertainty is the radius of the orbits involved, and it has been customary to adjust the size of the potential to reproduce the shape of the density. However, with the measurement of two densities a problem shows up. Whereas the charge density requires an increase in the well to reproduce the experiment, the current density requires a reduction

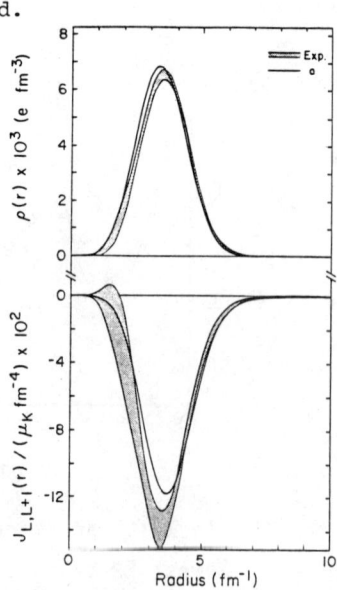

Fig.16: Transition charge and current density for the first E5 transition in ^{48}Ca.

of the well to get agreement. This apparent inconsistency is an
indication that significant effects have not yet been included in
the theoretical calculation. This is a case where the convection
current gives a significant contribution to the full current
measured. Thus we are sensitive to the effective mass correction.
This correction helps with the discrepancy, but it is not enough to
resolve it. Thus, it seems that there are significant
contributions from core polarization that are not included in this
calculation. Core polarization has been calculated without any
density dependence in the $\sigma\sigma'\tau\tau'$ channel of the interaction. The
recent G-matrix interaction of Nakayama et al.[3] does predict some
density dependence in this channel. Thus this discrepancy may be
taken as some evidence for such a density dependence. These
examples show the significant increase in information that a second
transition density can add.

If we look into the future, we can see results that again will
influence our understanding of nuclear structure. There has been a
recent paper by Pandharipande et al.[23] again emphasizing the role
of occupation probabilities, and presenting a calculation for
^{208}Pb. Their result is shown in Fig. 17. High resolution proton
or electron scattering can again be employed to verify features of this
picture. As mentioned before, the excitation of the 8^+ levels in the
Zr-region proceeds only via the recoupling of a $g_{9/2}$ proton pair
from the ground state. Thus the measured spectroscopic amplitude
sets a lower limit of the occupation probability of that orbit. One can
be more precise by using shell model calculations to predict the
fragmentation of the 8^+ strength and then correct for the missing
strength. In the Zr region this becomes only significant for ^{92}Mo.
Thus we get quite a good picture of the occupation probabilities from
the strength of the 8^+ excitation.[17] This is shown in Fig. 18. It is
quite obvious that the shell model calculations predict too much
occupation of the $g_{9/2}$ orbit. Thus it shows that there must be already
significant occupation of the $2d_{5/2}$ orbit, which was not included in the
shell model calculation. In the same realm, it would be interesting
to look in ^{208}Pb for the 14^+ state which corresponds to the
$(1j_{15/2})^2 (3p_{1/2})^{-2}$ configuration or

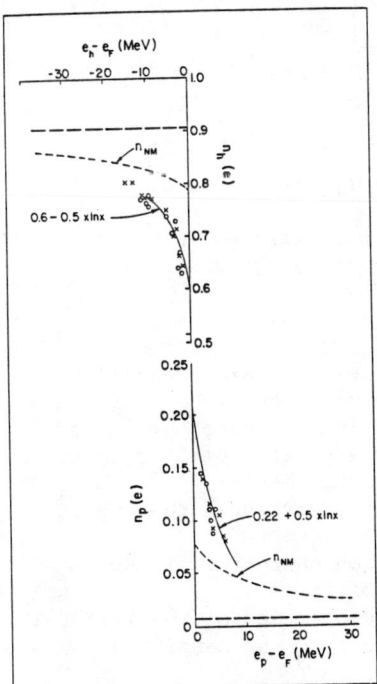

Fig.17: Occupation probabilities for orbits near the Fermi level in ^{208}Pb.

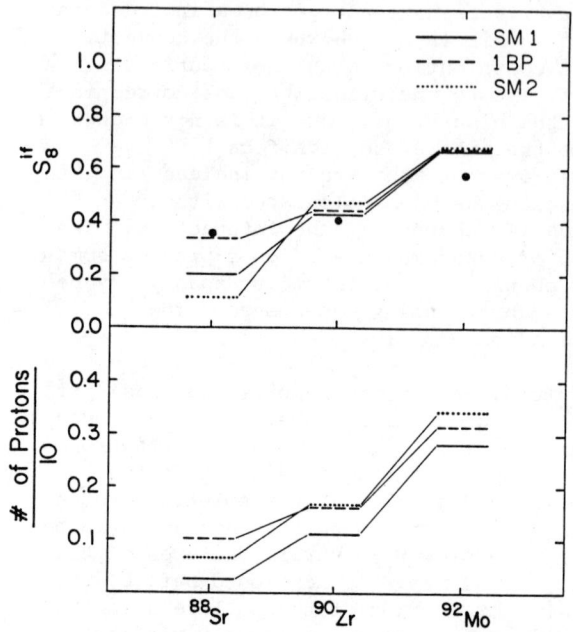

Fig.18: Spectroscopic amplitude for the 8⁺ levels and occupation probability of the proton $1g_{9/2}$ orbit.

some other states that can be excited mainly via the recoupling of a particle pair in a high spin orbit.

There are many additional exciting possibilities which I cannot present here. Instead let me present my conclusions:

CONCLUSIONS

(a) Measurements of different densities for the same transition strongly amplify our ability to interpret these measurements.
(b) There has been considerable progress in the understanding of proton scattering. In the near future we will be able to extract reliable densities from pp' experiments.
(c) Protons allow us to probe transition densities in the nuclear interior.
(d) Weaker transitions at higher excitation energy permit us to obtain ground state occupation probabilities.
(e) With multiple densities we can test properties of the residual interaction or higher order corrections to the density operator.

REFERENCES

1. B.A. Brown, R. Radhi and B.H. Wildenthal, Physics Reports 101 (1983), 315.

2. H. Sagawa and B.A. Brown, preprint MSUCL-476, July 1984.
3. K. Nakayama, S. Krewald, J. Speth and W.G. Love, submitted to Med. Phys. (1984).
4. T.T.S. Kuo and G.E. Brown, Nucl. Phys. $\underline{A114}$ (1968), 241.
5. J.B. McGrory and T.T.S. Kuo, Nucl. Phys. $\underline{A247}$ (1975), 283.
6. J.L. Friar in "Mesons in Nuclei" M. Rho and D.H. Wilkinson eds. (North Holland), Amsterdam (1973).
7. J.J. Kelly, W. Bertozzi, F. Petrovich: "Synthesis of electromagnetic and hadronic probes of nuclei" preprint 1984.
8. O. Schwentker et al., Phys. Rev. Lett. $\underline{50}$ (1983), 15.
9. M.M. Gazzaly et al., Phys. Rev. $\underline{C28}$ (1983), 294.
10. J. Heisenberg and J.F. Dawson, unpublished.
11. S. Krewald and J. Speth, Phys. Rev. Lett. $\underline{45}$ (1980), 417.
12. C.N. Papanicolas et al., Phys. Rev. Lett. $\underline{52}$ (1984), 247.
13. J. Heisenberg et al., Phys. Rev. $\underline{C29}$ (1984), 97.
14. T.E. Milliman et al., submitted to Phys. Rev. Lett.
15. M.M. Gazzaly et al., Phys. Rev. $\underline{C25}$ (1982), 408.
16. J.J. Kelly, private communication.
17. F.W. Hersman, private communication.
18. F.W. Hersman, private communication.
19. J.J. Kelly, preliminary analysis.
20. J. Heisenberg and H.P. Blok, Ann. Rev. Nuc. Part. Sci., Vol. $\underline{33}$ (1983), 569-609.
21. J. Heisenberg et al., to be published.
22. J. Wise, Ph.D. thesis, Charlottesville 1982, and J. Wise et al., submitted to Phys. Rev.
23. V.R. Pandharipande et al., Phys. Rev. Lett. $\underline{53}$ (1984), 1133.

RESOLUTION LIMITS WITH COOLED BEAMS

Heiner Herr
CERN, EP Division, Geneva, Switzerland

ABSTRACT

Storage rings for intense proton and ion beams are under construction which will work in the low and intermediate energy range, for experiments with internal targets. These machines will have a very good vacuum and will be equipped with efficient cooling devices which will give small beam dimensions and a small momentum spread. Under these conditions intrabeam scattering becomes a dominant effect. Calculations were made to estimate the resulting beam dimensions and momentum spread for different ions and energies.

INTRODUCTION

Storage rings are under construction at Uppsala, Sweden, (CELSIUS)[1] and at IUCF, Bloomington[2] for experiments with proton and ion beams. For good energy and spatial resolution in experiments with internal targets a circulating beam of small size and small energy spread is desirable. Therefore the phase space density of the injected beam will be increased by an efficient cooling device such as electron cooling with time constants of seconds. The stored beam cannot be cooled to infinitely small dimensions and momentum spread because there are other mechanisms which will "heat" the beam, so that the final dimensions are given by an equilibrium between heating and cooling. These mechanisms can be:

- Heating by machine imperfections:
  Resonances, ripple in power supplies, alignment errors etc.

- Heating by scattering:
  Residual gas (vacuum), internal targets, intrabeam scattering, electron cooling.

For the following discussions one assumes that the machine is well aligned, the ripple of the power supplies is low and the working point is not too close to a resonance so that heating due to machine imperfections is negligible.

The heating of the beam by scattering on the residual gas and on internal targets and the final equilibrium between heating and cooling can be easily estimated[3]. For example in CELSIUS one finds for 200 MeV protons and a cooling time of ~ 4 sec the following equilibrium emittances[4]:

Residual gas :   vacuum 10^{-12} Torr (NEP), no internal target
ε_{EQ} = < 10^{-3} π mm mrad

Internal target: perfect vacuum, target thickness 10^{-9} g/cm^2
ε_{EQ} = 0.12 π mm mrad

0094-243X/85/1280068-8 $3.00 Copyright 1985 American Institute of Physics

It will be shown that the equilibrium emittances due to intrabeam scattering, i.e. the scattering of the beam particles on each other, will be much bigger, so that the scattering on the residual gas and the internal target can be neglected at least for beams with more than 10^9 particles.

To calculate the equilibrium between heating by intrabeam scattering and electron cooling some assumptions are necessary which will be explained hereafter.

INTRABEAM SCATTERING

Scattering of the particles within the beam has the tendency to equalize the "temperature" of the beam in all three planes. Here the temperature is defined for one plane by

$$T = 1/2 \; mc^2 \beta^2 \gamma^2 \Theta^2 \tag{1}$$

where Θ is the beam divergence in one plane. For the longitudinal direction Θ is replaced by $1/\gamma \times \Delta p/p$. As the scattering probability depends on the particle density, intrabeam scattering therefore depends on:
- the beam dimensions, which are given by the focusing properties of the storage ring and the beam emittances
- the number of particles.

An additional problem occurs due to dispersion, the dependence of the radial position of the particle on its momentum. Particles which are changing their momentum by scattering start to make betatron oscillations around the orbit which corresponds to their new momentum. The energy for these oscillations is taken from the longitudinal motion of the particles, which is continuously renewed either by a RF-system or a cooling device. As a consequence the beam is heated continuously.

This heating is counteracted by electron cooling which leads to an equilibrium between heating and cooling. A numerical evaluation of this effect can be made with the help of the theory of A. Piwinsky[5], from which follows that the behaviour of the beam can be well described by the following invariant:

$$(\frac{1}{\gamma^2} - \frac{D^2}{\beta_x^2}) \; \langle \frac{\Delta^2 p}{p^2} \rangle + \langle \Theta_x^2 \rangle + \langle \Theta_z^2 \rangle = \text{const.} \tag{2}$$

with:

$\quad D$ = dispersion function
$\quad \beta_x$ = horizontal focusing function
$\quad \gamma$ = relativistic factor
$\quad \Theta_{x,z}$ = horizontal resp. vertical divergence of the beam
$\quad p, \Delta p$ = momentum and momentum deviation, respectively.

The three terms of the invariant correspond to the temperature of the beam in the two transversal and the longitudinal plane. Depending on the local value of D and β_x one can discriminate between the following cases:

1. $D^2/\beta_x^2 < 1/\gamma^2$

   In this case the temperatures in all 3 planes will equalize which means that the emittances and momentum spread will reach an equilibrium even without cooling (Note that one neglects here other effects such as residual gas scattering etc.). This is in particular always the case for $D = 0$.

2. $D^2/\beta_x^2 > 1/\gamma^2$

   In this case the first term is negative which means that all three "temperatures" can increase until the beam dimensions are so big that the scattering probability tends to 0.

3. $D^2/\beta_x^2 = 1/\gamma^2$

   As $< D^2/\beta_x^2 > \sim 1/\gamma_{tr}^2$ one finds for a well defined energy (transition energy) the longitudinal temperature equal to 0, which in so far has a meaning that at this energy all particles in the machine have independently of their energy the same revolution velocity. Longitudinal scattering cannot take place !

As the dispersion and the beta-functions change around the storage ring it is therefore evident that intrabeam scattering calculations have to be made in small steps along a superperiod using the local lattice functions. For a precise description it is also necessary to take D' and β_x' into account.

Furthermore intrabeam scattering depends strongly ($\beta^3 \gamma^4$) on the energy and (for different ions) on the square of the classical particle radius $r_i^2 = (Z^2/A)^2 \cdot r_p^2$. To calculate the equilibrium between intrabeam scattering and electron cooling a program written by F. Sacherer and D. Möhl in collaboration with A. Piwinski was modified. It is based on further development of Piwinski's original theory to include the deviations of D and β_x. In Fig. 1 the calculations are compared with the results from ICE[6]. They are in good agreement with the measurements.

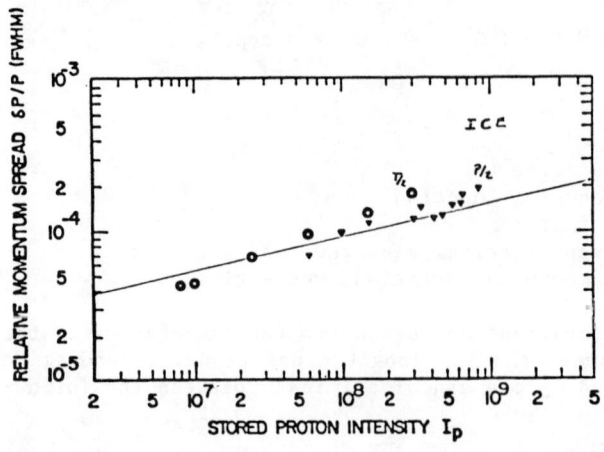

Fig. 1: Comparison between the measured equilibrium momentum spread in ICE and the calculations (straight line).

ELECTRON COOLING

Cooling means the reduction of the transverse velocities and the longitudinal velocity spread in the proton or ion beam. This is achieved by repetitive interactions with a "cold" electron beam. The amplitude cooling time is given by:

$$\tau = 0.16 \frac{\beta^4 \gamma^5 e (\theta_e^2 + \theta_i^2)^{3/2}}{r_i r_e j L_c \eta} \quad [\text{sec}] \quad (3)$$

where β, γ are the usual relativistic factors, e [Cb] the electric charge, $\theta_{e,i}$ the divergence of the electron resp. ion beam, j [A/m^2] the current density of the electron beam, L_c the Coulomb logarithm and η the ratio between the length of the cooling section and the circumference; r_e [m] is the classical electron radius and $r_i = Z^2/A \cdot r_p$ the "ion radius" given by the charge state Z, the atomic number A and the classical proton radius r_p.

Cooling stops when electrons and ions have the same temperature:

$$T = 1/2 \, m_e c^2 \, \beta^2 \, \gamma^2 \, \theta_e^2 \sim 1/2 \, M_i c^2 \, \beta^2 \, \gamma^2 \, \theta_i^2$$
$$\theta_i \sim \theta_e \sqrt{m_e/M_i} \quad (4)$$

which means that in the case of protons, for example, θ_p can become 1/43 of θ_e. Below this value the electron beam will heat the ion beam. However, as the divergence of the ion beam is proportional to $\sqrt{\varepsilon/\beta_c}$ with β_c being the local focusing function of the storage ring and ε the beam emittance, this can be avoided by a careful selection of the beta-function.

Another problem may occur at very low energies because of the space charge depression in the electron beam. Due to this effect there is a velocity difference between the outer and inner electrons of the beam. (Fig. 2)

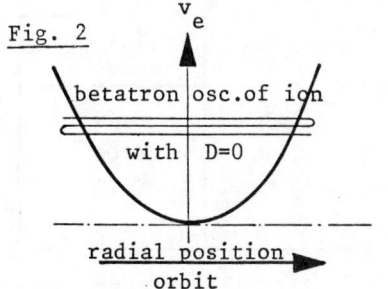

Fig. 2
velocity profile of e$^-$ beam

If this velocity difference becomes too big, the ions are alternatively accelerated or decelerated due to their betatron oscillations in the machine thus making the cooling under certain circumstances impossible. As the velocity difference is proportional to $\sqrt{j}$ with j being the current density, this can be avoided by reducing the beam current. However, this also prolongs the cooling time.

The electron cooling devices foreseen for CELSIUS and for the IUCF cooler ring are similar. They both use a Pierce configuration for the gun region accelerating the beam up to 20-50 keV and afterwards an acceleration tube to bring the beam to the desired energy between 250-300 keV. The beam may even be decelerated down to 10 keV by this tube. As a consequence the current density j, normally showing a $U^{3/2}$ behaviour, is constant.

In the following calculations a current density of 0.5 A/cm^2, a cathode diameter of 2 cm and a ratio η between the cooling length and the circumference of the ring of 3×10^{-2} (2.5m/80m) was assumed. For low ion energies of 10 MeV/n (corresponding to 5.45 KeV electron energy !) the current density was reduced by a factor 10 to lower the above mentioned space charge depression. Otherwise, the potential difference in the electron beam would have been 325 V≈6% !!

As shown before the amplitude cooling time τ of electron cooling depends strongly on the angular divergence of the ion beam Θ_i and of the electron beam Θ_e:

$$\tau \sim (\Theta_e^2 + \Theta_i^2)^{3/2} \quad (5)$$

The divergence of the electron beam is given by the construction of the cooling device and the transverse temperature of the cathode, while the divergence of the ion beam depends on the beta-function in the cooling straight section which can be adjusted within some limits.

Therefore the fastest cooling time can be achieved if the divergence of the ion beam in the cooling straight section is smaller than the divergence of the electron beam. If $\Theta_i < \Theta_e$, the cooling time is independent of the ion beam properties.

For calculating the cooling times for different energies and ions the above condition $\Theta_i < \Theta_e$ was assumed. For the temperature of the electron beam the value T ~ 0.2 eV was taken from the results from Novosibirsk and CERN.

Furthermore as the current density stays constant and the divergence of the electron beam varies with 1/βγ, the cooling times scale with $\beta \gamma^2$. With these assumptions the following amplitude cooling times were calculated:

Table I. Cooling times (in seconds)

	p	$^{12}C^{6+}$	$^{20}Ne^{10+}$	$^{238}U^{92+}$
10 MeV/n	3.8	1.26	0.76	0.10
200 MeV/n	2.2	0.74	0.44	0.06
500 MeV/n	4.7	1.56	0.94	0.13

EQUILIBRIUM CALCULATIONS

For the equilibrium calculations one of the possible lattices for CELSIUS[7] was taken. However, repeating the calculations with a lattice for the IUCF Cooler[8] gives only differences within a factor of 2.

The calculations were made for p, $^{12}C^{6+}$ and $^{238}U^{92+}$. Figs 3 and 4 show the dependence of the equilibrium emittance and momentum spread on energy for 10^{10} particles circulating in the machine; Figs 5 and 6 show the dependence from the number of circulating particles for 10 MeV/n and 500 MeV/n.

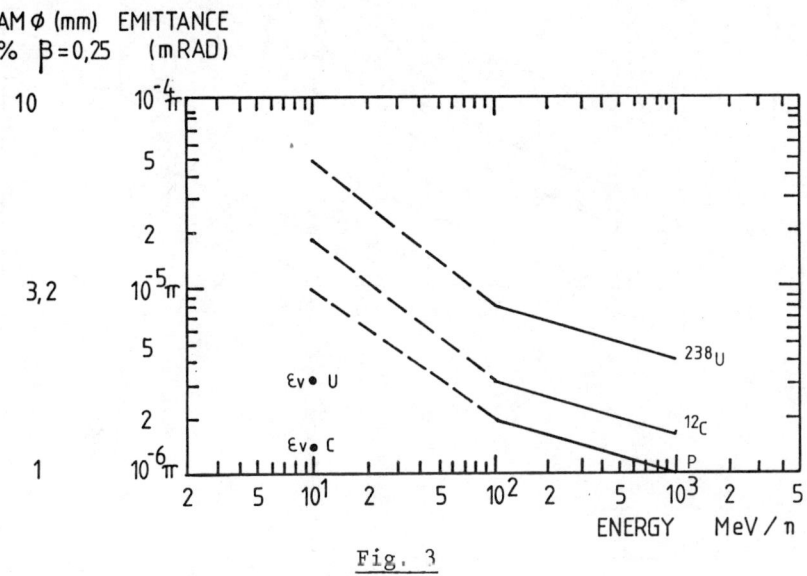

Fig. 3

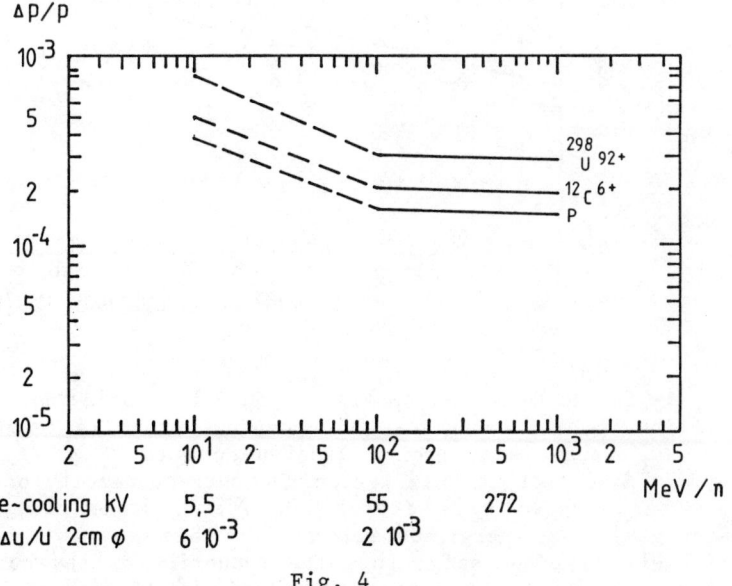

Fig. 4

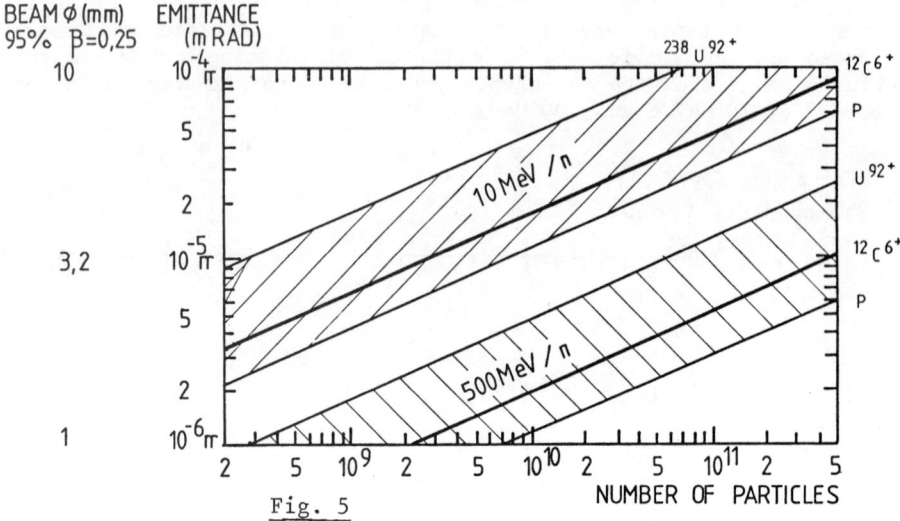

Fig. 5

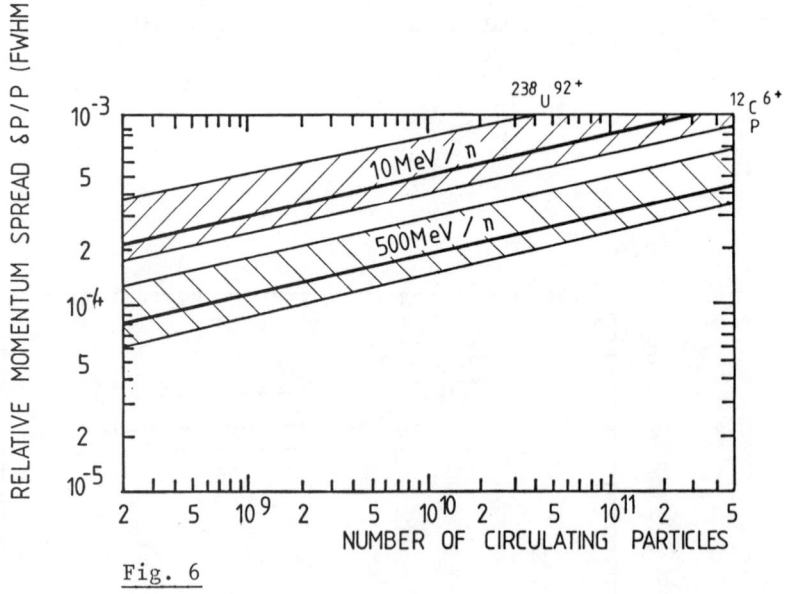

Fig. 6

As it can be seen from Fig. 3 and 4 the energy dependance of the equilibrium beam emittance and momentum spread is relatively small. For the energy range from 100 MeV/n down to 10 MeV/n it should be kept in mind that in this region the current density of the electron beam was reduced by a factor 10. Due to this change in cooling "strength" the energy dependence in this energy region is much stronger. As discussed before the reduction of the current density was assumed to avoid a too big velocity difference in the electron beam due to space charge depression. The factor 10 was chosen to keep the velocity difference within the same range as for the ICE gun. This choice however is somehow arbitrary and should depend on the individual set-up. Therefore the calculations below 100 MeV/n should be considered only as a guide-line.

Fig. 5 and 6 show that the equilibrium also does not depend strongly on the number of stored particles. Below $\sim 10^9$ particles it might be possible that, depending on the particular set-up the equilibrium emittances and momentum spreads are smaller due to the "magnetization" effect[9] of the electron beam, which will reduce the cooling times. Above $\sim 5 \times 10^{10}$ particles it may be desirable to have bigger equilibrium emittances and momentum spread to avoid instabilities for example caused by the resistive wall impedance or the Laslett tune shift.

Furthermore the calculations show that the equilibrium depends only weakly on the type of fully stripped ion: one finds a factor 2 between protons and $^{238}U^{92+}$ for the equilibrium momentum spread and a factor 4 for the equilibrium emittance.

CONCLUSION

Intrabeam scattering is an important effect in the discussed energy and beam intensity range. It may present a serious limit to the available energy resolution for experiments.

REFERENCES

1. D. Reistad, Proc. of this Workshop.
2. R. Pollock, Proc. of this Workshop.
3. K. Kilian, D. Mohl, J. Gspann, H. Poth, Proc. of the Workshop on Physics at LEAR with Low Energy Antiprotons, Erice 1982.
4. H. Herr, CELSIUS Note 83-2.
5. A. Piwinski, Proc. of the 9th Int. Conf. on High Energy Accelerators, Stanford 1974
6. M. Bell, J. Chaney, H. Herr, F. Krienen, P. Moller-Perterson, G. Petrucci, Nucl. Inst. Meth. 190, 237 (1981).
7. H. Herr, Proc. of the Workshop on the Physics Program at CELSIUS, Uppsala 1983.
8. R. Pollock, private communication.
9. V.Z. Kudelainen, V.A. Lebedev, I.N. Meshkov, V.V. Parkhomchuk, B.N. Sukhina, Novosibirsk, Inst. of Nucl. Physics, Preprint 82-78.

COOLED BEAMS WITH INTERNAL TARGETS

H.O. Meyer
Department of Physics, Indiana University, Bloomington, IN 47405 USA

ABSTRACT

Some aspects of the interaction of a stored, cooled beam with internal targets are discussed. Recent results of a computer simulation are given. In particular, the effect of the dependence of the velocity of cooling electrons on the distance from the center of the electron beam is investigated, limits in the use of fiber targets are listed, and the effect of the residual gas in the storage ring is studied.

INTRODUCTION

Storage rings for ion beams are currently under construction or are being designed as additions to accelerators at several laboratories. Confinement of the beam offers the possibility of duty factor manipulation. If a beam cooling device is added, a reduction in phase space volume can be achieved. Possible cooling mechanisms[1] include electron cooling, damping by the emission of synchrotron radiation (important for highly relativistic particles), or the active sensing and reducing of stochastic fluctuations of the beam ensemble.

The cooling method most appropriate in conjunction with internal targets is electron cooling. It makes use of the frictional force that acts on a charged particle in motion relative to an electron gas. In the practical cooling of ion beams, the electrons are provided by an intense e^- beam, matched in velocity to the stored ion beam and sharing a fraction of the ring circumference with this beam. Some practical details on electron cooling technology can be found, e.g., in Ref. 2, while theoretical aspects are given, e.g., in Ref. 3.

The time constant associated with electron cooling is of the order of one second. This sets the scale for the necessary duration of the confinement of the beam in the storage ring. In such a "batch-mode" processing of the beam, experiments with an <u>extracted</u> beam are limited to a low average beam intensity. In carrying out nuclear physics experiments with a cooler ring, the main interest will therefore be in the use of <u>internal targets</u>.

The performance of internal targets is characterized by a balance between the effect of electron <u>cooling</u> and the effect of <u>heating</u> of the beam by the passage through the target. It is clear that the presence of the target affects the stored beam in many ways; some aspects of this interplay are the subject of this paper.

0094-243X/85/1280076-13 $3.00 Copyright 1985 American Institute of Physics

EQUILIBRIUM DISTRIBUTIONS

The distribution of beam particles in phase space is given by an equilibrium which results from the opposing effects of the electron cooler and the internal target. The properties of these equilibrium distributions have been investigated by a computer program simulating the effects of the storage ring optics, the electron cooler and the target passage.[4] Simulations usually were carried out with a few hundred representative beam particles. The interaction with the target was treated on a microscopic level, using a Monte-Carlo technique. The cooling force in this investigation originally was a one-parameter function of the momentum, separable in the transverse and longitudinal directions. Recently, terms arising from the space charge of the electron beam have been included. Such terms introduce a dependence of the longitudinal equilibrium phase-space distribution on the (transverse) emittance of the stored beam.

The results from the computer simulation can be summarized as follows. The emittances of individual beam particles have nearly a Gaussian distribution. The beam emittances ε_x and ε_y in the horizontal and vertical direction, respectively, are here defined as to contain 90% of all beam particles and are therefore related to the width of these Gaussian distributions. They are proportional to the slope of the cooling force vs. transverse momentum and (for sufficiently thin targets) to the target thickness. As an example, the emittance of a 200-MeV proton beam in equilibrium with various targets is shown as a function of target thickness d in Fig. 1. It should be noted that these emittance values depend linearly on the radial aperture function, β_x or β_y, at the position of the target. For the case shown in Fig. 1, β_x = 0.35 m was chosen and the design cooling force for the Indiana Cooler was used.[4]

The longitudinal momentum distribution differs greatly from the transverse phase space distribution. Due to a peculiarity of the velocity distribution of the cooling electrons — it is much narrower longitudinally than transversely — the cooling force is a very non-linear function of the longitudinal beam particle momentum p in the electron rest frame. For a hypothetical beam of zero emittance, or infinitely small transverse dimensions, the equilibrium momentum distribution is characterized by a narrow spike, a few keV/c wide, in conjunction with a long tail towards lower momenta. The width of the narrow component is given completely by details of the cooling force and is independent of target thickness. The target thickness does, however, affect the number of particles in of the tail region as compared to the spike, since the tail is populated by δ-ray production in the target, leading to the possibly of relatively large, single-step momentum losses of the beam ions.

For a more realistic beam of non-zero emittance one has to consider an important aspect of the cooling electron beam. Due to the potential energy given by the space charge of the electron beam, there is a quadratic dependence of the electron velocity on the distance from the symmetry axis of the cooling section. A beam particle with a

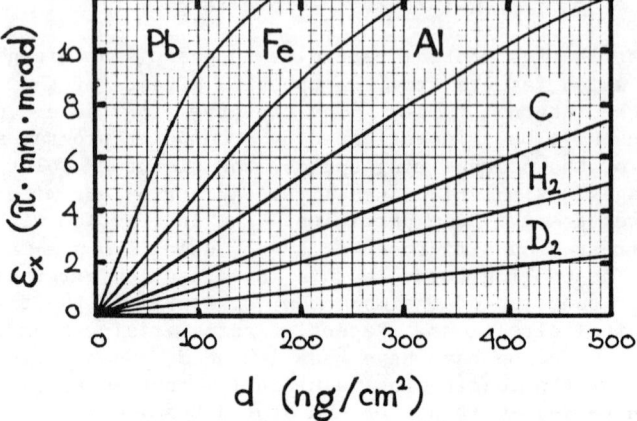

Fig. 1. Horizontal equilibrium emittance ε_x for a 200-MeV proton beam versus target thickness d for various target materials. The emittance ε_x is proportional to the radial aperture function β_x at the target; for this example $\beta_x = 0.35$ m. The same graph applies in the vertical direction as well. The emittance has been defined as the area of the smallest phase ellipse that contains 90% of all beam particles.

finite betatron amplitude thus samples a range of different cooling forces. This, in a beam with a distributed transverse emittance leads to a broadening of the narrow momentum spike, even in the absence of an internal target. The radial velocity dependence of the cooling electrons has recently been included in the computer simulation discussed in Ref. 4. It was found that the resulting momentum distributions depend on the sum $\varepsilon_s = \varepsilon_x + \varepsilon_y$ of horizontal and vertical emittance. Beam momentum distributions for a fixed value of ε_s, but without internal target, are shown in Fig. 2a. The full width Δp at half maximum is proportional to ε_s, as shown in Fig. 2b. In these calculations a 200-MeV proton beam was studied. The aperture function in the cooling region was taken as $\beta_c = 3.56$ m. The electron current was $I_e = 4$ A and the diameter of the e^- beam was 26 mm.

BEAM LOSS MECHANISMS

The stored beam intensity as a function of time decreases exponentially due to interactions in the internal target, the residual gas and the cooling electron beam. Possible loss mechanisms include the electron pickup by the beam ion,[5,6] either in the target material or in the electron cooling region, where velocity matched electrons are available. Nuclear reactions in the target also remove beam particles. These processes are, at least for light ions in the energy range of the Indiana Cooler, of minor importance compared with the losses due to the limitations in acceptance of the storage ring.

The momentum acceptance of the storage ring (for the Indiana Cooler, $\delta p/p$ = 0.2%) excludes part of the tail of the <u>longitudinal</u> momentum distribution, mentioned earlier. The resulting equilibrium beam loss is <u>not</u> proportional to the target thickness, therefore it can not be expressed in terms of a cross section. Instead, it is convenient to quote the 1/e lifetime τ_ℓ of the beam in the absence of all other loss mechanisms. The Monte-Carlo simulation of the stored beam[4] has also been used to calculate the longitudinal lifetime τ_ℓ for a number of cases; the results are given elsewhere.[4,6] The main reason for the longitudinal losses is the weakness of the cooling force for particles, which are badly mismatched in velocity with respect to the cooling electrons such as those near the momentum acceptance limit. If an additional mechanism is provided to restore those particles, such as continuous acceleration of the beam by, typically, the average energy loss in the target, then the longitudinal losses become indeed negligible.

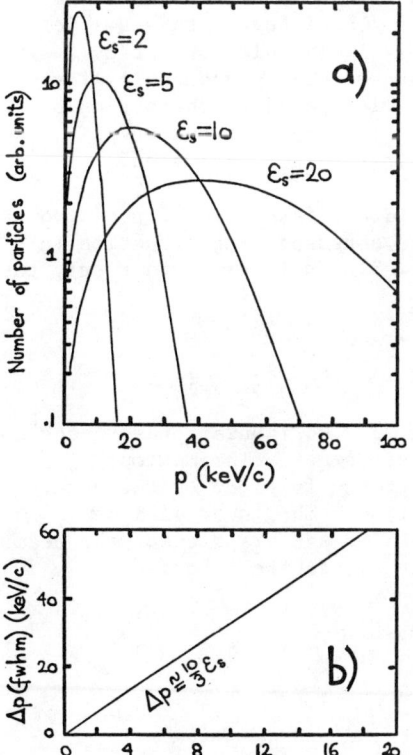

Fig. 2. Momentum distribution of a 200-MeV proton beam of fixed emittance $\varepsilon_s = \varepsilon_x + \varepsilon_y$. The figure illustrates the effect of the transverse variation of the velocity of the cooling electrons. The influence of the internal target is neglected.

a) Momentum distributions for various values of the sum emittance ε_s.

b) Full width at half maximum of the equilibrium momentum distribution.

In the transverse direction the acceptance, A_x or A_y, of the storage ring can be exceeded after a single Rutherford scattering event, which leads to the loss of the corresponding particle. Calculated transverse lifetimes τ_t for an equilibrium beam due to this mechanism are also given in Refs. 4 and 6. It is possible to find simple, semi-empirical expressions[6] for the dependence of both transverse and longitudinal lifetime on target material and target thickness.

FIBER TARGETS

Beam lifetime considerations and the effectiveness of the electron cooling limit the maximum thickness of the internal target.[4,6] For medium energy proton beams this limit is a few times 100 ng/cm^2. This rules out self supporting foils as targets. Gas jets or differential pumping arrangements are possible schemes, but, in view of the stringent vacuum requirements in the storage ring, such techniques are certain to require a major technical effort. It is therefore advisable to also investigate alternative ways to present target nuclei to the stored Cooler beam.

It has been suggested that a thin fiber exposed to the stored beam may constitute a possible scheme for an internal target. Here, the idea is that the traversal of a relatively thick object by a given beam particle can be tolerated if this particle subsequently misses the fiber often enough, while it is still exposed to the action of the cooler on every turn. There are, however, several fundamental limitations that come into play when fiber targets are used.

Assume that a vertical fiber of width w is placed at position x in a beam of horizontal radius x_0. The corresponding situation in horizontal phase space is shown in Fig. 3. It is clear that only the

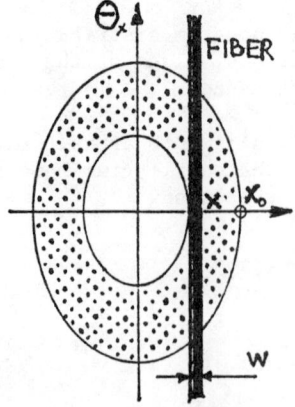

Fig. 3. Horizontal phase space and fiber target. The maximum emittance is given by the outer ellipse. The inner ellipse contains the fraction of the beam that misses the fiber.

beam particles with betatron amplitudes larger than x, i.e., those
located in the dotted phase space region, can interact with the fiber.
For these particles, the probability to interact is given by the
projected area of the fiber inside the beam divided by the dotted
area. The average target thickness $\langle d \rangle$ seen by those (interacting)
particles is then

$$\langle d \rangle = \frac{2}{\pi} w^2 \rho \bigg/ \sqrt{x_o^2 - x^2} , \qquad (1)$$

where ρ is the density of the target material. It is obvious that the
average target thickness, according to Eq. 1, has to be less than the
upper limit on thickness one would obtain for a homogeneous target.
Placing the fiber in the center (x = 0) minimizes $\langle d \rangle$ and thus
represents the most favorable case. In this case, Eq. 1 constitutes
an upper limit on the fiber thickness. As an example, using
$\langle d \rangle_{max}$ = 200 ng/cm^2, x_o = 0.5 mm, ρ = 2.2 g/cm^3 results in an upper
bound of w < 0.8 μm. For certain materials, fibers of this thickness,
or less, have been reported, but they constitute exceptions and will
be difficult to manufacture. One also expects problems with the
stable, continuous operation of very thin fibers in the extreme
environment of an intense ion beam.

It would seem that thicker fibers can be used if they are placed
at x ≠ 0, i.e., in the tail of the beam distribution. However, this
is not a feasible arrangement, because beam particles with betatron
amplitudes less than x, inside the smaller ellipse in Fig. 3, never
intercept the fiber and the ones that do, according to Eq. 1,
experience an average target thickness that is much too large, leading
to almost instantaneous loss of this fraction of the beam.

It is, however, possible to use thicker fibers if they are moved
across the entire profile of the beam, thus giving every beam particle
an equal chance to interact with the target. The sweeping can be
periodic and, instead of actually moving the target, the beam
may be wiggled. This introduces a time-dependent luminosity (probably
a disadvantage), whose time average can be adjusted to any desired
value (a definite advantage). This arrangement of a moving fiber has
been tested, by a Monte-Carlo simulation of the interaction of the
stored beam with the fiber and the cooler. It was found that sweep
velocities as low as 5 cm/s still provide sufficient mixing and that
the beam lifetime in such a case is very similar to that obtained with
a homogeneous target with the same average luminosity.

If a vertical fiber is used at a target station with dispersion,
a beneficial effect may arise from the localization of the luminous
volume. For instance, it would be possible to discriminate in an
experiment against interactions involving beam particles in the tail
region of the momentum distribution.

Another important limit that might restrict the use of fiber targets arises from the <u>thermal load</u> represented by the intercepting beam. It is obvious, that for a given luminosity L (in $cm^{-2}s^{-1}$) the thermal power delivered in a target material of mass number A in which the beam particles lose an energy of dE/dx (MeV cm^2/g) is completely determined. Electromagnetic radiation is the only mechanism by which this deposited power can be transported away. Assuming thermal equilibrium and black-body radiation and using the law of Stefan-Boltzmann one can determine the temperature T_F of the fiber:

$$T_F = 3.5 \cdot 10^{-7} \left[\frac{K^4 \, g \, cm^2 \, s}{MeV} \right]^{1/4} \left[\frac{L \cdot \frac{dE}{dx} \cdot R \cdot A}{w \cdot \ell} \right]^{1/4} \quad (2)$$

Here, w is the fiber diameter (in cm) and ℓ is the illuminated length of the fiber (in cm). R is the fraction of deposited energy that is thermalized, i.e., that does <u>not</u> leave the fiber in the form of δ electrons. A numerical estimate for μm size fibers gives typical values of R ~ 0.8. As an example, we take 150 MeV protons incident on a carbon fiber. From a table we find dE/dx = 4.89 MeV cm^2/g. For the illuminated length of the fiber we choose ℓ = 0.1 cm. For this situation, Fig. 4 shows the dependence of the fiber temperature on the

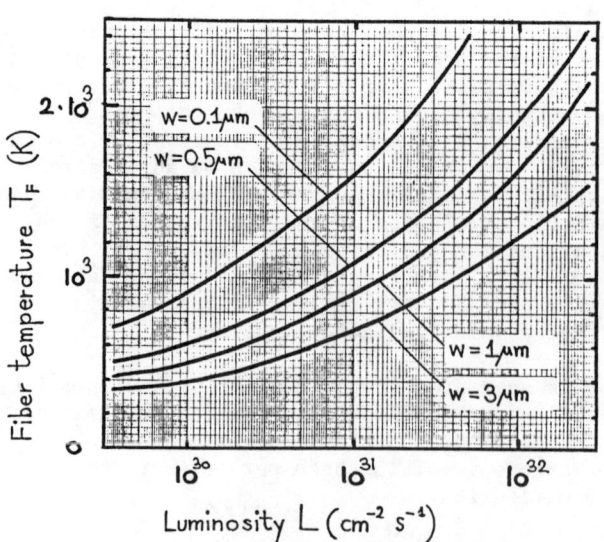

Fig. 4. Temperature T_F of a ^{12}C fiber used as internal target in an experiment with luminosity L. A proton beam of 150 MeV is assumed. The length of fiber illuminated by the beam is 1 mm.

luminosity, according to Eq. 2, for a number of different fiber thicknesses. It can be seen that temperatures beyond 1800 K can be expected in some cases. At these elevated temperatures the sublimation of many target materials presents a serious problem in the continuous operation of such a target.

Another consequence of high target temperatures concerns the experimental energy resolution. Because of the thermal motion of the target nuclei, one expects an uncertainty of the total center-of-mass energy s of the reacting nuclear system consisting of an incident projectile of mass m and total energy $E = \gamma mc^2$ and a target nucleus of mass M. For a target material well above the Debye temperature one finds for the full width Δs at half maximum of the reaction energy s

$$\Delta s = 2.35 \frac{mc^2 \cdot \beta\gamma}{s} \sqrt{Mc^2 \, k \, T_F} \quad (3)$$

where

$$s^2 = m^2c^4 + M^2c^4 + 2EMc^2,$$

$k = 0.86 \cdot 10^{-10}$ MeV/K is Boltzmann's constant, and β and γ are the usual relativistic parameters. As an example, for a 150 MeV proton beam on a graphite target (Debye temperature: 420 K) at a temperature $T_F = 2000$ K one finds $\Delta s = 4.6$ keV. For temperatures T_F near or below the Debye temperature collective excitations of the lattice in the target material may become relevant; on the other hand, Debye temperatures beyond 1000 K are quite rare (beryllium and diamond).

VACUUM LIMITS

The residual gas in the storage ring influences the stored beam in much the same way as an internal target distributed over the whole ring circumference. In particular, the loss mechanisms discussed in section 3 of this paper set a practical upper limit on the residual gas pressure in the storage ring. In the following, we calculate the acceptance related losses in the rest gas. In this calculation we ignore the influence of the (longitudinal) momentum acceptance, since by auxiliary acceleration, it is possible to make these losses negligibly small, as pointed out earlier.

We first assume that the emittance of the stored beam is zero; in Fig. 5a where the transverse acceptance ellipse is shown, all beam particles are therefore at the origin. A single Rutherford scattering

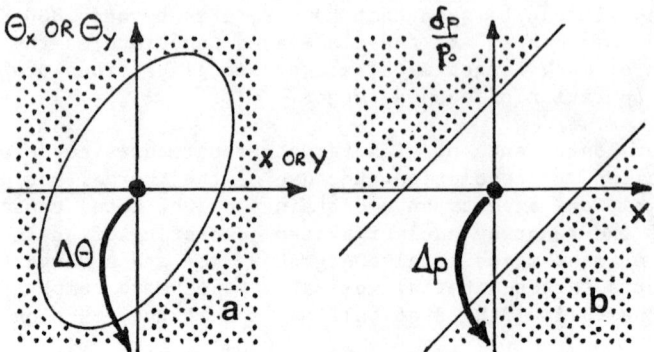

Fig. 5. Zero-emittance beam is shown at the orgin of (a) the transverse phase space and (b) the subspace of longitudinal momentum versus horizontal coordinate x. The dotted region is excluded by the transverse acceptance criterion. The tilt of the allowed region in (b) is given by the dispersion.

event displaces a beam particle parallel to the θ axis. If this displacement, e.g., in the x direction is larger than

$$\Delta\theta_x = \sqrt{\frac{A_x}{\pi \beta_x}}, \qquad (4)$$

where A_x is the ring acceptance and β_x is the local, radial aperture function, then the beam particle is lost. The Rutherford cross section for scattering by any angle larger than $\Delta\theta$ is

$$\sigma_R(>\Delta\theta) = \pi(\alpha\hbar c)^2 \left(\frac{zZ}{T}\right)^2 \left(1 + \frac{T}{mc^2}\right) \Delta\theta^{-2}, \qquad (5)$$

where z, m, T refers to charge number, mass and kinetic energy of the projectile and Z is the charge of the target nuclei. Using Eqs. 4 and 5, it is straightforward to calculate the sum of all transverse acceptance losses along the circumference of the ring. For a 200-MeV proton beam, using a ring acceptance $A_x = A_y = 25\pi$ mm·mrad, leads to a lifetime of

$$\frac{1}{\tau_t} = 19.1 \cdot m_n \cdot Z_R^2 \cdot p_R \; (\langle\beta_x\rangle + \langle\beta_y\rangle). \qquad (6)$$

Here, τ_t is in seconds and p_R, the residual pressure, is in torr. The atomic number is Z_R and m_n is the number of atoms per molecule of rest gas. The averages $\langle\beta_x\rangle$ and $\langle\beta_y\rangle$ of the radial aperture functions are taken over the whole circumference of the ring.

Dispersion provides a coupling of longitudinal changes into the transverse particle motion. From Fig. 5b it is obvious that a step Δp in longitudinal momentum in a dispersed section of the ring can lead to a loss caused by the transverse acceptance limit. This loss occurs if the step Δp is larger than

$$S_p = \left| \frac{p_o}{\eta} \sqrt{\frac{\beta_x A_x}{\pi}} \right|, \qquad (7)$$

where p_o is the beam momentum and η the dispersion, $\eta \equiv p_o(dx/dp)$. The probability Q for a decrease of the momentum by any value larger than S_p follows from a treatment of collisions of the beam ions with electrons of the rest gas and is given by

$$Q(>S_p) = 153.5 \text{ keV cm}^2/\text{g} \; \frac{z^2}{\beta^4} \; \frac{Z}{A} \cdot d \left\{ \frac{1}{S_p} - \frac{1}{p_H} - \frac{\beta^2}{p_H} \ln \frac{p_H}{S_p} \right\}, \qquad (8)$$

where d is the target thickness in g/cm^2 and $p_H = 2m_e c^2 \beta\gamma^2$ is the maximum momentum of a δ electron. Again, it is straightforward to calculate the associated beam lifetime. For a 200-MeV proton beam one approximately finds for the lifetime due to dispersion-coupled losses

$$\frac{1}{\tau_D} = 530. \cdot m_n \cdot Z_R \cdot p_R \left\{ \left\langle \frac{|\eta|}{\beta_x} \right\rangle + \left\langle \ln \frac{\beta_x}{|\eta|} \right\rangle - 2 \right\} \qquad (9)$$

Again, the averages $\langle...\rangle$ are over the ring circumference.

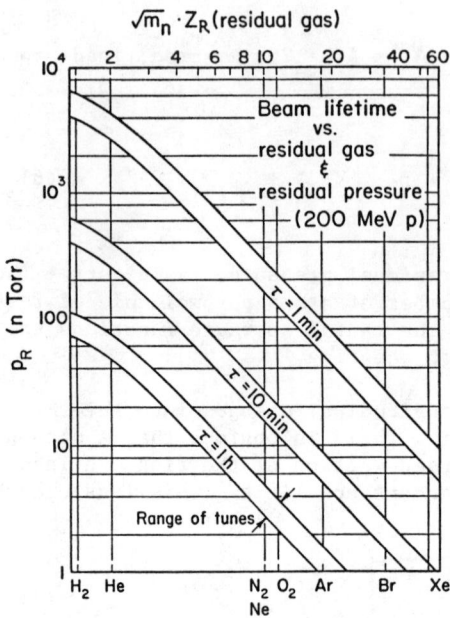

Fig. 6. Lifetime of a zero-emittance, 200-MeV proton beam due to residual gas in the storage ring as function of residual pressure in 10^{-9} torr. The bands indicate the dependence of the lifetime on the ring optics. The number of atoms per molecule is denoted by m_n.

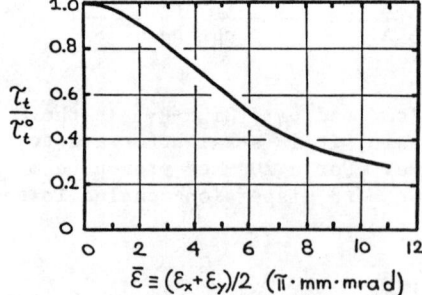

Fig. 7. Reduction of the transverse lifetime due to finite beam emittance. Lifetimes τ_t for different emittances $\bar{\varepsilon}$ have been obtained in a Monte-Carlo simulation by varying the target thickness. τ_t° is the corresponding zero-emittance lifetime. The emittance $\bar{\varepsilon}$ given is the average between horizontal and vertical emittance, ε_x and ε_y.

For the Indiana Cooler, the combined transverse and dispersion-coupled effect is shown in Fig. 6 as curves of constant lifetime in a graph of residual pressure p_R vs. target material. Since the lifetime due to the residual gas depends on the ring optics, Eqs. 6 and 9 have been evaluated for a number of different tunes of the storage ring.[7] In comparing Eqs. 6 and 9 it was found that transverse losses are more important than dispersion-coupled losses except for the lightest elements ($Z < 3$). For a mixture of gases of given partial pressures the corresponding individual lifetimes can be combined by adding their reciprocals.

Since for this study the assumption of a zero-emittance beam was made, it still remains to be investigated how the above results change if a beam of non-zero emittance is considered. For this purpose, a Monte-Carlo simulation[4] was used to evaluate the equilibrium emittances, ε_x and ε_y, and the lifetime of a 200-MeV proton beam traversing carbon targets of a number of different thicknesses. The ratios of the resulting realistic, transverse lifetimes τ_t and the corresponding zero-emittance values τ_t° are shown in Fig. 7 as a function of average beam emittance $\bar{\varepsilon} = (\varepsilon_x + \varepsilon_y)/2$. Since the calculational effort in obtaining this result is considerable, this specific example must suffice as a qualitative assessment of the importance of the effect.

CONCLUSION

We expect that in 1987, less than three years from now, Cooler rings will be in the service of nuclear physics research at IUCF, at CELCIUS in Uppsala and and TARN in Tokyo. Planning for early experiments is made difficult by uncertainties in the properties of a completely novel research instrument and its performance. The questions addressed in this paper, for instance, are typical of questions prospective users of the Cooler are certain to ask. Answers are needed in the development of experimental equipment as well as in in the late stages of the machine design.

ACKNOWLEDGEMENT

The author would like to thank G.T. Emery for his willingness to exchange ideas and his helpful advice and R.E. Pollock and W. Schaich for many stimulating discussions about various topics of this paper.

REFERENCES

1. F.T. Cole and F.E. Mills, Ann. Rev. Nucl. Part. Sci. $\underline{31}$, 295 (1981).

2. M. Bell et al., Nucl. Instr. Meth. $\underline{190}$, 237 (1981).

3. T. Ogino and A.G. Ruggiero, J.Phys. Soc. Japan $\underline{49}$, 1654 (1980); A.H. Sorensen and E. Bonderup, Nucl. Instr. Meth. $\underline{215}$, 27 (1983).

4. H.O. Meyer, Proc. 8th Int. Conf. on Applications of Accelerators in Research and Industry, Denton, TX, 1984, Nucl. Instr. Meth., to be published.

5. R.E. Pollock, Proc. Workshop on Polarized Targets in Storage Rings, Argonne 1984, ANL-84-50, (1984) p. 1.

6. H.O. Meyer, Proc. Symposium on Nuclear Shell Models, Philadelphia 1984, World Scientific, Singapore, to be published.

7. J. Dreisbach, private communication

GAS JET INTERNAL TARGETS

M. Macri

CERN, Geneva, Switzerland
Istituto Nazionale di Fisica Nucleare, Sezione di Genova, Italy

1. INTRODUCTION

The use of gaseous internal targets in antiproton storage rings [the CERN Intersecting Storage Rings (ISR), the Super Proton Synchrotron (SPS), and the Low-Energy Antiproton Ring (LEAR)] can provide sources of p$\bar{\text{p}}$ interactions which have unique characteristics. The internal gas target in ring 2 of the ISR[1-5] was successfully used to perform an experimental study of the formation of charmonium (c$\bar{\text{c}}$) in proton-antiproton annihilation. The internal target for the SPS[6] has been installed in the machine, and tests are in preparation.

The main problem created by the use of an internal gas target in a storage ring is the effect of the target gas on the coasting beam lifetime and characteristics (e.g. size, momentum). The most efficient solution to this problem is to apply horizontal and vertical betatron stochastic cooling and momentum stochastic cooling[7]. Also of great interest is the electron cooling foreseen in LEAR, which will extend the antiproton momenta range for experiments.

The features of the system, a coasting antiproton beam interacting on a gas internal target, are the following:

i) it has high luminosity ($\geq 10^{31}$ cm^{-2}·s^{-1});
ii) the interaction region has small dimensions (of the order of 1 cm³ or less); this is fundamental in order to simplify the design of the experimental apparatus and to facilitate the analysis of the experimental data;
iii) the value of the momentum resolution $\Delta p/p$ for the cooled antiproton beam is not spoiled by the interactions in the gas target;
iv) operation is continuous;
v) there is efficient use of antiprotons;
vi) there is a possibility of polarizing the atomic gas jet target;
vii) the effect of multiple scattering on the target is negligible for very low energy recoil protons or secondary particles.

The target which gives the largest luminosity is a type of molecular beam[8] called a 'cluster' beam, which provides a flow of gas at supersonic speed (hence the name of gas 'jet' target) due to the expansion of gas from a vessel at high pressure and low temperature into the vacuum through an injector (nozzle) of very small aperture (10-150 μm) and special geometry. The flow of gas is directional in the axis of the expansion and it is absorbed after having traversed the storage ring vacuum pipe (Fig. 1).

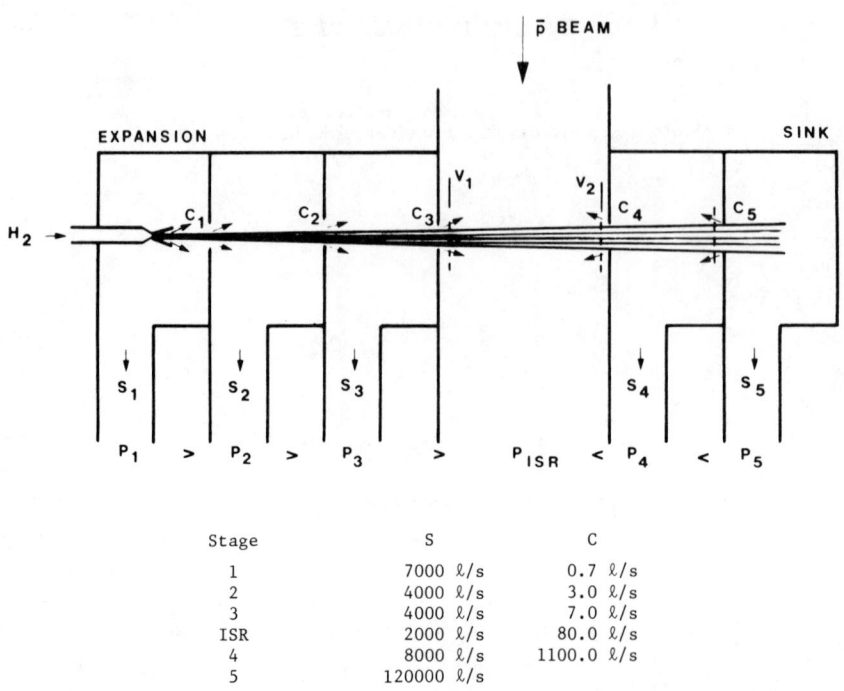

Fig. 1 Scheme of the use of molecular beam as an internal gas target. $s_{1,5}$ and s_{ISR} are the speeds of the pumping systems acting on the different chambers and $p_{1,5}$ and p_{ISR} the respective pressures.

Section 2 describes the operations of molecular and atomic beams. Section 3 deals with the problems of the use of a jet target in a storage ring. To outline problems and solutions, I will describe many of the characteristics of the jet target of the ISR and that of the SPS.

2. MOLECULAR AND POLARIZED ATOMIC BEAMS

2.1 Effusive gas source

The simplest source one can consider is the effusive source. Gas at temperature T_0 and pressure p_0 is kept in vessel A (Fig. 2), which has a small hole with diameter d allowing the gas to flow towards vessel B ($p_B \ll p_A$). This establishes molecular flow conditions. Molecules in the forward direction can be selected with a screen between vessels B and C. The luminosity possible from this source is very low. For example: using H_2 in a source with $T_0 = 100$ K, $p_0 = 1$ Torr, $d = 50$ μm, $\lambda = 25.0$ μm, we obtain a gas flux with intensity $I(\theta) = 2.5 \times 10^{17} \times \cos\theta$ atoms per steradian per second. The luminosity of the interaction of a beam of 3×10^{11} antiprotons, coasting at a frequency of 3.3×10^5 Hz on an internal target built from this gas source, will be $\sim 2.5 \times 10^{26}$ cm$^{-2} \cdot$s^{-1} taking into account the dimensions of the antiproton beam at the crossing with the H_2 beam (the H_2 beam diameter at the crossing point with the antiproton beam is ~ 1 cm at a distance of 26 cm from the H_2 beam origin).

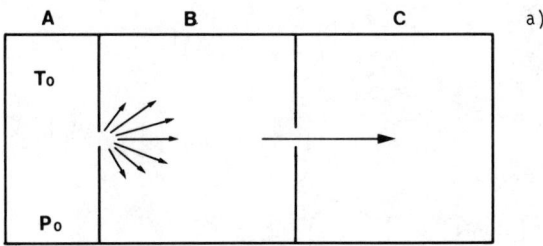

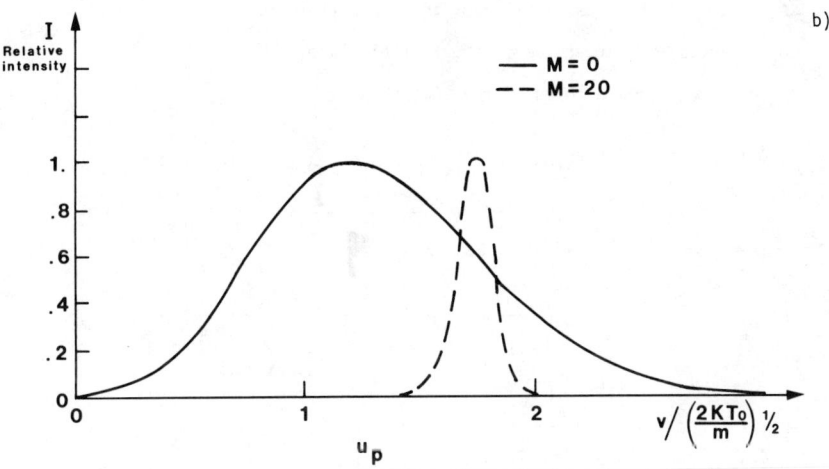

Fig. 2 Effusive gas source scheme and velocity distribution of molecules.

2.2 Sonic gas source

To increase the density of the gas source we have to consider the free expansion under continuum conditions. The gain with respect to the effusive source is that the final flow will be molecular, but the pressure (hence the density in the jet) can be increased in the initial state. In this case the gas in vessel A can escape to vessel B via an injector (nozzle) of the type shown in Fig. 3a. The equations that describe the expansion of the gas are derived from those for the conservation of energy and mass of a fluid flow: if we use the Mach number they can be expressed as[8]:

$$T = T_0 \left[1 + (\alpha-1) \left(\frac{M^2}{2} \right) \right]^{-1} \tag{1}$$

$$p = p_0 \left[1 + (\alpha-1) \left(\frac{M^2}{2} \right) \right]^{-\alpha/(\alpha-1)} \tag{2}$$

$$u = u_0 M \left[1 + (\alpha-1) \left(\frac{M^2}{2} \right) \right]^{-\frac{1}{2}} \tag{3}$$

$$n = n_0\left[1 + (\alpha-1)\left(\frac{M^2}{2}\right)\right]^{-1/(\alpha-1)} \quad (4)$$

$$A = \left(\frac{\alpha+1}{2}\right)^{-B}\left[1 + (\alpha-1)\left(\frac{M^2}{2}\right)\right]^{B}\left(\frac{A_0}{M}\right) \quad (5)$$

where
- $B = (\alpha+1)/[2(\alpha-1)]$
- u = local flow velocity
- $\alpha = C_p/C_v$
- A_0 = area of the cross-section of the nozzle throat
- u_0 = velocity of molecules in the initial state
- n_0 = density of molecules in the initial state.

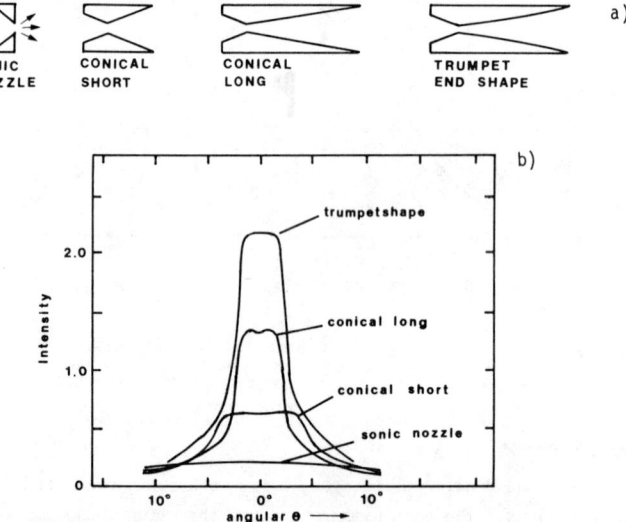

Fig. 3 a) Various types of nozzles. b) Mass flux distribution for nozzles with the same throat diameter and different geometries. For all distributions the stagnation condition and thus the nozzle mass flux are the same. The intensity scale is in units of 10^{22} atoms per steradian per second. (From Ref. 9.)

The nozzle flow field is similar to that from a point source close to the nozzle throat. At a short distance x from the nozzle throat, the streamlines are almost straight, the density decreases as $1/x^2$, and the flow velocity has approached its final value

$$v_{max} = u_0\left(\frac{\alpha-1}{2}\right)^{-0.5}. \quad (6)$$

We note that the velocity distribution (see Fig. 2b) has completely narrowed approaching the end of expansion. As for the effusive source we can fix the parameter for the gas source and calculate the luminosity of interactions of an antiproton beam with an internal target built from this gas source. Suppose we can use $\Delta\Omega = 1.2 \times 10^{-3}$ sr of solid angle and a total flow I from the nozzle of 13 cm$^3 \cdot s^{-1} \approx$ 10 Torr$\cdot \ell \cdot$s^{-1}. We get $I_0 = I\Delta\Omega = 4.2 \times 10^{17}$ molecules per second; the density expressed in molecules per cubic centimetre at the crossing point with the antiproton beam is given by

$$\rho = \frac{I_0}{Av_{max}}, \qquad (7)$$

where A is the area of the cross-section of the gas flow at the crossing with the antiproton beam. For a 1 cm diameter beam with v_{max} = 1290 m$\cdot$s^{-1} we get a luminosity of $\sim 8 \times 10^{29}$ cm$^{-2}\cdot$s^{-1} if the beam intensity is 3×10^{11} antiprotons at a frequency of 3.3×10^5 Hz.

2.3 Cluster gas source

The use of a converging diverging nozzle[9] increases the intensity in the axial direction of expansion of the gas by a large factor. Figure 3b displays the intensity of the gas source from different nozzles. All the nozzles have the same throat diameter, hence the same quantity of expanding gas. This is due to the fact that with a suitable geometry (Fig. 4a) of the nozzle (converging/diverging), the rate of expansion can be slowed down. Then the description of the expansion given above is only valid up to the cross-section of the diverging part of the nozzle, where the values of p and T of the expanding gas reach the corresponding vapour-pressure values (Fig. 4a). At this time the gas is saturated and as the expansion

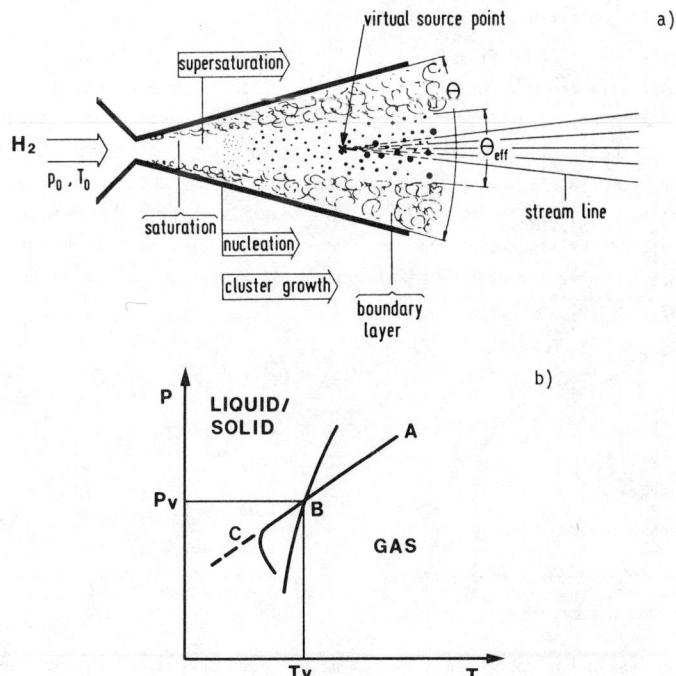

Fig. 4 a) Converging/diverging nozzles; the relevant phases of the generation of the gas beam are indicated. (From Ref. 9.) b) log p, log T diagram for H_2. p_v: vapour pressure. T_v: vapour temperature (see text).

goes on it will undergo a supersaturation condition in which, if the density and the pressure are large enough -- or in other words if the number of collisions is large -- the molecules start to form clusters. This nucleation phase will result in a central part of the beam being composed of clusters of molecules of a size which, depending on p_0, T_0, and nozzle geometry, can reach 10^5-10^6 molecules per cluster.

Figure 4b shows schematically the diagram of the expansion in a log p, log T plot. Expansion from point A to point B is described by Eqs. (1) to (5). The gas becomes supersaturated from B to C, and in C the formation and growth of clusters begin, that is to say p and T tend to move towards the equilibrium curve. The effects which are relevant for our applications are the following:

i) there is a formation of clusters;
ii) there is an increase in the density of the central region;
iii) clusters can fly straight for long distances in high-vacuum regions without disturbance from the residual gas.

A complete theory of condensation in beams of clusters has still not been formulated. The understanding of the thermodynamic properties of small clusters and the kinetics by which they are formed and destroyed is not complete. A complete condensation theory should be able to:

i) find the equilibrium concentration for a given cluster size in a supersaturated state;
ii) calculate the nucleation rate;
iii) calculate the further growth of clusters.

The main difficulties come from the fact that given a cluster size N, these clusters will not only interact with monomers but also with other clusters of any size.

A phenomenological approach is to try to formulate models which can find, from the properties of the beams of different gases, scaling laws which correlate the effects of initial pressure and temperature and the nozzle geometry. Let us consider as an example the corresponding jet model[8]. The main result of this model is to find the reduced values of p_0 and T_0 such that, on the assumption that the basic process (interaction of a cluster with a monomer) has the same strength for both gases, the properties of the beam of gas type 1 are the same as those of the beam of gas type 2.

$$p_0 = \frac{p_{0,1}}{\varepsilon_1/\sigma_1^3} = \frac{p_{0,2}}{\varepsilon_2/\sigma_2^3} = \text{const} \tag{8}$$

$$T_0 = \frac{T_{0,1}}{\varepsilon_1/K_1} = \frac{T_{0,2}}{\varepsilon_2/K_2} = \text{const}, \tag{9}$$

where for gas i,

ε_i: characteristic energy

σ_i: characteristic length

K_i: Knudsen number of expansion.

This is most valid for rare gases, but is also a good approximation for other gases (CO_2, N_2, etc). For the H_2 cluster beam a large effort has been devoted to formulating, from extensive measurements with different nozzle parameters[9], scaling laws for cluster size and intensity.

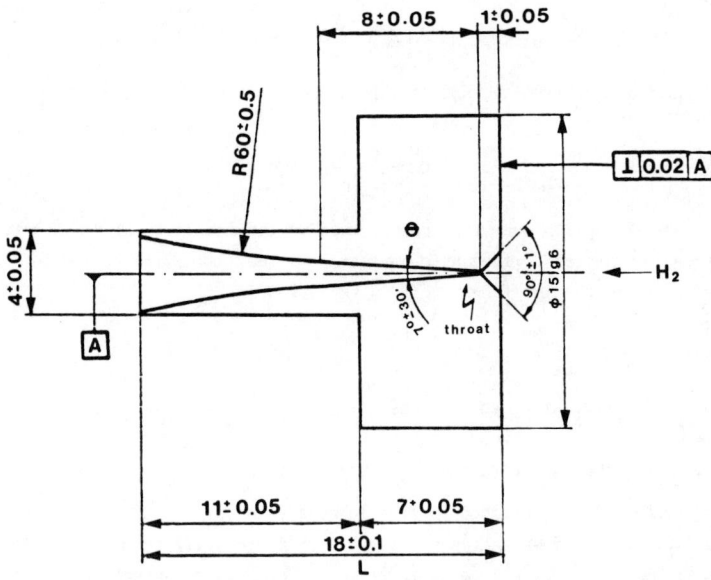

Fig. 5 Converging/diverging nozzle with special trumpet-shaped end part.

If the characteristics of the nozzle (Fig. 5) are semi-aperture θ, throat diameter D, length of diverging part L, one finds[9] that the cluster size N is

$$N \propto p_0 T_0^{-2.4} \left(\frac{D}{tg\,\theta}\right)^{1.5} L^{0.2}, \qquad (10)$$

with p_0 in pascals, T_0 in kelvins, and D and L in millimetres. The following observations are important:

i) N is limited: in fact p_0 and T_0 should be on the gaseous side of the vapour-pressure curve (Fig. 4b): at fixed T_0, p_0 cannot be raised above $p_v(T_0)$; and at fixed p_0, T_0 cannot be lowered below $T_v(p_0)$.

ii) The gas flux from the throat is $\propto p_0 d^2$: so a smaller nozzle throat implies a smaller gas flux for the same stagnation pressure p_0.

iii) The pressure in the expansion chamber should be $< 10^{-1}$ Torr to avoid beam attenuation, thus implying that a powerful pumping system has to be employed for the expansion chamber.

The scaling law (10) can be applied if the characteristics of the injector are also scaled. For example, to get the same intensity from two nozzles, one with D_1 and θ_1, the other with D_2 and θ_2, we should use pressures $p_{0,1}$ and $p_{0,2}$ such that:

$$p_{0,1}\left(\frac{D_1}{tg\,\theta_1}\right)^{1.5} = p_{0,2}\left(\frac{D_2}{tg\,\theta_2}\right)^{1.5}. \qquad (11)$$

For example, if $D_2 = D_1/3$ and $\theta_1 = 5°$, $\theta_2 = 3.5°$ we should use a pressure $p_{0,2} = 3p_{0,1}$. But the injector geometry to get the same expansion parameters is also different in the two cases and it should be calculated, according to Ref. 8, as follows:

$$\left(\frac{D_2}{2} + \ell \text{ tg } \theta_2\right)^2 = \left(\frac{D^2}{2}\right)^2 \left(\frac{\alpha+1}{2}\right)^{-B} \frac{\left[1 + (\alpha-1)(M^2/2)\right]^B}{M}, \tag{12}$$

where ℓ is the distance from the nozzle throat and $B = (\alpha+1)/[2(\alpha-1)]$. The length L_2 of the nozzle is:

$$L_2 = \frac{D_2 \text{ tg } \theta_1 L_1}{D_1 \text{ tg } \theta_2}. \tag{13}$$

The choice of the jet beam that can be used as the internal target in a storage ring depends on many parameters. The basic parameter is the luminosity of the interactions of the coasting beam on the jet target. It is expressed as:

$$L = I\omega\rho \text{ (cm}^{-2} \cdot \text{s}^{-1}), \tag{14}$$

where

I: number of circulating antiprotons,
ω: their frequency of revolution,
ρ: density of the target (atoms per square centimetre).

The maximum luminosity is obtained for the largest value of ρ such that the beam lifetime and size and the momentum spread Δp/p at equilibrium are still compatible with useful running conditions (a few days per antiproton injection). Fixing the density we can choose the other characteristics of the target, namely:

 i) the geometry of the pumping systems to absorb the jet (to be built in such a way that apart from the gas in the jet, no other large gas quantity enters the vacuum pipe),
 ii) the geometry of the production system,
iii) the distance from the nozzle throat to the interaction region, and the collimation geometry which determines the jet dimensions at the interaction region.

In the case of the ISR, we fixed the diameter of the H_2 beam at the crossing point with $\bar{p}$ = = 10 mm to match the particle beam equilibrium dimensions. The jet density had to be $\rho = 10^{14}$ atoms per square centimetre to obtain a luminosity of 10^{31} cm$^{-2} \cdot$ s^{-1} with 3×10^{11} antiprotons at 3.3×10^5 Hz. To determine the actual values of the jet parameters p_0, T_0, D, and the consequent nozzle geometry, the velocity of the clusters had to be fixed. Since we were working at T_0 = 77 K (liquid N_2 temperature) then v_{max} = 1290 m$\cdot$s^{-1}. The jet intensity was $I \simeq 5 \times 10^{18}$ molecules per square centimetre per second $\simeq 4.4 \times 10^{21}$ molecules per steradian per second. This could be obtained with $p_0 \simeq 10^6$ Pa and a trumpet-shaped nozzle (Fig. 5) with the following characteristics: D = 30 μm, θ = 3.5°, and L = 18 mm.

2.4 Polarized target

Atomic beams which can be obtained with the technique described above can be polarized with the selection of pure spin states[6,10]. In this case the formation of clusters has to be avoided and the resulting intensity will depend on i) the maximum pressure that can be used in the stagnation region without recombination (biatomic gas), ii) the value of T_0 of the source, and iii) the geometrical acceptance of the six-pole magnets used to focus the atoms in the interaction region. The speed of the atoms, which is a function of T_0, will influence the geometrical acceptance of the six-pole magnets. Polarized target of this type can reach densities of the order of 10^{11}-10^{12} atoms per square centimetre in the interaction region.

Stable atomic hydrogen can be produced at very low temperatures (0.3 K or below) with the use of very high fields (5-10 T) and all surfaces coated with a superfluid ^{4}He film[10]. An internal jet target could be built starting with a source of this type, but the technology is still in a development phase.

3. GASEOUS TARGETS IN STORAGE RINGS

3.1 Introduction

One of the characteristics of jet molecular or atomic beams is to be able to fly over long distances in high vacuum without absorption or diffusion by the rest gas. This, together with their high intensities, makes their use very attractive as internal targets. The problems are the following:

i) blow-up of the coasting beam dimensions and momentum degradation of the beam,
ii) increase of the pressure in the vacuum pipe of the ring.

These problems influence the lifetime of the beam and the luminosity that can be integrated for physics data taking. Other problems are connected with the operation of the target itself. Possible nozzle blockage or the contamination of the vacuum of the storage ring by the target system could lead to interruption of the operation for repairs, leading to loss of the coasting beam.

3.2 Target-system configuration

Figure 1 shows the schematic organization of an internal target. The jet of gas is produced in the gas expansion which takes place in chamber 1, and is absorbed in chamber 5 after having passed through chamber 2, chamber 3, the interaction region with the circulating particles, and chamber 4.

Our aim is to keep the value of the pressure in the interaction region (p_{int}) as near as possible to that in the case of operation of the storage ring without target. The clusters of the jet can enter and leave the vacuum pipe of the storage ring without perturbation of the pressure condition if the aperture towards the beam sink is large enough. So they will not produce a pressure rise in the interaction region. A rise of the pressure in the interaction region will only be provoked by the gas diffused in that region if $p_3 > p_{int}$ and $p_4 > p_{int}$. The amount of gas entering the storage ring vacuum pipe is $\dot{Q} = (P_3 C_3 + P_4 C_4)$, where C_3 and C_4 are the conductances between chambers 3 and 4 and the ISR vacuum pipe (see Fig. 1). And if $C_3 + C_4 \ll s_I$, where s_I is the pumping speed applied to the interaction region, we obtain $p_{int} = \dot{Q}/s_I$. Since C_3 and C_4 cannot be reduced below the values corresponding to the aperture needed to get the correct target dimensions at the interaction region with the circulating beam, one has to minimize p_3 and p_4, and maximize s_I; hence, the interest of interposing the maximum number of differential pumping stages compatible with the geometry of the system, in order to minimize the quantity of gas diffusing in the vacuum pipe. The reduction in pressure from chamber i to chamber i+1 is given by $p_{i+1}/p_i \simeq C_{(i \to i+1)}/s_{(i+1)}$ if $C_{(i \to i+1)} \ll s_{(i+1)}$.

3.3 Interactions of stored beam on the jet target

Losses of particles of the coasting beam can be caused by small-angle scattering, which adds up at each crossing. This multiple scattering blow-up can be compensated by a cooling system which achieves transverse cooling time smaller than the multiple scattering blow-up time. Then particles of the beam are lost if they undergo (i) elastic scattering at angles larger than the machine acceptance or (ii) inelastic reactions.

$$I = I_0 \, e^{-\sigma \rho \omega t} \,, \tag{15}$$

where

I_0: initial intensity
ω : revolution frequency
ρ : density of the jet (atoms/cm^2),

$$\sigma = \sigma_{el}(> \theta_0) + \sigma_{inel} \tag{16}$$

θ_0 being the storage-ring acceptance angle.

3.4 Beam blow-up

The blow-up rate is

$$\frac{1}{\sigma}\frac{d\sigma}{dt} = 0.698 \, \frac{\bar{\beta} p_{ms}}{p^2 E} \,, \quad E = (\alpha^2 - 1)^{-\frac{1}{2}} \varepsilon \,, \tag{17}$$

where

$\alpha = \frac{1}{2}\left[(E\bar{\beta})/\pi\right]^{\frac{1}{2}}$,

p_{ms} = equivalent distributed N_2 pressure.

To counteract multiple scattering the cooling rate should be of the order of or larger than the blow-up rate.

The beam size at equilibrium is found from the emittance balance. If $\dot{E}_{ms}$ is the rate of emittance increase due to the multiple scattering and $\dot{E}_c$ is the rate of emittance due to cooling, then (in real life corrections have to be applied to take into account electronic noise):

$$\frac{dE}{dt} = \dot{E}_{ms} + \dot{E}_c \tag{18}$$

$$\dot{E}_{ms} = 0.4 (\bar{\beta})^{3/2} \, \frac{p_{ms}}{p^2} \tag{19}$$

$$\dot{E}_c = \frac{2}{\tau} E \tag{20}$$

where τ is the amplitude cooling time

$$E(t) = \left(1 - e^{-2t/\tau}\right)\tau \, \frac{\dot{E}_{ms}}{2} + E_0 \, e^{-2t/\tau} \,, \tag{21}$$

where E_0 is the initial beam emittance.

3.5 Beam loss by target system breakdown

Nozzle blockage can occur owing to mechanical blockage (remember the nozzle throat is very small) or condensation of gas on the cold nozzle. The pumping systems should be oil-free and fast acting ultra-high vacuum (UHV) valves should separate the target system from the vacuum pipe to avoid contamination or sudden pressure bump.

3.6 ISR internal target

The density of the target[11] was chosen as the maximum compatible with a reasonable beam lifetime, and the beam characteristics were determined by the horizontal and vertical cooling systems of the ISR[7]. The gas jet dimensions were determined by a series of skimmers, which also separate the production system into three different stages. Each stage had an individual 'oil-free' pumping system. The nozzle was mounted on a moving mechanism to optimize its position with respect to the axis defined by the skimmers and thus the intensity. The jet was absorbed in the sink system, which was built of three cryogenic pumps dimensioned to minimize the flow of gas streaming back towards the interaction region (Figs. 6, 7).

A system to monitor the jet intensity was mounted before the last cryopump, which was very large (120,000 $\ell \cdot s^{-1}$). The monitor gives a measurement of the vertical projection of the jet intensity. The dimensions of the interaction region were $6 \times 8 \times 10$ mm^3. This is very helpful in the design of a compact experimental apparatus and enhances its capability. Also, analysis of data is easier (pattern recognition, etc.).

Particular care was devoted to the H_2 injection line in order to avoid nozzle blockages by gas impurities. The following points are very important:

  i) the use of pure H_2 (10 ppm maximum),
 ii) the use of mechanical filters (2 μm),
iii) the use of a condensation trap (LN_2 temperature),
 iv) the use of an active charcoal trap (LN_2 temperature).

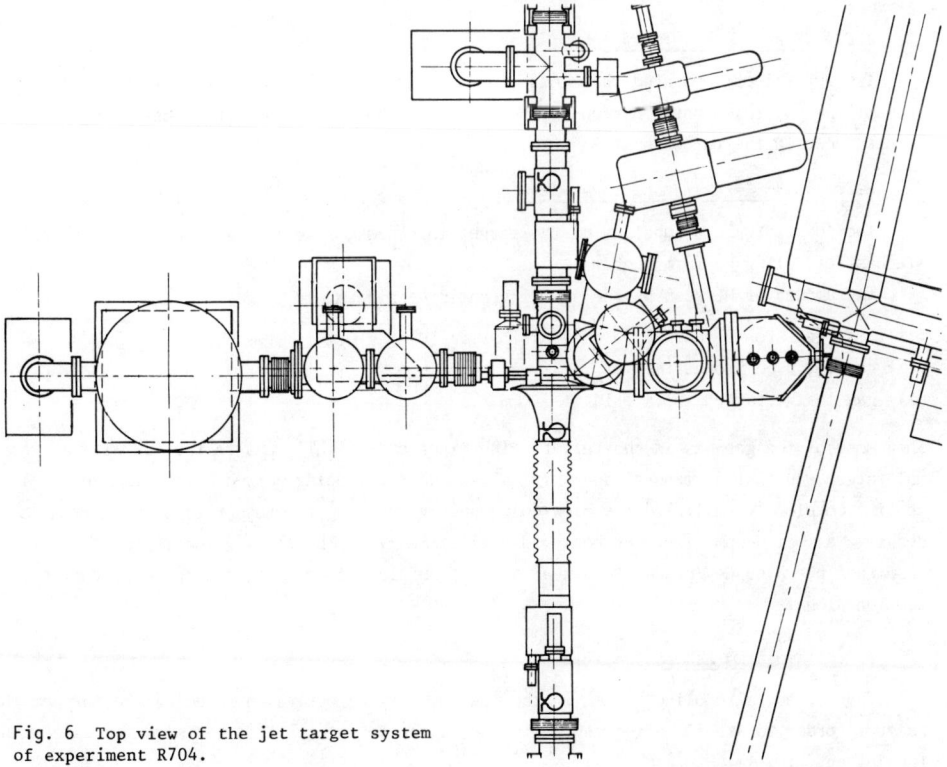

Fig. 6 Top view of the jet target system of experiment R704.

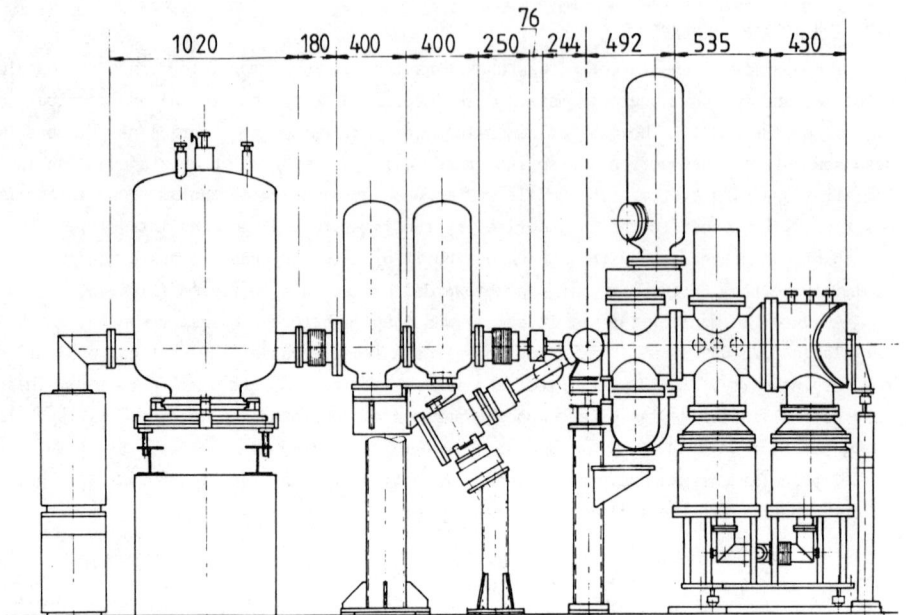

Fig. 7 Side view of the jet target system of experiment R704.

3.6.1 Ease of on/off jet conditions

Two UHV valves separated the production and sink from the vacuum pipe. This was to prevent the possible contamination of ISR ring 2 in the case of a large pressure bump due to breakdown in the target system.

3.6.2 Beam lifetime and loss rate

The loss rate is a function of the target thickness, rest gas, machine aperture, and stochastic cooling performances.

The amount of H_2 seen by the coasting antiproton beam was

i) rest gas around the ISR (p = 10^{-11} Torr) $\approx 10^{-12}$ g·cm^{-2}
ii) gas diffused around the jet $\sim 2 \times 10^{-12}$ g·cm^{-2}
iii) gas in the target $\sim 1.5 \times 10^{-10}$ g·cm^{-2}.

Then most of the gas was in the target. The experience of all runs in the ISR with an internal target was that horizontal and vertical stochastic cooling worked well. The vertical cooling completely nullified the blow-up caused by the target. Loss rate at a maximum target density was ~ 170 ppm, i.e. the beam had a lifetime of 100 h. This value is for the worst case when p = 3.68 GeV/c and it is near to that obtained from the total p$\bar{\text{p}}$ nuclear cross-section alone.

3.6.3 Momentum resolution

The horizontal cooling system in the ISR was very effective in reducing the $\Delta p/p$ of the beam, in order to stabilize the central value of the momentum. This was also the case with the jet on (Fig. 8).

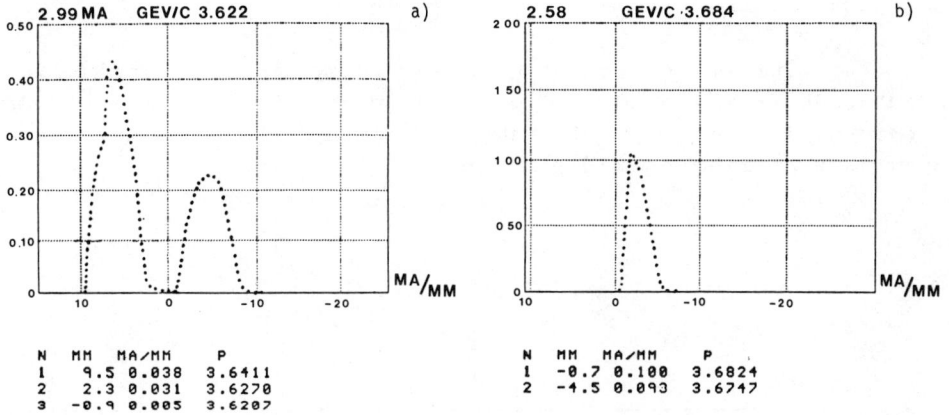

Fig. 8 a) Two pulses stacked in run 1366 during initial momentum cooling. b) Final stack after 12 hours of physics and continued momentum cooling. (Note change of vertical scale.)

The value of $\Delta p/p$ can be as low as 4×10^{-4} when the antiprotons traverse the jet target. In this case, in 'formation' experiments the width of the state will be measured using the machine parameters rather than the resolution given by the experimental apparatus. This leads to a much better precision of the measurement, for example:

$$\Delta M \simeq \frac{0.938}{\sqrt{s}} \Delta p \,, \quad \Delta M(\eta_c) = 460 \text{ keV} \,. \tag{22}$$

Final-state particles had to traverse a thickness of 0.3 mm of stainless steel. This does not introduce a large multiple scattering angle. If the detection takes place inside the vacuum pipe, particles of very low momentum can be detected: the luminosity monitor is, for example, obtained through the measurement of recoil protons of momentum. It should be mentioned here that there have been applications of an internal target in Saturn at Saclay[12] (Fig. 9) and at FLAB[13] to measure pp elastic scattering at very small angles. This last application is only suitable for an accelerating cycle.

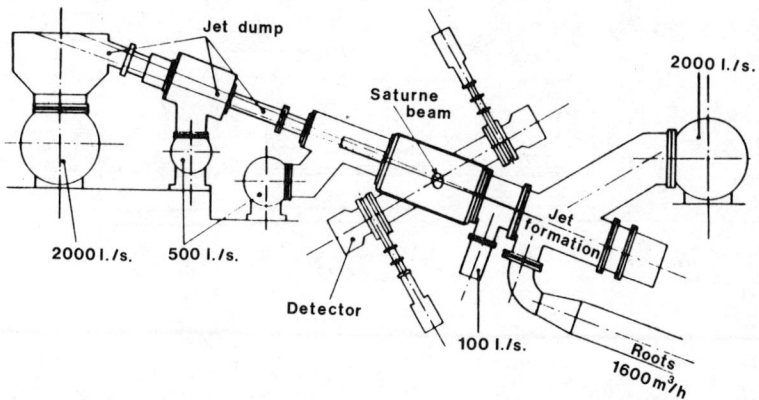

Fig. 9 Saclay jet target system.

3.7 SPS jet target

The working conditions are $T_0 = 30$ K, $p_0 = 1 \times 10^5$ Pa. The nozzle is trumpet-shaped with a throat diameter of 100 µm (Figs. 10 and 11). The thickness is 4×10^{14} atoms per square centimetre, in order to compensate for the reduced revolution frequency. With 6×10^{11} antiprotons in the coating beam the luminosity is

$$L = 10^{31} \text{ cm}^{-2} \cdot \text{s}^{-1} \ . \tag{23}$$

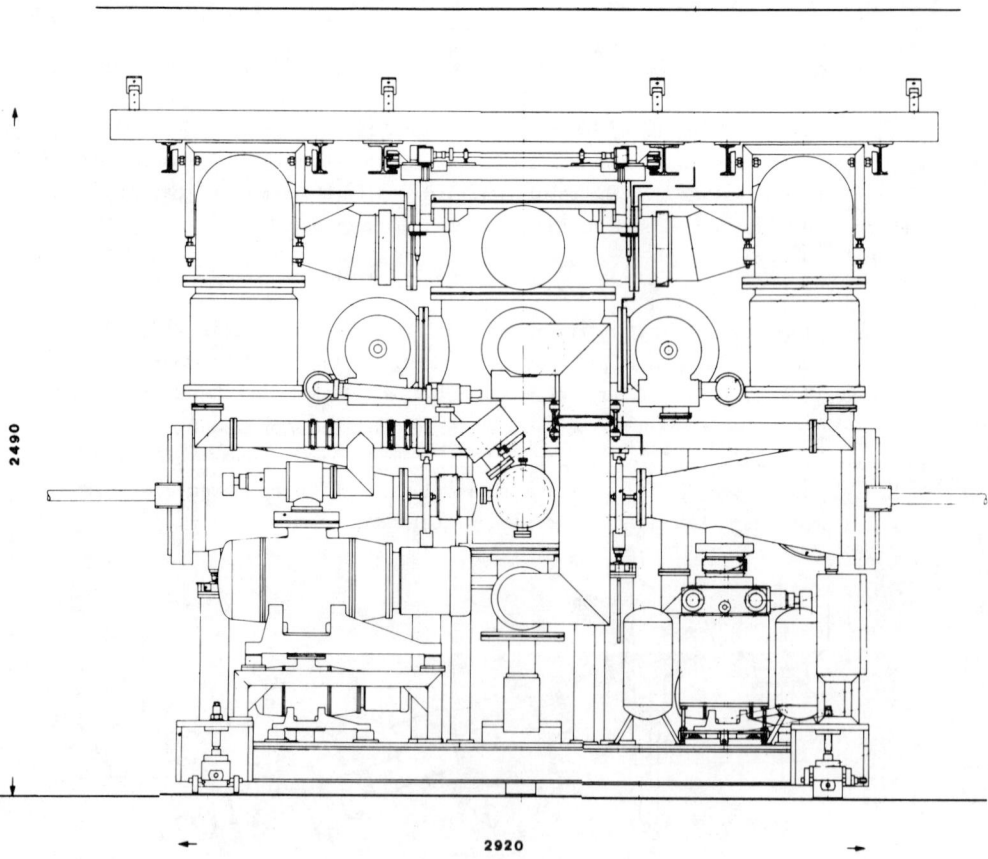

Fig. 10 SPS jet target system.

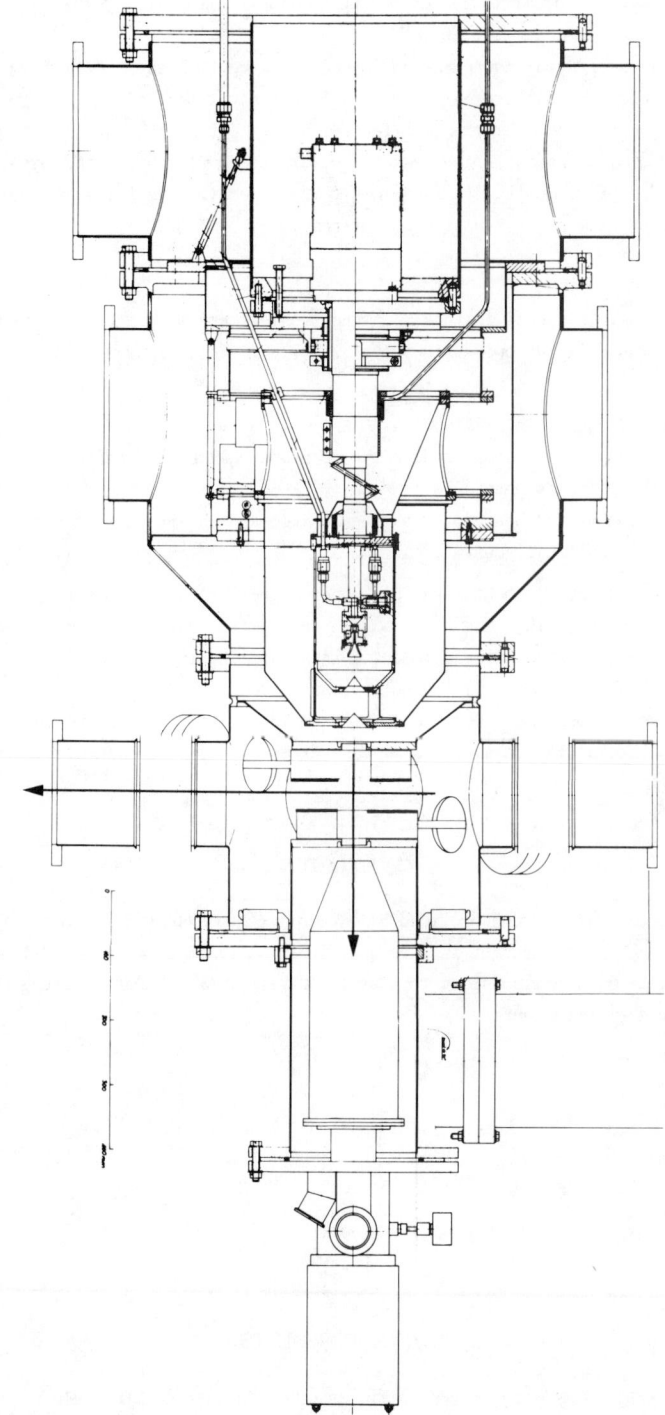

Fig. 11 SPS jet target system with details of expansion-chamber arrangement.

This is a compromise in order to reduce the p̄p Collider luminosity by 10% per day. In this case the momentum of the beam is 270 GeV/c.

The multiple-scattering contribution to beam blow-up is less important, but no stochastic cooling is applied. The calculated losses are 3% per day. The calculated losses due to interaction luminosity (both beams are affected) are 6% per day. Tests are in preparation.

Also of great importance is the test to measure the perturbation induced by the target on the UA1 and UA2 experiments. In fact this target operation is foreseen together with the UA1 and UA2 experiments (parasitic operations).

3.8 A jet target for LEAR

The third possible use is in LEAR[14]. The problems are much the same as in the ISR, but the momenta range is in lower values (0.1-2 GeV/c). The following conditions are needed:

- stochastic cooling (this exists);
- electron cooling, which would reduce the $\Delta p/p$ and enlarge the range of momenta for a jet operation (this is envisaged);
- low-β insertion at the beam-jet crossing to increase the machine acceptance (this would reduce loss by multiple scattering).

Depending on the beam momentum various modes of operation can be foreseen. A thickness of the target as large as 10^{14} atoms per square centimetre can be obtained and used. The actual thickness will depend on beam momentum, availability of antiprotons from the Antiproton Accumulator (AA), and the intensity of pulse transferred from it[14]. This would provide a very powerful tool to perform experiments which would benefit from the performances of the antiproton beam interacting on an internal gas target that I have been discussing.

4. CONCLUSION

The source of interactions provided by a coasting beam (to which cooling is applied) hitting a gas jet internal target is a very efficient tool to study in detail nuclear and particle physics. It makes possible the detection of narrow resonant states and the study of low cross-section phenomena.

ACKNOWLEDGEMENTS

I wish to express my warm thanks to Professor U. Valbusa for the helpful conversations I have had with him.

REFERENCES

1) C. Baglin et al., Charmonium spectroscopy at the ISR using an antiproton beam and a hydrogen jet target, CERN/ISRC/80-14 (1980).

2) C. Baglin et al., First evidence of J/ψ formation in antiproton-proton annihilation, *in* Proc. 1983 Int. Symposium on Lepton and Photon Interactions at High Energies, Cornell, 1983 (Cornell University, Ithaca, 1983), p. 906.

3) R. Cester, Formation of charmonium states in antiproton-proton annihilation, *in* Proc. Workshop on Heavy Flavours, Como, 1983 (Ed. Frontières, Paris, 1984) p. 327.

4) C. Baglin et al., Antiproton collisions with a gas H_2 jet and formation of charmonium states, *in* Proc. 12th Int. Conf. on High Energy Accelerators, Fermilab, Batavia, 1983 (Fermilab, Batavia, 1984), p. 251.

5) C. Baglin et al., Formation of charmonium states in antiproton-proton annihilation to be published in Proc. 22nd Int. Conf. on High-Energy Physics, Leipzig, 1984.

6) J. Antille et al., CERN/SPSC/80-63 (1980).

7) A. Peschard and M. Studer, Stochastic cooling in the CERN ISR during p$\bar{\text{p}}$ colliding beam physics, CERN/LEP/ISR/RF/83-15 (1983).

8) O.F. Hagena and W. Obert, Cluster formation in expanding supersonic jets, J. Chem. Phys. 56, 1793 (1972).
   O.F. Hagena, Surface Science, 106, 101 (1981).

9) W. Obert, Properties of clusters beams formed with supersonic nozzles, 11th Symposium on Rarefied Gas Dynamics, Cannes, 1978 (CEA, Paris, 1978), p. 1181.

10) T.O. Niinikoski, CERN-EP/83-124 (1983) and references therein.

11) M. Macri, A clustered H_2 beam, *in* Physics with Low Energy Cooled Antiprotons (Plenum, New York, 1983) p. 432.

12) R. Burgel et al., Nucl. Instrum. Methods 204, 53 (1982).

13) V.A. Nikitin et al., Adventures in Experimental Physics 8, 263 (1972).

14) LEAR Design Study, CERN/PS/DL/80-7 (1980).
    P. Lefevre, D. Möhl and G. Plass, *in* Proc. 11th Int. Conf. on High Energy Accelerators, Geneva, 1980 (Birkhäuser, Basel, 1980), p. 819.
    K. Kilian and D. Möhl, CERN p̄ LEAR note 44 (1979).

TARGET DEVELOPMENTS FOR CELSIUS

C. Ekström, B. Holmqvist and H. Sterner
The Studsvik Science Research Laboratory,
S-61182 Nyköping, Sweden

ABSTRACT

The target developments for the CELSIUS storage ring project at Uppsala is concentrated to a gas-jet target, primarily for cluster beams of hydrogen and deuterium, and different fiber targets of refractory materials. Following a general discussion of internal targets possible in storage ring experiments, details of the gas-jet and fiber targets, under development at Studsvik, will be presented.

GENERAL

There is a large choice of internal targets to be used in storage ring experiments. A summary is given in Table I. In addition, more exotic ones, like dust targets and pellet targets, should be considered.

The target thicknesses of the atomic-beam and molecular-beam targets in Table I are given for a distance of 25 cm between the beam source and the crossing with the circulating beam, and a target width of 1 cm at the crossing. The higher values of the target thicknesses, found in the literature, refer to systems where the ion beam intercepts the target beam very close to the source. This, however, gives such a pressure rise that storage ring operation is excluded. A strong differential pumping is required between the target-beam source and the crossing, thus a distance of typically 25 cm, and as a result, relatively low values of the target thicknesses.

Atomic-beam targets may be obtained by evaporation of a metal in some kind of oven, or through dissociation of gas molecules. Atomic beams are often used in connection with polarized targets, and they reach target thicknesses of 10^{12} atoms/cm^2. The polarization may be obtained either by state selection in sextupole magnets or by optical pumping induced by laser light. The possibility of using polarized atomic-beam targets in storage ring experiments has been discussed in Ref. 1.

The molecular-beam targets are essentially of three types (Cf. Table I). At low input pressures, there is a molecular flow through some kind of nozzle with target thicknesses of 10^{12} atoms/cm^2, comparable with those of the atomic-beam sources. At higher pressures we get a supersonic gas jet, which at room temperatures reaches 10^{13} atoms/cm^2. When it is cooled, a so-called cluster beam is formed with molecules of $10^4 - 10^5$ atoms. This target gives at present the highest thicknesses - 10^{14} atoms/cm^2. They have been used in high-energy physics experiments in the Saturne-ring at Saclay [2]

Table I   Internal Targets

	Target thickness atoms/cm^2
Atomic-beam targets	
Metallic evaporation Molecular dissociation (+polarization)	$<10^{12}$
Molecular-beam targets	
Molecular effusion	$<10^{12}$
Supersonic, room temp.; Gas jet	$<10^{13}$
Supersonic, cooled; Cluster beam	$<10^{14}$
Ion-beam target	
ISBIT	est. $10^{12} - 10^{14}$
Fiber targets	eff. $<10^{16}$

and at ISR [3]) and SPS [4]) at CERN. This type of target, under construction for CELSIUS, will be discussed in more detail below.

A new target concept, introduced by G. Andersson [5]), is based on a small target storage ring, ISBIT, intercepting the main ring. The ion beams may be injected into the target ring either directly from an ion source or after mass separation. There is thus a wide choice of target material, and the estimated target thicknesses range between 10^{12} and 10^{14} atoms/cm^2.

The fiber targets are substantially thicker than the gas targets. The effective target thickness, obtained by taking into account the overlap between the circulating beam and the fiber, is however limited to about 10^{16} atoms/cm^2 by the capacity of the electron cooling and by the heating and evaporation of the fiber. Different types of fibers are now being tested at Studsvik.

THE GAS-JET TARGET

The hydrogen gas-jet target, under construction at Studsvik, is to a large extent based on the design of the SPS-target [4]) at CERN. The central parts, shown in Fig. 1, consist of a beam source with a nozzle, skimmer and collimators; the experimental region where the ion-beam in the main ring intercepts the target beam; and finally a target-beam dump. Hydrogen gas of a few atm is pressed through the 0.1 mm aperture of the nozzle which is cooled to 20 K by a two-stage cryogenerator. The hydrogen cluster beam thus formed shows a well bounded intensity profile. The flow through the nozzle is typically 50 torr l/s, of which about 1% is used in the target beam. This gives a thickness of 10^{14} atoms/cm^2 or 2×10^{-10} g/cm^2. At the interception, the cross-section of the target beam is 5 mm x 10 mm, with possibilities to move the beam ± 5 mm in the horizontal plane. Assuming 10^{10} protons of 200 MeV in the

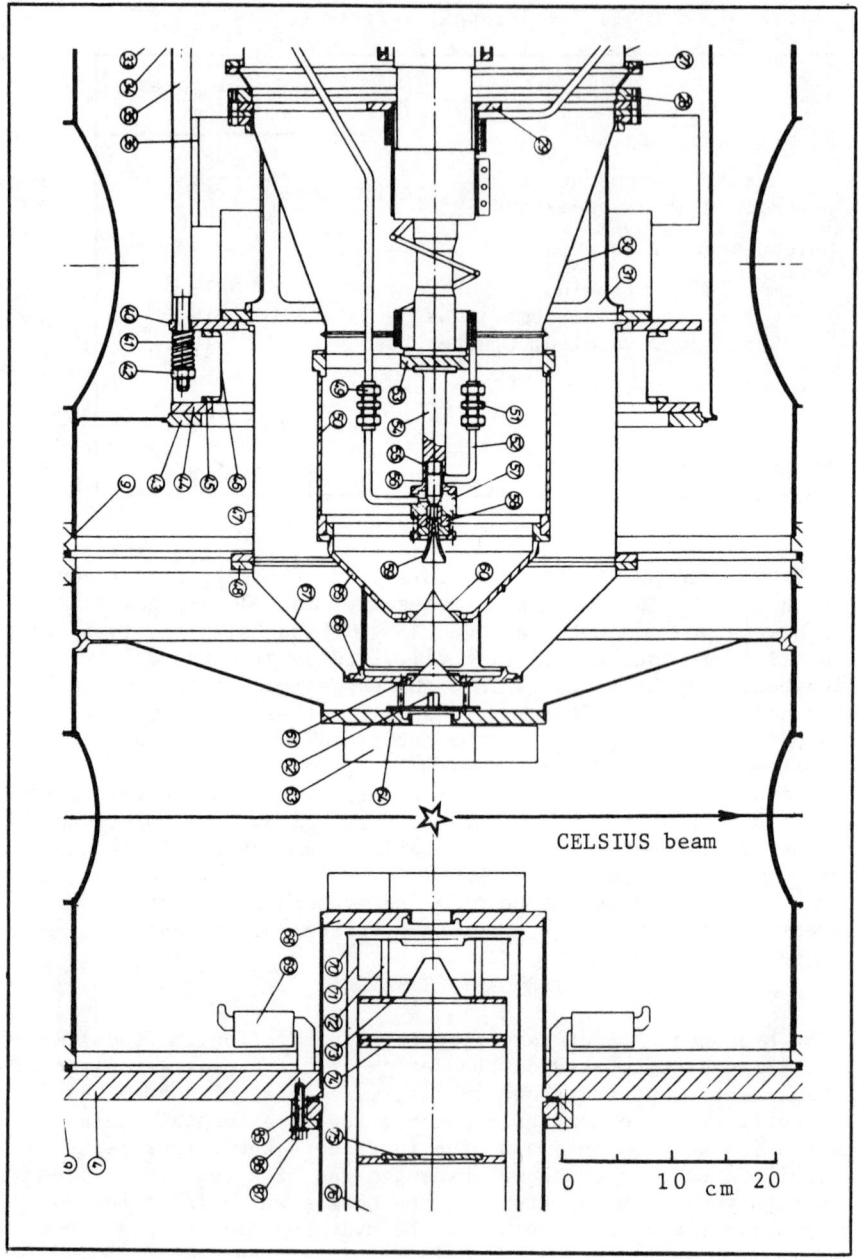

Fig. 1 Layout drawing of the central part of the gas-jet target showing the beam source with the cooled nozzle, skimmer and collimators; the experimental region where the CELSIUS beam intercepts the target beam; and the target beam dump.

CELSIUS ring, overlapping the target beam, the luminosity amounts to 2.3×10^{30} cm^{-2} s^{-1}.

The strong differential pumping of the target-beam source is made in three steps. The main part of the hydrogen gas is pumped off in the first stage around the nozzle by a set of four turbo-molecular pumps backed by a roots and a rotary pump. The two following steps have a turbo-molecular pump each and a common roots and rotary pump. The target-beam dump is evacuated by a specially made cryopump. A strong pumping of the experimental chamber will be required to improve on the vacuum of 10^{-7} torr reported in previous gas-target experiments [2-4]. Differential pumping towards the ring will also be needed.

As shown in Fig. 1, the target beam will intercept the CELSIUS beam vertically, leaving the horizontal plane free for detectors and connections to spectrometers. In the construction, attempts have been made to free as much as possible the forward and inward directions from pumps and vacuum lines. The final design of the experimental chamber is still awaiting. Instead of a general purpose chamber, different ones probably have to be used depending on the experiment to be done.

The gas-jet target, although designed primarily for cluster beams of hydrogen and deuterium, may also be used for other gases by adapting the input parameters for the gas and the nozzle.

The cost of the gas target is estimated to 0.4 million dollars the main part going into the extensive pumping system. Concerning the time schedule, the stand for the target is now being mounted in the new experimental hall at Studsvik, and bids for the manifacture of the target with a simple test chamber have been asked for. We expect at least a two-year period for the manifacture, mounting and testing of the target before it will be moved to the CELSIUS ring at Uppsala.

FIBER TARGETS

The second type of targets being developed at Studsvik is the so-called fiber targets. They are expected to be simpler than the gas target both in construction and operation. In particular, they will not require an extensive pumping system. The problems are instead connected with the manifacture and handling of the thin fibers, diameter of about 1 μm, and that one is limited to a few refractory materials like carbon, quartz and tungsten.

To simulate the conditions in the CELSIUS ring, a number of tests are being prepared at the Van de Graaff accelerator at Studsvik. In a first test experiment, a tungsten fiber with a diameter of 4.5 μm was irradiated with 3.5 MeV protons of 25 μA. The beam cross-section was 5 mm x 5 mm, i.e. a current of 23 nA onto the fiber. The energy deposition under these conditions is 250 keV [6] and the power deposition 5.75 mW. The corresponding energy deposition with 200 MeV protons in CELSIUS is 20 keV which means a current of 288 nA onto the fiber to get the same power deposition. If the cross-section of the CELSIUS beam is 1 mm x 5 mm, the total current amounts to 64 μA. This, in turn, means 1.7×10^8 protons in the ring and a luminosity of 4×10^{31} cm^{-2} s^{-1}.

The tungsten fiber was irradiated during 5.5 h, and showed after that period an uneven surface structure and a high degree of brittleness at the irradiated part. The ratio between the resistance of the fiber before and during the irradiation indicates a temperature of about 1500 K. This value is different from the 1100 K obtained from the Stefan-Boltzmann´s law and the assumption that all power was carried away by radiation. The difference has to be looked into in further experiments. This first test has shown, however, that it is possible to handle a thin tungsten fiber and that it can stand a power deposition which corresponds to a rather high CELSIUS luminosity.

The evaporation of the fiber will put a limit on the temperature, power deposition and thus the luminosity in this kind of experiments. Tungsten represents one of the most favourable cases with a very low vapour pressure and a luminosity limit of about 10^{33} cm^{-2} s^{-1}. Almost all other fibers will evaporate at lower temperatures, leading to lower luminosity limits.

We will continue with the tests at the Van de Graaff accelerator to learn more on the handling of these thin fibers and to study their properties in simulated CELSIUS experiments.

We would like to thank Dr. W. Kubishta, CERN, for several clarifying discussions about the gas-jet target.

REFERENCES

1. C. Ekström, CELSIUS-Note 83-6, Uppsala, 1983.
2. R. Burgei, M. Garçon, M. Grand, B. Gonel, R. Maillard, A. Malthiery and J. Martin, Nucl. Instr. and Meth. 204, 53 (1982).
3. C. Baglin et al., CERN/ISRC/80-14 (1980), and M. Macri (private communication).
4. J. Antille et al., CERN/SPSC/80-63 (1980), and W. Kubishta (private communication).
5. G. Andersson, CELSIUS-Note 83-18, Uppsala 1983.
6. H.H. Andersen and J.F. Ziegler, Stopping Powers and Ranges in All Elements, Vol. 3 (Pergamon Press Inc. 1977).

SESSION C

SHORT-RANGE AND

OFF-SHELL STUDIES

Short Range Effects
in Nucleon-Nucleus Scattering

Edward F. Redish

Department of Physics and Astronomy
University of Maryland
College Park, Maryland 20742

ABSTRACT

The character of the two-nucleon wave function within the range of the interaction potential is manifested in intermediate energy nucleon-nucleus scattering via off-shell effects and a resulting non-locality of the T operator. I present here a study of the off-shell and non-local behavior of the free nucleon-nucleon T matrix for the RSC and Paris potentials in the energy regime between 100 and 350 MeV. We find the following results: (1) The off-shell amplitudes for the two potentials are strikingly similar, both for the individual partial waves and for the central and spin-orbit parts of the amplitude, despite large differences in the momentum-space matrix elements of the potential. (2) The amplitudes are strongly non-local. (3) Although they are non-local, they are nearly independent of the angle between the vectors $\mathbf{k}-\mathbf{k}'$ and $\mathbf{k}+\mathbf{k}'$, and they vary very slowly with energy in the regime considered. This reduces a function of 4 variables to a function of 2. (4) Elastic scattering calculations using the free non-local T operator in an approximate way give results comparable to the Love-Franey Ansatz.

INTRODUCTION

In traditional nuclear physics, the nucleus is considered as being made up of nucleons moving non-relativistically and interacting with effective two-body potentials. These potentials are obtained phenomenologically by choosing some forms, perhaps motivated by a more general theory (such as relativistic quantum field theory or dispersion relations), and adjusting the parameters of these forms to fit two-body data. These potentials are then used in many nucleon Schrödinger equations which are solved using some approximation scheme. I refer to this approach as the *standard model* of nuclear physics.*

Nuclear physics is currently in a state of turmoil. The standard model appears to fail to give a number of detailed properties of nuclei, though it has many qualitative successes. These failures have led nuclear physicists to seek to develop new fundamental assumptions, some of which are spectacularly successful in narrow areas.

None of these models has, however, attempted to explain the wide variety of phenomena to which the standard model has been applied. The standard model, when applied carefully with an attempt to control errors, can serve as a benchmark against which new approaches can be measured. Since none of the revolutionary approaches can yet be considered as approximations to a complete theory, even in principle, sharp identifications of the failures of the standard model may provide guidance in formulating the next generation's standard model for nuclear physics. Furthermore, the standard model provides an excellent testing ground - one in which a more complete treatment of the dynamics can be considered than in other, more complex cases. The results of the standard model can be used to provide guidance as to what physics is important to include in building more sophisticated models.

Part of the problem with the standard model is that the potential which serves as its foundation is badly determined. Potentials which fit two-body data equally well may differ substantially when compared in detail. We do not yet know in what way the output of standard-model calculations depend on these uncertainties. The essential question is: How do two nucleons behave when they are within the range of their potential? The long-range part of the nucleon-nucleon wave function (outside of 3 f) is well known. The phase shift and the one-pion exchange potential dominate. Between 1 and 3 f (the mid-range) most realistic phenomenological potentials agree reasonably well and this is usually used as an argument to say we understand the nucleon-nucleon system in to separations of about 1 f.

However, some recent models of the nucleon's internal structure predict that sub-nucleon degrees of freedom come in to play at two-nucleon separations of as much as 1.5–2 f. Since there is no experimental evidence confirming the prejudices based on phenomenological potentials, I will consider the region inside of 2 f as being short range. Any experimental information that can be extracted about this region could be of considerable importance.

In this talk, I present a progress report on a project to attempt to probe the short range character of the effective nucleon-nucleon interaction inside a nucleus by an analysis of elastic and inelastic nucleon-nucleus scattering at the low end of the intermediate energy (IE) regime (100–350 MeV). At higher energies (800–1000 MeV), the impulse approximation (= single scattering + on-shell) works well [1], especially if supplemented by the inclusion of some nucleon resonances [2]. At the higher energies on-shell approximations suffice and we could imagine many theories leading to the same or similar results. Therefore, we expect to learn little either about the validity of the standard model or about the behavior of two nucleons at short ranges. Below about 300 MeV correlations become more important and some interesting sensitivities are observed. Inelastic scattering in particular appears to be sensitive to the

---

* As a result of the extensive information obtained over the past decade from intermediate energy accelerators, a strong need has been demonstrated to include at least some additional degrees of freedom: a small number of nucleon resonances seem to be required and an occasional meson-exchange current. I include these developments under the rubric "standard model" when they can be included perturbatively.

off-shell and density dependence of the effective interaction. The analyses currently used are based either on phenomenological approaches to the free T matrix [3] or to the G matrix in infinite nuclear matter [4]. The connection with the standard model is weak or non-existent, as the off-shell or off-diagonal matrix elements are extrapolated rather than calculated.

I report here on some early results of studies Karen Stricker-Bauer and I are carrying out on the off-shell structure of the free nucleon-nucleon interaction. I will also discuss some preliminary implications of these results for the analysis of nucleon-nucleus elastic and inelastic scattering.

The structure of the paper is as follows. In the next section I consider the state of the standard model with respect to some new proposals, and the non-local and off-shell character of the effective interaction. In section III I discuss briefly the theoretical basis for the single scattering approximations and look at their non-locality structure in detail. I then explain how we carry out the off-shell calculation. In section IV I display a sample of the off-shell results and analyze the dependence on the various variables. In section V, I present some of the implications of these results and discuss my conclusions.

BACKGROUND AND MOTIVATION

The State of the Standard Model

The standard model does reasonably well in describing nuclear properties to the level of a few percent (if one's quantities are appropriately chosen). Worse failures occur when sensitive quantities are predicted (*e.g.*, ones which are the difference of two large numbers, or which are much smaller than leading terms). The theory

- has been unable to correctly give the binding energy and radii of nuclei, whether this latter is reflected in the density of nuclear matter [5], or in the position of the dip of the charge form-factor for ^{3}He [6];
- has been unable to give correctly the details of the spin dependence for nucleon-nucleus scattering at IE where the model is "well understood" [7];
- has been unable to predict the properties of nucleons in nuclei as measured by the EMC effect [8].

Some of these failures may be ameliorated by the extension of the standard model to include energy dependent or non-local two-body potentials, or by introducing three-body forces. We don't want to carry this too far, however. Even were sub-nucleon or relativistic degrees of freedom of great importance, we could "in principle" formally reduce the problem to one with nucleons [9]. The impact of choosing inappropriate degrees of freedom would be that the effective potentials and effective operators would be exceedingly complex —the two-body potentials strongly energy dependent and strongly non-local, and complicated many-body forces would be required. If the standard model requires extreme modifications, we may take it as a signal that we are using the wrong degrees of freedom and must start again.

New Degrees of Freedom

Some attempts have been made to build models of nuclear systems using other degrees of freedom than those available to non-relativistic nucleons (with an occasional delta or pion). The most important challenges to the standard model are:

- Relativistic (Dirac) degrees of freedom: small components for nucleons are included dynamically [10].
- Relativistic mean-field degrees of freedom: meson fields are added to the Dirac nucleons [11].
- Internal degrees of freedom: Quark models are now very diverse and wildly speculative. For each nucleon one usually includes three particle degrees of freedom which may be Pauli (non-relativistic) [12] or Dirac (relativistic) [13]. One may also include fields to represent the effects of large numbers of soft gluons [14].

These models are fundamentally different from the standard model, but at the present moment, none is part of a complete theory in the sense that a wide variety of phenomena can be addressed in a low order approximation, and systematic corrections sketched out. On the other hand, the standard model can address a wide variety of phenomena systematically and corrections can (in principle) be calculated, but so far the best calculations give results which seem to be irreducibly in error by a few percent.

There have been two sharp tests of the standard model: the properties of nuclear matter and of ^{3}He. These are essentially the only cases where a full calculation as prescribed by the standard model has been carried out and higher order terms calculated to the point that errors are controlled [5][6]. The results in nuclear matter miss the binding energy by about 3–4 MeV/nucleon, while that in ^{3}He miss by ½–1 MeV. These are 5–10% effects when the errors are compared separately to the kinetic and potential energies (two large terms which nearly cancel). The trinucleon case is particularly clean since the three-body Schrödinger equation can now be solved directly by Faddeev methods to almost any desired accuracy. There is no problem of developing a convergent scheme of approximations as one has in nuclear matter.

It seems that in both cases a moderately small amount of three-body force could solve the problems. It is difficult to include this in the framework of the standard model, however. A three-body force can be extremely complicated in its dependence on spin, isospin, and three continuous variables. If the two-nucleon force provides any guidance, the three-nucleon force will be as complicated as possible. At present, three-body experiments are insufficient to pin these dependences down accurately, though there have been some useful theoretically developed models [15].

The uncertainties in which degrees of freedom must be used in nuclear physics may be probed in two ways. We can either ask about the free N-N wave function inside the range of the potential or we can ask about the effect of other nucleons on a pair. The former will affect the off-shell behavior of the nucleon-nucleon amplitude; the latter will modify on and off-shell matrix elements. We may try to measure both experimentally via a study of the effective interaction of pairs in nuclei.

The off-shell behavior of the free N-N amplitude tells about the short range part of the two-body wave function. This is of particular importance in developing quark models of nucleons. Some models [12] indicate that the transition from a quark picture to a meson-nucleon picture may take place between 1 and 2 f. Since the average spacing between nucleons in a nucleus is between these two numbers, the specification of that distance is crucial. One does not have to go very far off-shell to probe these distances.

Any information that can be obtained about the three-nucleon force would be of great interest, especially if some constraints could be placed on its spin and isospin dependence. This may be probed through studying the density dependence of the effective interaction. (A three-body force leads to a density dependence in the effective interaction if one extracts it only assuming two-body forces.) This fact is often overlooked by workers who compare calculated density dependences with experiment. It is inconsistent to compare a nuclear matter calculation of density dependence with data, since these calculations do not predict the correct properties of nuclei. It is perhaps more logical to have a phenomenological form for the density dependence and to extract this empirically first [16]. Then we may try to disentangle how much of it is due to Pauli and how much to three-body effects.

A careful calculation of the non-relativistic model is also important in deciding to what extent relativistic degrees of freedom must be considered. It is very common in the history of nuclear physics that similar corrections or phenomena can be obtained from very different (but non-orthogonal) models. A relevant example is the fact that at 100 MeV, a preliminary non-relativistic calculation of p + ^{4}He elastic scattering including off-shell effects produces an optical potential with a "double WS" shape - *i.e.*, a weak short range attraction and stronger surface attraction. This is very similar to the result obtained in the effective Schrödinger equation one obtains using strong scalar and vector potentials in a relativistic Dirac model [17].

All of these facts suggest that we make a careful study of the standard model in considering the elastic and inelastic scattering of nucleons from nuclei in the 100–350 MeV energy range. There is already strong evidence that at these energies the effective interaction can be studied in detail [3] [18] and that density dependence can be extracted [16].

Non-Localities in the Effective Interaction

The analysis of intermediate energy elastic and inelastic scattering has been developed in a very detailed manner over the past few years. Most calculations of both processes rely on a multiple scattering theory framework [19] and apply some form of single scattering approximation. The multiple scattering series is arranged in a manner appropriate for strongly interacting pairs in a low density environment. The interactions between a given pair is summed to all orders, while an expansion is made in the number of target particles the projectile interacts with.

As a result, the effective interaction is a complicated many-body operator, and further approximations are needed in order to evaluate it. Two approaches are commonly made:

- The effective interaction is assumed to be a free nucleon-nucleon scattering amplitude. A local pseudo-potential is fit to the on-shell two-body data in Born approximation including non-localities arising from exchange. This pseudo-potential is then used as an effective interaction and provides an off-shell continuation for use in the single scattering term of the MSS [3].

The effective interaction is assumed to be the Brueckner G matrix in nuclear matter. A local pseudo-potential is fit to the diagonal forward scattering amplitude in nuclear matter with the strengths adjusted as a function of density. The interaction is then used in the single scattering term of the MSS using a local density approximation (LDA) [4].

Neither of these methods can be considered as being approximations to the single scattering term of a multiple scattering series. No plausible expansion has ever been written down which yields the kind of off-shell continuation provided by the prescription "fit a local + exchange pseudopotential to the on-shell data and then use this off shell" used by Picklesimer-Walker (PW) or Love-Franey (LF) [3]. A similar statement is true for the fitting prescription used by Geramb in nuclear matter [4]. What may be worse in the nuclear matter case is that there is also no justification for a local density approximation of the type used, beyond the insensitivity of the results to the detailed choice of point at which the density is measured. A self-consistency criterion has been proposed to test the validity of the LDA, namely, that the strength of the effective interaction should change by only a small fraction when one moves a distance of the range of the force [20]. This provides a small parameter, but unfortunately, one which has not been integrated into the theory in such a way as to suggest corrections. However, it is at least a plausible test. Unfortunately, applying this to the results of Geramb's calculations (see IUCF ref. in refs [4] and [21]) shows that even this condition is not satisfied for many of the force components.

Both methods work reasonably well and should be considered as a good starting point for evaluating the correctness of the standard model. It would certainly be much more convenient for calculations if one or the other approaches were valid. In the case where the free interaction is used, if we require a more sophisticated analysis of off-shell behavior to construct an optical potential, it appears as if we would have to carry out a multi-dimensional quadrature. Inside nuclear matter, failure of the LDA might imply the need for a distinct G matrix calculation for each individual nucleus. Either of these results would increase the difficulty of the calculation by an order of magnitude.

Nonetheless, it is important to consider these issues if we are to obtain a sharp test of the standard model. Since the experiments analyzed in conjunction with these models have shown significant sensitivities and since the quantity being measured is of such fundamental importance, it is essential to try to improve the quality of the approximations being made. This is the case even if the result turns out to be that the approaches stated above are accurate. Note that we are not at this point claiming that either method is wrong - only that they are poorly justified with respect to a foundation in the standard model.

We focus on the question of the non-locality of the free nucleon-nucleon effective interaction as this can be considered as the starting point for either approach. A primary failure of both methods is the assumption that the effective interaction is local. The off-shell scattering amplitude that is prescribed by the MS Theory has non-localities arising from three sources:

1. exchange (as applied to the two interacting particles);
2. the non-locality of the model potential;
3. dynamical effects (like dispersive corrections in electron scattering but stronger).

Only exchange is included in the two approaches discussed above. The on-shell matrix elements of the scattering amplitude are exchange symmetric, so if only on-shell momenta are required this effect is automatically included. However, if a model is constructed to extend the on-shell amplitudes off-shell, then the exchange symmetry must be explicitly included. Since there is no unique way to do this directly from on-shell data without postuling a model for the potential, substantial ambiguities may arise. These are most obvious in considering the contribution of "direct" and "exchange" terms to inelastic scattering [18].

Two-body potentials used in the standard model differ widely in their short range character. The Hamada-Johnston potential is local except for the presence of spin-orbit terms. The Reid potentials have some angular non-localities (*i.e.*, the radial potentials are local, but the potentials in different partial waves differ from each other in arbitrary ways), and the Paris potential has very strong short-range radial non-localities in the form of a velocity dependence. Any way to determine the non-local character of the short range part of the interaction could be of considerable importance. For example, we may think of the nucleon-nucleon interaction as an effective interaction arising from a lagrangian field theory after the elimination of meson and/or quark degrees of freedom. If the short range repulsion in the nucleon-nucleon sector arises from the exchange of a heavy vector meson, then the potential will be nearly local. If, on the other hand, it arises from quark exchange between extended nucleons, then it will be strongly non-local.

The dynamical non-localities can also be expected to be important. At low and intermediate energies, the Born approximation is considerably larger than the full T matrix. The higher Born terms are also large (indeed, the series diverges for some partial waves), and their sum almost cancels the Born approximation on shell. Because the higher order terms in the Born series allow the particles to propagate between interactions, their sum should be strongly non-local. Due to the different character of the Born term and the higher Born terms, the cancellation should fail off shell, and the full T matrix should have a substantial non-locality.

The effect non-local effective interactions will have on the analysis of elastic and inelastic scattering data is two-fold. First, any non-localities in T will lead to non-localities in the optical potential. The elastic scattering will be similar to that produced by a local potential, but the interior wave function will be reduced [22]. Leaving out this effect in a DWIA analysis of inelastic scattering would lead to the conclusion that the interior contributes less than expected and therefore that the effective transition operator must be reduced. Thus one would infer an incorrect density dependence.

A second effect would be on the separation of the effective interaction into spin dependent parts. Non-localities in the transition operator can affect both the magnitude and selection rules in inelastic scattering [23]. This could be particularly troublesome if one were attempting to use the nucleus as a filter and suppress large terms by selection rules in order to learn something about smaller terms.

The reader should note that the problems discussed above that arise from non-localities are not restricted to the non-relativistic standard model. If one considers nuclear physics as described by a full quantum field theory of mesons and nucleons, the four-point function corresponding to the T matrix will still be non-local, for many of the same physical reasons as in the standard model (propagation in between interactions, elimination of quark effects, and exchange). If these non-localities turn out to have important effects on elastic and inelastic scattering non-relativistically, then it is likely that they will have to be considered in the relativistic analysis as well.

CALCULATIONS

The Multiple Scattering Series

We consider the scattering of a nucleon from an A-particle nucleus in the framework of the standard model of nuclei: non-relativistic nucleons interacting via energy independent two-body potentials. The hamiltonian for this system is:

(1) $\quad H = \sum_j K_j + \frac{1}{2} \sum_{ij} V_{ij}$

where K_j is the kinetic energy of the j-th nucleon and V_{ij} is the potential between nucleons i and j, which is determined by fitting some model parameters to experimental two-nucleon data. We will label the projectile nucleon as 0. Unmarked sums will run from 0 to A. Sums marked with a prime will run only over the target nucleons, $i = 1,..A$.

The many-particle scattering operator for nucleon-nucleus scattering without rearrangement is:

(2) $\quad T = T^{00} = V^{00} + V^{00} G V^{00}$

where

(3) $\quad V^{00} = \sum'_j V_{0j}$

(*i.e.*, the sum only runs over the target nucleons), and G is the exact system Green's operator

(4) $\quad G = 1/(E+i\eta-H)$.

Since H is the Hamiltonian for the full $A+1$-particle system, G is a complicated many-body operator. In order to calculate it we require an approximation scheme. The appropriate scheme is the multiple-scattering series originally formulated by Lax and Watson, extended by Kerman, MacManus, and Thaler (KMT) and recently reanalyzed from a more rigorous point of view by Picklesimer, Tandy, and Thaler (PTT) [19].

The basic physical ideas underlying the scheme are the observations that the density of nucleons in the nucleus is low (*i.e.*, the average spacing of nucleons in the nucleus is larger than the range of the strong part of the nucleon-nucleon force*), and the binding energies of the nucleons are small compared to the potential energies involved. As a result, the calculation is organized into scatterings of the projectile by nucleon clusters of increasing size (see Siciliano and Thaler, ref. [19]). Scattering from a single nucleon (the single scattering approximation (SSA)) is the first term, scattering from a correlated pair of nucleons is the second, *etc.* Within each term of the series, the interactions of all particles in the cluster must be summed to all

---

* Note that the average spacing between nucleons in nuclear matter is about 1.8 f, if we take the reciprocal of the cube root of the density, 0.17 nucleons/f^3 as our spacing. This is not so much greater than the range of the force, about 1.4 f.

orders.

One begins by using the resolvent equation for the full Green's operator to rewrite (2) as a Lippmann-Schwinger (L-S) equation:

(5) $\quad T = V^{00} + V^{00} G_0 T$

where

(6) $\quad G_0 = 1/(E+i\eta-H_0)$

and H_0 is the operator in which the interactions between the projectile and the target nucleons, V_{0j}, have been turned off:

(7) $\quad H_0 = \sum_j K_j + \frac{1}{2} \sum'_i \sum'_j V_{ij}$.

We use (3) in (5) to break T up into parts which end with the interaction with a specified pair:

(8) $\quad T_j = V_{0j} + V_{0j} G_0 T$.

Then we have that the full T operator is the sum of these parts:

(9) $\quad T = \sum_j T_j$.

We now introduce the operator which sums the pair interactions between the projectile and the j-th nucleon,

(10) $\quad \tau_j = V_{0j} + V_{0j} G_0 V_{0j} + \ldots$

$\quad\quad \tau_j = V_{0j} + V_{0j} G_0 \tau_j$.

Notice that between the interations, the target nucleus propagates as a fully interacting A-body system since H_0 has only the projectile-target interactions turned off. This means that τ is still a many-particle operator and must be approximated further. We can do this by introducing a shell-model approximation for the target nucleus. In the denominator of G_0 we add and subtract a shell-model or self-consistent mean-field potential for each target nucleon, U_i, and expand in $V-U$. See the discussion in Tandy, Redish, and Bollé (TRB) [24] for more details.

The resulting operator may be related to the two-body T matrix, t_j, defined at an energy shifted by the kinetic energy of the center of mass of the interacting pair plus the single particle energy. The relevant two-body operator is:

(11) $\quad t_j(e) = V_{0j} + V_{0j} g_{0j}(e) t_j(e)$

where $g_0(e)$ is the free propagator for the $0j$ pair:

(12) $\quad g_{0j}(e) = 1/(e+i\eta-k_{0j})$.

k_{0j} is the kinetic energy of relative motion between the projectile and the j-th target nucleon. The off-shell character of this operator when imbedded in the many-body problem arises from the fact that the scattering is being treated as a three-body problem: a pair of nucleons interacts in the presence of (but not interacting with) the nuclear core. This affects the kinematics in an essential way. Both the way the momentum is balanced and the choice of energy for the interacting pair changes from what is found in the two-body case. The fact that the operator is off-shell is an essential representation of the kinematics of the reaction and contains the physical information about the reactive content of the model [24].

The Optical Potential

In the energy range we are considering, the elastic scattering amplitude is too strong to be treated directly by a multiple scattering series. One must instead first perform a Feshbach separation of the Green's operator G_0 into a part in which the nucleus propagates in its ground state and the rest [9]. Solving the rest formally leads to the introduction of an effective projectile-nucleus interaction which leaves the nucleus in its ground state and can therefore be treated as a two-body operator (*i.e.*, only the projectile-nucleus relative coordinate is affected) or optical potential. A multiple scattering series can then be made for the optical potential and its effects summed to all order by solving the L-S equation:

(13) $P_0 T P_0 = P_0 U P_0 + P_0 U P_0 G_0 P_0 T P_0$

where P_0 is the projection on the nuclear ground state. In the SSA the optical potential is given by (see KMT and PTT):

(14) $<k\ \phi_0|U|k'\ \phi_0> = \sum'_j <k\ \phi_0|t_j(E_0 - K_{0j} - \epsilon_j)|k'\ \phi_0>$.

In this equation the initial and final states $|k'\ \phi_0>$ and $<k\ \phi_0|$ describe the projectile in a plane wave (initially with 3-momentum k, finally with 3-momentum k') and the nucleus in its ground state ϕ_0. The energy appearing in the two-body operator t_j is the projectile's kinetic energy in the overall CM frame, E_0, reduced by the kinetic energy of the center of mass of the interacting pair, K_{0j}, and the shell-model energy of the j-th particle, ϵ_j.

If we insert a complete set of plane waves for the struck particle before and after t_j, and apply the conservation of momentum delta function between the projectile and the struck particle, we obtain the explicit result for the SSA optical potential:

(15) $<k\ \phi_0|U|k'\ \phi_0> = \sum'_j \int dq\ dq'\ \delta(k+q-k'-q')$

   $\otimes\ <(k-q)/2|t_j(E_0-(k+q) - \epsilon_j)|(k'-q')/2> \rho_j(q,q')$

In this expression, ρ_j is the density matrix for the nucleon in the j-th shell model state. Recoil corrections of order $1/A$ have been suppressed to simplify the typography. If $t_j(e)$ were independent of energy, the sum over j could be carried out and ρ_j would become the density matrix for the full nucleus. If we want to use the full nuclear density rather than the shell model density, ρ_j is replaced by a spectral density function of a continuous energy variable and the sum over j is replaced by an integral over this variable.

For the rest of this paper I will suppress the energy dependence of the two-body T operator and concentrate on its non-locality, *i.e.*, on its dependence on its momentum arguments. We will see later that the slow energy dependence of the off-shell T matrix elements justifies this approximation somewhat *ex post facto*. With this restriction the integral (15) reduces to:

(16) $<k\ \phi_0|U|k'\ \phi_0> = \int dp\ <(k-p)/2|t|k'-(k+p)/2>\ \rho(p,k'-(k+p)/2)$

where now ρ is the full density matrix of the nucleus.

If the two-body T matrix were a local operator, *i.e.*, depended only on the difference of its momentum arguments, then the matrix element of t in (16) would reduce to $t(k-k')$, independent of p. The integral over p can then be carried out and the density matrix is reduced to the fourier transform of the density. The result is:

(17) $<k\ \phi_0|U|k'\ \phi_0> = t(k-k')\ \rho(k-k')$,

the well-known "$t\rho$" approximation.

Equation (17) is a considerable simplification over (16) for three reasons. First, the construction of the optical potential in (16) requires integration over 3 variables while (17) requires none. Second, the optical potential in (17) is local while that in (16) is non-local. This means that finding the elastic cross sections and wave functions for use in distorted wave calculations will be more difficult if carried out in configuration space (requiring solution of integro-differential rather than differential equations) or unfamiliar if carried out in momentum space. Third, if t is assumed to be local, then it can be taken from on-shell data (at least part of it) and one doesn't have to solve a complicated integral equation for it (equation (11)).

Unfortunately, at the present time there are no known justifications for this approximation. Indeed, there are many reasons for believing it to be poor. The other approximations leading up to this point have been more or less justified on the basis of physical arguments and the smallness of certain dimensionless parameters. If the approximation is poor we may not get a good representation of the scattering. Since neutron densities are sometime adjusted in this approximation, we may still be able to get good fits to the data as a result of compensating errors. If we are to take the entire structure seriously, we must ask ourselves whether the factorization approximation (17) is at all sensible. To do this, we need to investigate the non-locality structure of the off-shell T matrix appearing in (16).

If we have to use (16) rather than (17), there could be substantial changes in standard analyses arising from new physics. One of the most important uses of the optical potential is to construct distorted waves for use in analyzing inelastic scattering and transfer reactions. If the optical potential (16) is strongly non-local, then the distorted waves constructed from it will be different in the interior from those produced by (17), even if the elastic scattering results are nearly identical. This is a well-known phenomena in low-energy nuclear physics and is referred to as the Perey effect [22]. *

---

* We refer here to the fact that at a particular fixed energy, the Buck-Perey non-local potential can closely reproduce the elastic scattering results obtained from a local Woods-Saxon potential. The interior wave functions are, however, substantially reduced in magnitude. The fact that some of the energy dependence observed when data is fit with local potentials is not required if the Buck-Perey non-local potential is used is not relevant here.

If this effect is important but not included in the analysis of experimental data, then any information extracted about the interior of the nucleus will be wrong. *It is therefore essential to know whether we expect a significant non-locality in the optical potential. The only way to determine whether this should be present is to consider the full calculation of Equation (17).*

Inelastic Scattering

The derivation of a SSA for inelastic scattering is a bit more complex. The presence of initial and final distorted waves implies that certain scatterings are summed to all orders while others (the "impulse") are treated to first order. This requires additional assumptions. An approximate way of handling this is given in KMT where the optical potential operator is broken into diagonal and off-diagonal parts. An alternative analysis is given in PTT.

The result takes the form:

(18) $M_{fi} = \sum_j <\chi_f \phi_f|t_j|\chi_i \phi_i>$,

where χ_i and χ_f are the initial and final distorted waves of the projectile with initial and final momenta κ' and κ respectively. ϕ_i and ϕ_f are the initial and final nuclear bound states. The t is the same two body T matrix as in the discussion of elastic scattering.

If we put in a complete set of momentum states for the projectile and struck nucleon, we get the following expression:

(19) $M_{fi} = \int dq\, dq'\, dk\, dk'\, \delta(k+q-k'-q') <(k-q)/2|t|(k'-q')/2> \rho_{fi}(q,q')$.

We have again used the assumption that $t(e)$ does not vary rapidly with e. The sum on j then produces the transition density matrix, ρ_{fi}. If the transition corresponds to moving a single nucleon from an orbital ϕ_1 to ϕ_2 then the transition density is simply:

(20) $\rho_{fi}(q,q') = \phi_2(q)^* \phi_1(q)$.

As expressed in (19), there are 9 integrals (4x3 − 3 done by the delta function). If t is local, this will simplify further. Three of the integrals can be moved through t which is now only a function of $k-k'$. These convert the transition density matrix to a transition density and leave us with 6 integrals. Changing to relative and CM variables allows the integral to be separated into parts: an interaction kernel (integral of $t\rho$) integrated with the distorted waves.

Structure of the Off-Shell T Matrix

The off-shell T matrix element used in both the SSA for elastic and inelastic scattering (Eqs. (15) and (19)) is a function of 4 variables:

(21) $<p|t(e)|p'> = f(p,p',\theta,e)$,

that is, t is a function of the magnitudes of p and p', of the angle θ between them, and of the energy e. The complexity is increased by the fact that this is an operator in the (4 component) spin space and in the (4 component) isospin space. These variables are suppressed here.

A more convenient representation of the matrix element is to change variables to the sum and difference of the momenta:

(22) $\quad Q = p - p' \quad\quad K = \frac{1}{2}(p+p')$.

In these variables, the matrix element may still be expressed as a function of 4 variables:

(23) $\quad <p|t(e)|p'> = t(Q,K,e) = t(Q,K,\phi,e)$

where K and Q are the magnitudes of K and Q and ϕ is the angle between them. This is a more convenient representation in which to study the non-localities, since a local operator will be independent of K.

Changing to these variables also clarifies the structure of the SSA integrals and their relation to the local approximations. The optical potential SSA in terms of these variables becomes:

(24) $\quad <k|U|k'> = U(Q,K) = \int dP\, t(Q,K+P)\, R(Q,P)$.

where we have written the density matrix as a function of its transformed variables in a way similar to (23):

(25) $\quad \rho(q,q') = R(q-q', \frac{1}{2}(q+q'))$.

The reduction of Eq. (24) to local form becomes obvious if we note that

(26) $\quad \int dP\, R(Q,P) = \rho(Q)$.

If $t(Q,K)$ is independent of its non-local (second) variable, the $t\rho$ form immediately appears.

The inelastic equation becomes more transparent in coordinate space. The full integral takes the form:

(27) $\quad M_{fi} = \int dx_0\, dx_1\, dr\, \chi_f(x_0+r/4)^*\, \phi_f(x_1-r/4)^*$

$\quad\quad\quad \otimes\, t(x_0-x_1, r)\, \chi_i^+(x_0-r/4)\, \phi_i(x_1+r/4)$

Here, the function t is the Fourier transform of $t(Q,K)$. Its dependence on the second argument is a measure of its non-locality. To say that t is local in coordinate space means that

(28) $\quad t(x,r) = t'(x)\, \delta(r)$.

Using this reduces (27) to the familiar local finite-range form:

(29) $\quad M_{fi} = \int dx_0\, dx_1\, \chi_f(x_0)^*\, \phi_f(x_1)^*\, t'(x_0-x_1)\, \chi_i^+(x_0)\, \phi_i(x_1)$

The integral over x_1 yields the interaction kernel.

Calculation of the Off-Shell T Matrix

We calculate the off-shell T matrix explicitly by expanding in partial waves and solving the L-S equation in momentum space by matrix inversion. The equation for uncoupled partial waves has the form:

(30) $\quad t_{slj}(k,k',e) = v_{slj}(k,k') + 2/\pi \int dq \, (q^2/(e+i\eta - q^2/m)) \, v_{slj}(k,q) \, t_{slj}(q,k',e).$

In practice we solve this equation by breaking up the q integration into 4 regions. This is done in order to handle the pole at the on-shell momentum, $p = \sqrt{(me)}$, and the fact that the integration has an infinite range. The pole is handled by solving the equation with a principal value rather than a pole to obtain the K matrix and then transforming it into a T matrix using the Heitler equation. The principal value is handled by taking a symmetric interval about the on-shell momentum, $p-\Delta$ to $p+\Delta$, and using an even number of Gauss points in this interval. Our 4 regions are then: (I) 0 to $p-\Delta$, (II) $p-\Delta$ to $p+\Delta$, (III) $p+\Delta$ to some cutoff Λ, and (IV) the region from the cutoff Λ to infinity.

The equation is solved by discretizing the integral with linear Gaussian quadrature in regions (I), (II) (even number of points), and (III). We use Gauss-Laguerre quadrature in region (IV). For the Reid potential, 1% accuracy is obtained for the phase shift using a total of about a dozen quadrature points. For the Paris potential somewhat more are needed - about 20. The calculations reported here are done with 24 quadrature points for both potentials.

RESULTS

In this section I present a brief survey of some of the results found from calculating fully-off-shell T matrices with the RSC and Paris potentials. I will discuss three results: (1) a comparison of the momentum space matrix elements of the two potentials; (2) a comparison of the off-shell partial wave amplitudes they produce; (3) an analysis of the role played by high momentum components in the two cases, and (4) a study of the dependence of the central and spin-orbit parts of the off-shell T operator on some of the four variables involved.

The Potentials in Momentum Space

The Reid and Paris potentials represent two different approaches to the problem of finding a nucleon-nucleon potential. Reid's potential was found by taking sums of Yukawa potentials with masses corresponding approximately to those of known mesons, and adjusting the strengths of terms independently in each partial wave [25].

The Paris potential [26] was obtained by a more complex procedure. First the Paris group used dispersion relations to determine on-shell OPEP + TPEP amplitudes. Second, they postulated an off-shell continuation. This yielded a long and mid-range potential. Third, they added a flat, short-ranged repulsion and made it join smoothly on to the mid-range. They adjusted the height as a function of energy to fit the experimental phase shifts. It was

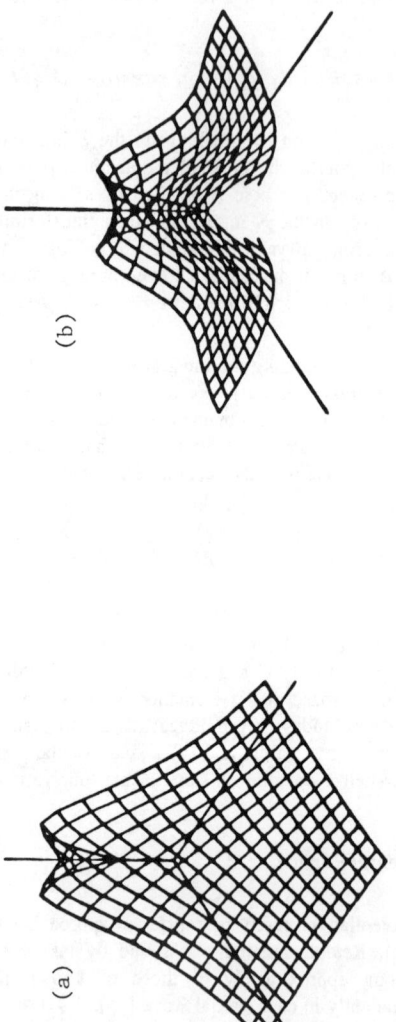

Fig. 1: Momentum space matrix elements for the $^1S_0$ potential, $v_{000}(k,k')$ for k, k' between 0 and 15 f^{-1}. (a) RSC, (b) Paris.

empirically determined that a linear function of the energy sufficed. This fact permitted them to transform the energy dependence into a momentum dependence. The result is a potential with a strong short range non-locality. Finally, they fit a sum of many Yukawas to the local functions and the coefficients of the non-local functions. The result is a potential which is reasonably convenient to use*, and which fits the two-body data better than any other potential (indeed, better than phase shift analyses).

These potentials are very different in character. The shortest ranged terms for the Reid potential (and therefore the ones which extend to the highest momenta in momenta space) are Yukawas with a range of about 0.2 f. The Paris potential is dominated at high momenta by the non-local terms. These arise from Fourier transforming coordinate space potentials of the form $p^2 V(r) + V(r) p^2$ where the V's are sums of Yukawas similar to Reid's. The extra powers of momentum push these terms to much larger values of momenta than in Reid's potential. In Fig. 1. (at the end of the paper), I have plotted the momentum space matrix elements $v_{slj}(k,k')$ for the Reid and Paris potentials for momenta between zero and 15 f^{-1} for the important $^1S_0$ state. It is clear from the figure that the potentials have a significantly different character. Indeed, even for an initial momentum as low as $k' = 1.0$ f^{-1}, the function $v_{000}(k,k')$ for the Paris potential peaks at a momentum of about $k = 20$ f^{-1}.

Off-Shell Partial Wave Amplitudes

The basic idea of the standard model is that by using a little bit of elementary particle physics supplemented by the requirement of fitting on-shell (and bound state) two-body data, that a reasonably unique two-body potential can be obtained —at least one with a sufficiently small uncertainty to be of use in low-energy nuclear physics. The pictures shown in Fig. 1 seem to suggest that this is exceedingly unlikely. The results, however, show that the situation is a bit more subtle.

When the off-shell partial wave amplitudes, t_{slj}, are compared, a startling result is observed. For the strong partial waves, the off-shell behavior of the Reid and Paris amplitudes are strikingly similar. Fig. 2 shows the real part of the $^1S_0$ amplitudes and the imaginary part of the $^3S_1$-$^3D_1$ coupling amplitudes for the RSC and Paris potentials for the range k and k' from 0 to 3.5 f^{-1}. These results are typical. †

The strong similarity of the fully-off-shell behaviors is reminiscent of results found in analyzing the off-shell sensitivities of experiments designed to probe the half-off-shell T matrix, such as bremsstrahlung or knockout [28]. The half shell results can be summarized as follows: *Although the half-shell amplitudes required are significantly different from the on-shell ones, variations from one realistic potential to another can be attributed largely to differences in the fits to the on-shell data. Potentials which have similar fits to on-shell data have similar half-shell behavior.*

This result can be explained by means of physical arguments [29]. The half-shell amplitude

---

* There are some technical difficulties in both coordinate and momentum space because of the non-local term and the delicate cancellations in the parametrization.
† The off-shell amplitudes for the RSC $^1P_1$ state, long known to be anomalous in Reid's parametrization [27], shows substantial differences from the off-shell Paris results.

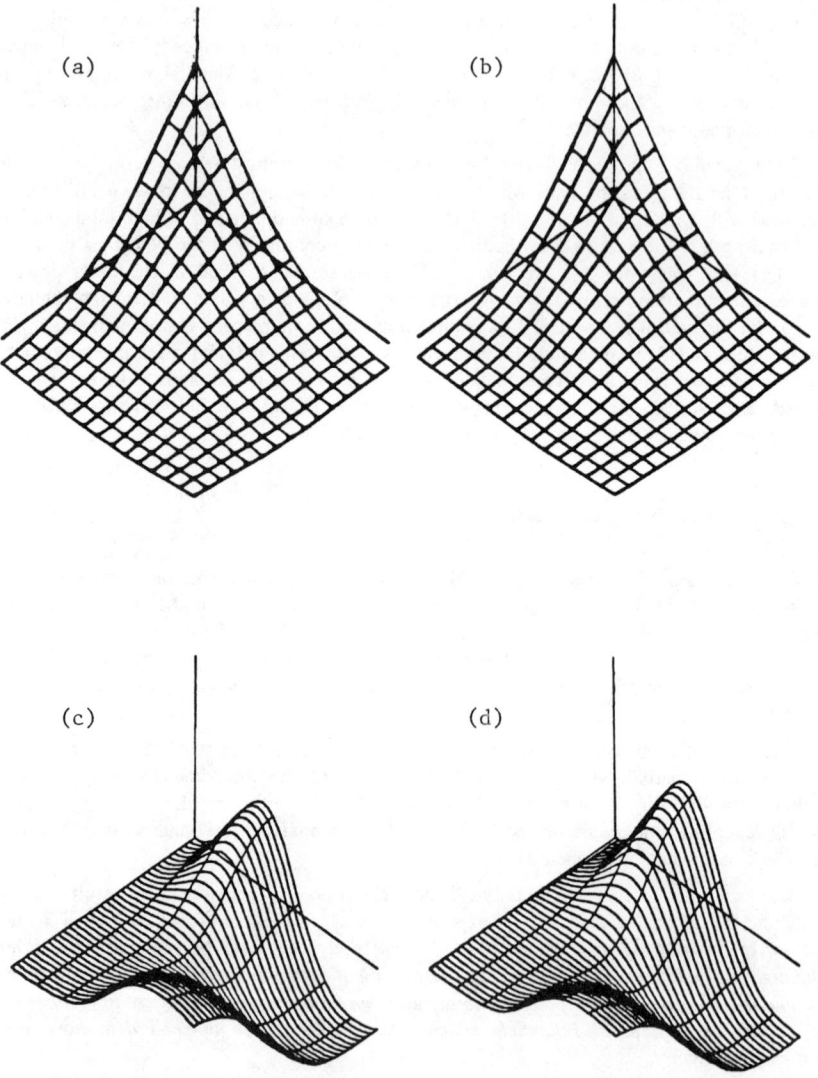

Fig. 2: Off-shell partial wave amplitudes $t_{slj}(k,k')$ for k, k' between 0 and 3.5 f^{-1}. $E_{lab} = 200\ MeV$. (a) Real part, RSC potential, $^1S_0$, (b) Real Part, Paris potential, $^1S_0$, (c) Imaginary part, RSC potential, $^3S_1$-$^3D_1$ coupling, (d) Imaginary part, Paris potential, $^3S_1$-$^3D_1$ coupling.

can be expressed in terms of the phase shift and the scattering wave function (with explicit reference to the potential eliminated). Most realistic potentials considered are similar outside of about 1.5 f and have some mechanism to suppress the wave function inside of about 0.5 f. Consider the case at a particular scattering energy. Outside the range of the potential, if the phase shifts are the same, then the wave functions are the same. By integrating the Schrödinger equation inward from infinity, we see that the wave functions for the two potentials must be similar in to about 1.5 f. If both wave functions are suppressed in the interior and must match smoothly to a small value near 0.5 f, the only uncertainty is what the wave function looks like between 0.5 and 1.5. If the wave function is required to be smooth with no extra wiggles (as is the case for most potential models), the resulting wave functions are very much alike. Since on-shell amplitude and wave function are both close, so are the off-shell amplitudes.

No analogous argument is known to hold for fully off-shell amplitudes. Nonetheless, a similar result seems to be true. Despite the substantial differences in the qualitative structure of their momentum space matrix elements, the Paris and RSC potentials yield almost identical off-shell results for momenta below about 5 f^{-1}. (Looking at higher momenta, the Paris amplitudes develop a small trough along the diagonal not present in the RSC amplitudes.)

The fact that the low momentum amplitudes are very similar does **not** imply that low energy calculations will yield the same answers. Even binding energy and low energy scattering processes depend to some extent on high-momentum amplitudes (as we will see below). It does suggest a new way of looking at nuclear physics. We should consider output calculations as a function of the off-shell amplitudes as input rather than as a function of a potential. If we could then determine experiments which only were sensitive to low momentum components, a failure of the standard model would be critical, and couldn't be blamed on "not using exactly the right potential".

The Role of High Momentum Components

High momentum components can play a role even in low energy processes and the different character of the potentials can thereby have an effect, despite the strong similarity of the low momentum amplitudes. Let us consider the effect of the high momentum components of the potentials on the phase shift. We can do this by setting the momentum space matrix elements of the potential to zero above some upper cutoff. We can then investigate how the phase shift changes as the cutoff is brought down.

The results for the 100 MeV (lab energy) phase shifts are shown in Fig. 3. The fractional change in the phase shift is displayed for the RSC and Paris potentials as a function of the upper cutoff. We see a substantial difference in the result for the two potentials. Low energy phase shifts are insensitive to matrix elements of the RSC above about 5 f^{-1} while for Paris, they are required up to about 20 f^{-1}. This is due to the strong off-diagonal character of the Paris potential and its failure to fall off at high momenta.

Note that it is the potential that cuts off the integration in the L-S equation (30), not the propagator. The integrand has the structure $(q^2/(p^2 - q^2))\, v(k,q)\, t(q,k')$ (integrated over q; p equals on-shell momentum). For large q the density of states times the propagator goes to -1.

The implications of this for currently favored procedures are disturbing. A standard approach

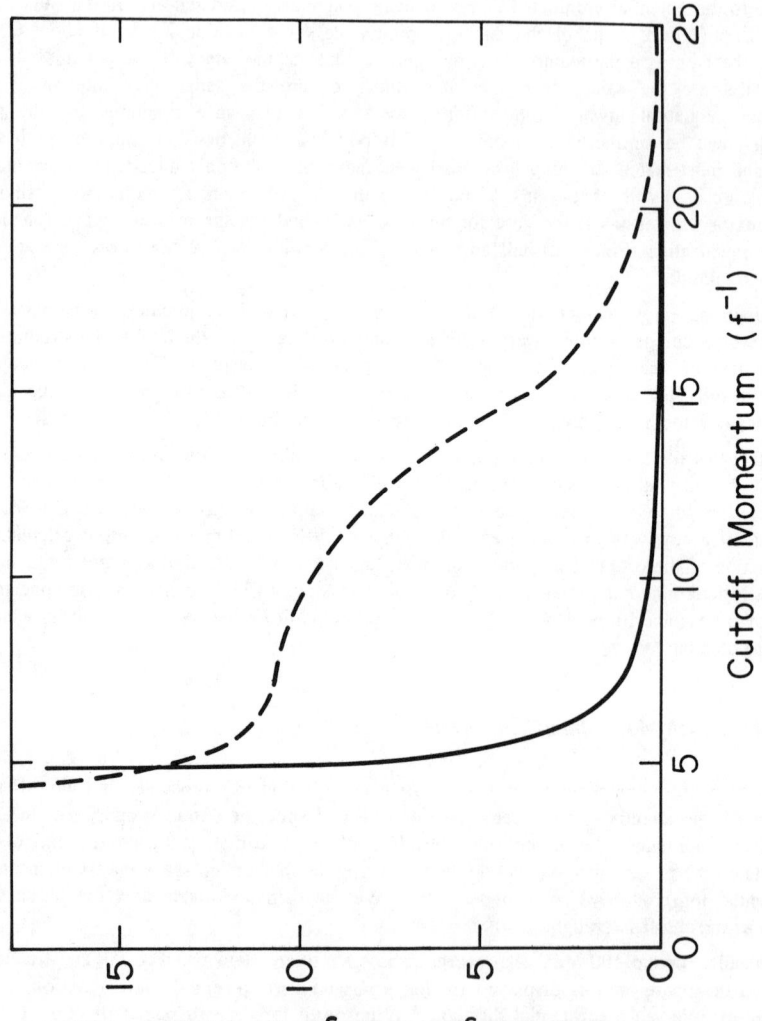

Fig. 3: Fractional change in 100 MeV $^1S_0$ phase shift as a function of upper momentum cutoff used in the L-S equation. Solid line = RSC potential dashed line = Paris potential.

[4] is to solve for the forward (diagonal) matrix elements of a two-body interaction in nuclear matter with low-momentum states suppressed by the Pauli operator. The result is fit with a sum of Yukawas which is then used as an effective (density dependent) interaction to generate off-diagonal matrix elements.

Because the Paris potential is so strongly off-diagonal, one might expect that suppression of low momentum intermediate states would have less of an effect than for the RSC. Indeed, less density dependence is found for Paris than for Reid [4][16]. This is, however, largely due to the strong off-diagonal character of Paris. When one then fits the result with an interaction that does *not* have this character (a sum of Yukawas), the result may misrepresent the true effective interaction. A more careful analysis of the nuclear matter equation is clearly required.

Central and Spin-Orbit Parts

The full T operator can be decomposed on the spin-space into parts having various spin structures (see, *e.g.*, KMT). The spin-independent term (A), and the term proportional to the total spin of the pair dotted into the normal to the scattering plane (C) give rise in the SSA to the central and spin-orbit parts of the optical potential for spin zero nuclei. I will therefore concentrate on the combinations of the partial wave amplitudes that correspond to A and C. Note that it is impossible to tell whether an operator is local from its partial wave components. It is only when it is reassembled into the full operator that we can investigate its dependence on Q ($= k - k'$) rather than on the magnitudes, k and k'.

Non-localities are present in the T operator for a number of reasons. If we look at the structure of the definition of T in Equation (2), they can easily be identified. T has the form $V+VGV$. If V were local, and the Born approximation held, then T would be approximately local. For nucleon-nucleon scattering V is non-local both because of exchange and because the potential form used to fit the data is non-local.

The exchange non-locality occurs because the potential is not really just $V(Q)$ but rather $PV(Q)P$ where P is the operator that exchanges the two interacting nucleons. This replaces a function of Q by a function of Q plus a function of K (modulo details of the spin and isospin decomposition). The forms used for both potentials are non-local, but that of the Paris potential is explicit in each partial was and is very strong. The VGV term can be expected to be non-local because of the finite range of the full propagator. Calculations show that in the energy region below 350 MeV, the most important partial waves are not Born dominated. In general, V is large and is substantially cancelled by the real part of the VGV term.

The results for the real and imaginary parts of the central and spin-orbit parts of the T matrix generated from the RSC potential are shown in Fig. 4. (The results from the Paris potential are very similar, as may be expected from the discussion above.) The plots are made as a function of the variables K and Q (Eq. (22)) for fixed values of the energy, e, and the angle ϕ. The most interesting result is that *the amplitudes A and C' are nearly independent of the angle ϕ for all energies considered and for both potentials*. (C' is equal to C with a factor of $\sin\phi$ exctracted. See KMT.)

The dependence of the amplitudes on the energy is also slow in this energy regime. When e gets below 50 MeV, the imaginary part of A begins to develop a peak near the origin.

These slow dependences mean that a full operator need not be considered as a function of all

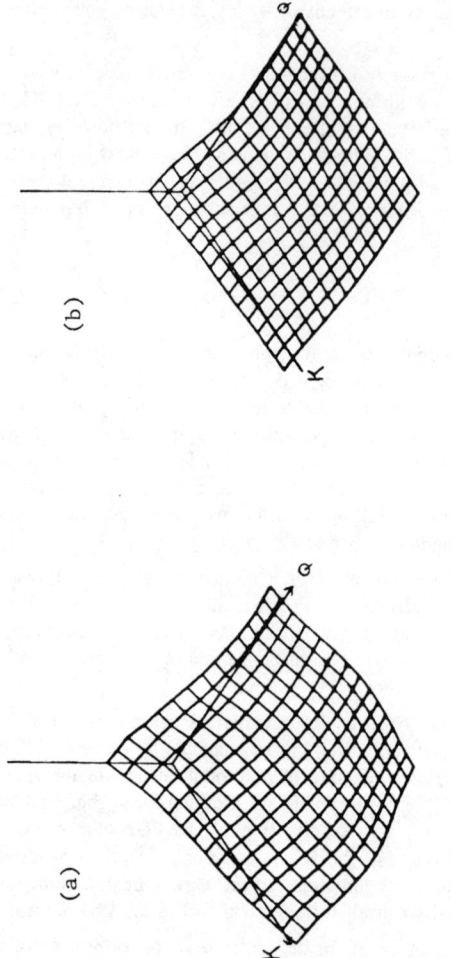

Fig. 4: Central part of the off-shell T matrix, $A(Q,K,\phi,e)$ for $\phi = 90°$ $e = 100$ MeV (note this is the pair's CM energy). Q varies from 0 to 3.5 f^{-1}, K from 0 to 1.75 f^{-1}, RSC potential. (a) Real part, (b) Imaginary part.

4 variables: 2 may be enough. This implies important simplifications.

Both A and C have substantial non-localities, though the imaginary part of A is nearly local (I was very much surprised by this), and the real part of C is small. An examination of the small K behavior of $A(K,Q)$ for $Q = 0$ shows that it is reasonably well approximated by $exp(-Kr_0)$ where the range of the non-locality, r_0, is about 0.6 f.

IMPLICATIONS AND CONCLUSIONS

Calculation of the Full-Folding Optical Potential

The empirical fact that the off-shell A and C amplitudes for both our test potentials are nearly independent of the angle, ϕ between Q and K has some important implications for the construction of the full folding integral. If we consider Eq. (24) this becomes clear. Because of the way we have chosen the integration variable, if t is independent of ϕ then it does not depend on the angles of P. The two angular integrals may be moved onto the density matrix and done explicitly. Karen Stricker-Bauer has pointed out that if the density matrix is expressed in terms of harmonic oscillator states, then the integrations can be done analytically. This would mean that the full folding integral can be reduced to a single quadrature. She has written a program to do this calculation [30], and we hope to be able to calculate elastic scattering with it in the next few months.

Elastic Scattering Results

Even though we have not yet completed the full folding calculation, we test our amplitudes in an approximate calculation of elastic scattering. If one assumes that the P dependence of the density matrix in Eq. (24) is sharply peaked compared to the rate of variation of the T matrix, then one can evaluate t at $P = 0$ and pull it out of the integral. This yields the approximation:

(31) $<k|U|k'> = t(Q,K) \, \rho(Q).$

The scattering from this non-local potential can be calculated using the momentum-space optical-model code WIZARD developed by Hynes, Picklesimer, Tandy, and Thaler (HPTT). In collaboration with this group, we have carried out the calculation of elastic scattering of 100–350 MeV protons scattering from ^{16}O and ^{40}Ca using fully-off-shell A and C amplitudes from the RSC potential in the approximation (31).

The RSC potential is rather old and does not fit the on-shell data as well as either the current phase shift analyses or the Paris potential does. Therefore, in order to test whether this potential provides an adequate on-shell starting point for an off-shell extrapolation, we have taken the on-shell RSC amplitudes and calculated the elastic scattering produced in an impulse approximation [1]. The predicted measurable quantities agree almost exactly with those using Arndt's recent amplitudes.

When used in approximation (31), the off-shell amplitudes yield results in the forward hemisphere at 135 and 180 MeV which are very close to those using the L-F amplitudes in the same approximation. Backward observables are substantially different, but since they are both small, and since both begin to deviate from experimental spin observables in the region when they still agree, the backward comparison may not be meaningful. The differential cross-section and the analyzing power for scattering of protons from ^{16}O are shown in Fig. 5.

We will repeat all calculations with Paris amplitudes as well as with the RSC, but since they are so close I do not believe this will make much difference.

Conclusions and Prospectus

Our study of the fully off-shell scattering amplitude has produced a number of conclusions that are highly promising. The similarity of the off-shell amplitudes from widely different, realistic phenomenological potentials should make it possible to reduce the ambiguities that arise in the standard model due to the lack of a unique potential. The systematic independence of the results on two of the four variables leads to substantial simplifications in dealing with the fully-off-shell amplitudes. These two results together lead to a situation much better than we had any right to expect. They strongly suggest that the off-shell amplitude is the correct starting point for an analysis of the effective interaction.

In the next few months we hope to use these amplitudes in elastic scattering calculations to answer some of the important questions underlying the analysis of elastic and inelastic scattering in the IUCF energy range. Some of these are:

- Do the elastic scattering wave functions inside the nucleus differ when RSC amplitudes are used rather than LF? Is there a substantial Perey effect?
- Is the approximation (31) a good approximation to the full folding integral? If not, is some other factorized approximation adequate?
- Is the LF amplitude a good representation of potential model amplitudes off shell?
- What are the off-shell systematics for the other amplitudes in the spin decomposition of the T matrix?

In the longer term, we should be able to address additional important questions:

- What are the effects of non-localities on the inelastic transition operator? (This requires a DWIA code in momentum space, able to handle general non-local transition operators.)
- Does the nuclear matter calculation simplify when viewed from the starting point of the off-shell free T matrix?
- What physics is responsible for the insensitivity of the off-shell amplitudes to the details of the potential?

Once these questions are answered, we can begin the delicate disentangling of the short range information from the experimental determinations of the effective nucleon-nucleon interaction.

I would like to thank my collaborator, Karen Stricker-Bauer, for reviewing this manuscript and for her extensive and constructive suggestions. Most of the work discussed here will be

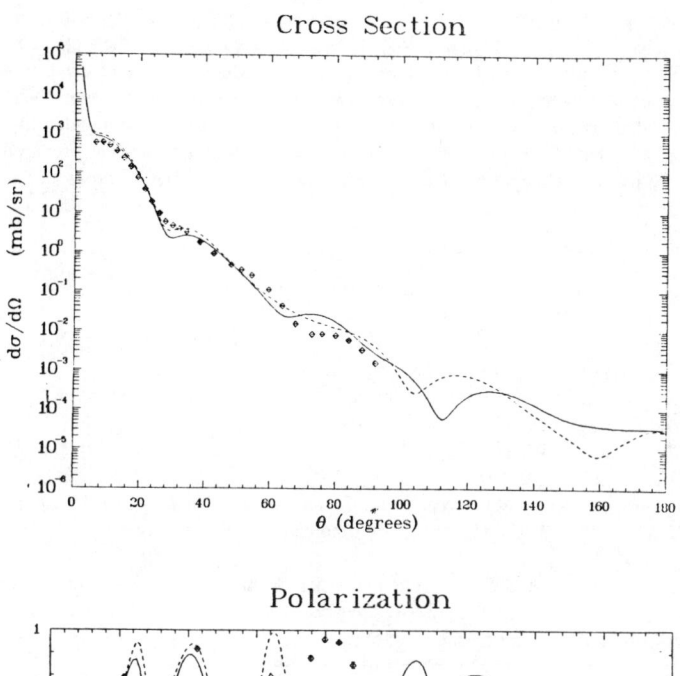

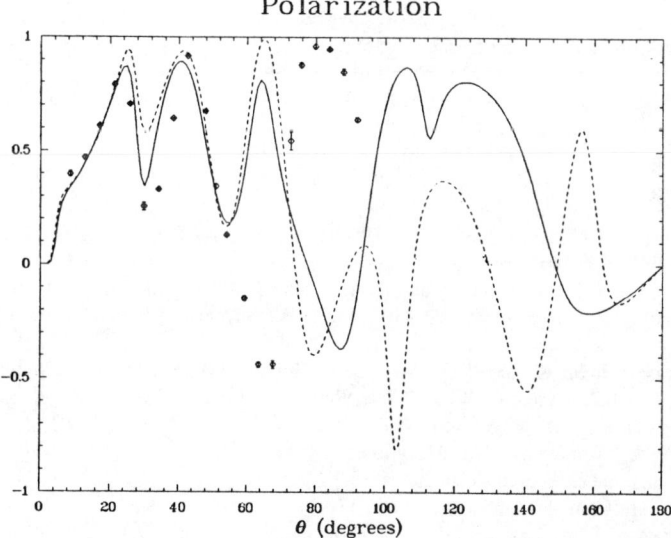

Fig. 5: Differential cross section and polarization for scattering of 135 MeV protons from ^{16}O. Solid line = RSC off-shell amplitude in approximation (31), dashed line = Love-Franey amplitude in same approximation.

published in a series of joint papers with her that are now in preparation. I would also like to thank Alan Pickelsimer, Peter Tandy, and Roy Thaler for discussions on the optical potential and for the use of the code WIZARD. I am very grateful to Mike Hynes and the Los Alamos National Laboratory for their hospitality at Los Alamos where some of this work was done. Some of the calculations reported here were carried out under a grant of computer time from the University of Maryland Computer Center. The manuscript for this paper was prepared at the University of Maryland Computing Center using the Center's DPS program and the Merganthaler Linotron 202.

References

[1] L. Ray et al., Phys. Rev. **23** (1981) 828.
[2] S. J. Wallace, "High Energy Proton Scattering", in Advances in Nuclear Physics, vol. 12, J. W. Negele and E. Vogt, eds. (Plenum, NY, 1984).
[3] A. Picklesimer and G. Walker, Phys. Rev. **C17** (197 8) 237; W. G. Love and M. A. Franey, Phys. Rev. **C24** (1981) 1073.
[4] H. V. von Geramb, AIP Conf. Proc. No. 97: The Interaction Between Medium Energy Nucleons in Nuclei - 1982, H. O. Meyer, ed. (American Institute of Physics, NY, 1983), pp. 44–77.
[5] B. D. Day, "Brueckner-Bethe Calculations of Nuclear Matter", in From Nuclei to Particles, Proceedings of the International School of Physics "Enrico Fermi", Course LXXIX, A. Molinari, ed. (North Holland, Amsterdam, 1981) pp.1–72.
[6] M. I. Haftel, Phys. Rev. **C7** (1970) 80; **C14** (1976) 698.
[7] B. C. Clark, S. Hama, R. L. Mercer, AIP Conf. Proc. No. 97, op. cit., p.260.
[8] J. J. Aubert et al., Phys. Lett. **123B** (1983) 275.
[9] H. Feshbach, Ann. Phys. (NY) **5** (1958) 357; **19** (1962) 287.
[10] J. A. MacNeil, J. Shepard, and S. J. Wallace, Phys. Rev. Lett. **50** (1983) 1439, 1443.
[11] B. Serot and J. D. Walecka, to be published.
[12] K. Maltman and N. Isgur, Phys. Rev. **D29** (1984) 952; A. Faessler and F. Fernandez, Phys. Lett. **124B** (1983) 145.
[13] A. W. Thomas, "Chiral Symmetry and the Bag Model: A New Starting Point for Nuclear Physics", in Advances in Nuclear Physics, vol. 13, J. W. Negele and E. Vogt, eds. (Plenum, NY, 1984).
[14] M. A. Birse and M. K. Banerjee, Phys. Lett. **136B** (1984) 284, and to be published in Phys. Rev. D; R. Goldflam and L. Wilets, Phys. Rev. **D25** (1982) 1951.
[15] S. A. Coon et al., Nucl. Phys. **A317** (1979) 242.
[16] J. J. Kelly, AIP Conf. Proc. No. 97, op. cit., p. 153.
[17] G. M. Lerner and E. F. Redish, unpublished.
[18] C. Olmer, AIP Conf. Proc. 97, op. cit., p. 176.
[19] K. M. Watson, Phys. Rev. **89** (1953) 575; M. Goldberger and K. M. Watson, Collision Theory, (Wiley, NY,1964); A. K. Kerman, H. McManus, and R. M. Thaler, Ann. Phys. (NY) **8** (1959) 551; E. Siciliano and R. M. Thaler, Phys. Rev., **C16** (1977) 1322; A. Picklesimer, P. C. Tandy, and R. M. Thaler, Phys. Rev. **C25** (1982) 1215, 1233.
[20] C. Mahaux, AIP Conf. Proc. 97, op. cit., p.20.
[21] E. F. Redish, in U. of Alberta/TRIUMF Workshop, Edmonton, Alberta, 11–13 July 1983,

J. M. Greben, ed.; TRIUMF preprint TRI-83-3, p.441.
[22] F. G. Perey, in *Direct Interactions and Nuclear Reaction Mechanisms*, E. Clementel and C. Villi, eds. (Gordon and Breach, NY, 1963); N. Austern, *Direct Nuclear Reaction Theories* (Wiley, NY, 1970).
[23] E. F. Redish and K. Stricker-Bauer, Phys. Lett. **133B** (1983) 1.
[24] P. C. Tandy, E. F. Redish, and D. Bollé, Phys. Rev. Lett. **35** (1975) 921; Phys. Rev. **C16** (1977) 1924.
[25] R. V. Reid, Ann. Phys. (NY) **50** (1968) 411.
[26] M. Lacombe et al., Phys. Rev. **D12** (1975) 1495; Phys. Rev. **C21** (1980) 861.
[27] M. Reiner, Ph.D. Thesis, U. of Maryland, 1974, unpublished; H. Bethe, private communication.
[28] D. Marker and P. Signell, Phys. Rev. **185** (1969)1286; E. F. Redish, G. J. Stephenson Jr., and G. M. Lerner, Phys. Rev. **C2** (1970) 1665; G. J. Stephenson Jr., E. F. Redish, G. M. Lerner, and M. I. Haftel, Phys. Rev. **C6** (1972) 1559.
[29] E. F. Redish, G. J. Stephenson Jr., and H. S. Picker, Phys. Rev. **C5** (1972) 707.
[30] K. Stricker-Bauer and E. F. Redish, in preparation.

BREMSSTRAHLUNG AND RADIATIVE PROCESSES
AT MEDIUM ENERGIES

Harold W. Fearing
TRIUMF, 4004 Wesbrook Mall, Vancouver, B.C. Canada V6T 2A3

ABSTRACT

Some general features of the electromagnetic interaction are discussed, features which make it useful as a probe of nuclear and few body systems. Then some specific reactions are reviewed, including pp → ppγ, np → npγ, pd → ^{3}Heγ, dd → ^{4}Heγ, and A(p,γ)A+1, with the aim of focusing on some of the interesting physics which perhaps could be investigated via such reactions at a medium energy proton facility with stored cooled beam as is being developed at IUCF.

I. INTRODUCTION

We want to discuss radiative processes at medium energies with particular emphasis on some of those nucleon initiated reactions which involve interesting and exciting physics and which at the same time may exploit some of the novel features of the proton cooler being built at IUCF. An immediate reaction to this aim might be that electrons rather than protons are the primary electromagnetic probes relevant for medium energy nuclear physics studies. It is true that much has been learned using electrons and we can anticipate learning a lot more with ongoing improvements of existing machines and with the development of high intensity, high duty factor machines. However there are also a number of proton or light ion induced electromagnetic processes which promise interesting physics and perhaps different kinds of information than that available from electron scattering.

To begin, we want to talk about some of the general features of the electromagnetic interaction used as a probe of hadronic systems. We then look at some specific radiative processes including proton-proton bremsstrahlung, some neutron induced processes such as np → npγ and np → dγ, some simple few body reactions such as pd → ^{3}Heγ and dd → ^{4}Heγ and finally the (p,γ) reaction on "real" nuclei. In each case we want to ask why the reaction is interesting, what is already known both theoretically and experimentally, what are some of the important theoretical ingredients, and if possible what could be measured to give additional insight.

It is hard for a theorist to fully appreciate the technical advances which have gone into the development of a machine such as the cooler. However there are some anticipated features which may make possible entirely new kinds of experiments. In particular we note the possibility of polarized atomic beam targets of a variety of nuclei, the possibility of measuring recoil nuclei, the good energy resolution, and the general cleanliness of the beam and lack of background, all of which could be used to advantage for some of these radiative processes.

0094-243X/85/1280138-19 $3.00 Copyright 1985 American Institute of Physics

II. GENERAL FEATURES OF ELECTROMAGNETIC PROCESSES

The electromagnetic interaction has several features which make it very useful as a probe of hadronic systems. In the first place it is "weak". Thus it can almost always be treated just to first order. One does not have to worry, for the photon, about the uncertainties introduced by distortion effects which are crucial for reactions induced by, say, protons or pions. One also does not have to worry about the double counting problems which enter in purely strong processes as for example in (p,π) where the basic reaction mechanism involves deltas and pion rescattering effects very closely entwined with the rescattering interactions or distortion of both proton and pion as they traverse the nucleus. The electromagnetic interaction is also "known" so that the photon becomes in principle a useful probe of the less well known strong interactions.

However, the electromagnetic interaction, while known, is also "complicated". This is not because the primary interaction itself is complicated. The coupling of a photon to a free elementary particle is in fact very simple. Its form is determined by general principles and there is usually a great deal of experimental information available to pin down form factors. The interaction becomes complicated because the hadronic interaction being probed is complicated, involving nucleons and clouds of pions or other mesons being exchanged, or at another level many quarks and gluons. The photons must couple to each charged particle. Thus even fairly simple microscopic descriptions of a strong process will lead to many contributions when a photon is coupled to all charges. We see this for example in the relatively complicated meson exchange correction calculations necessary for even simple systems or in the many diagrams required in diagramatic approaches to simple (p,γ) processes.

Another feature of the electromagnetic interaction is that it couples to a "conserved current". This is both a blessing and a curse. Current conservation allows one to use very general relations which often combine a lot of unknown and complicated physics into a simple result. The classic example of this is the use of Siegert's theorem[1] in reactions like np $\rightarrow$ dγ. In this case current conservation is used to express the relatively complicated current density in terms of the charge density, which is presumably less sensitive to the details of the process. Soft photon theorems[2] are another example of the use of current conservation to obtain a simple result, in this case a result which expresses the amplitude for a radiative process in terms of that for the non radiative process through the first two orders in the photon momentum k. However, current conservation implies constraints among many different contributions, and a photon coupling to a particular diagram may often imply coupling to many other diagrams so as to enforce the conservation. Thus it is often hard to obtain a "simple" conserved current.

An essentially equivalent property of the electromagnetic interaction is that of "gauge invariance", really just another way of stating current conservation. The full theory must be gauge invariant and in practice this both requires the existence of more complicated contributions and enforces cancellations among the various contributions. For example consider bremsstrahlung in the scattering process, depicted in Fig. 1, of a spinless particle of mass m on one of mass M. The amplitude corresponding to the first two diagrams,

$$M_1 + M_2 \sim g^2 e \left\{ -\frac{\varepsilon \cdot p_1}{k \cdot p_1} \frac{1}{(p_3+p_4)^2 - M^2} + \frac{\varepsilon \cdot p_3}{k \cdot p_3} \frac{1}{(p_1+p_2)^2 - M^2} \right\}, \quad (1)$$

is not gauge invariant. We must add the contribution

$$M_3 = g^2 e \, \varepsilon \cdot (p_1+p_2+p_3+p_4) \left\{ \frac{1}{(p_1+p_2)^2 - M^2} \cdot \frac{1}{(p_3+p_4)^2 - M^2} \right\} \quad (2)$$

corresponding to radiation from the interior of the "strong" interaction in order to make the result gauge invariant. Observe that an ad hoc term $\tilde{M}_3$, of the form

$$\tilde{M}_3 = \frac{\varepsilon \cdot X}{k \cdot X} \left\{ \frac{1}{(p_3+p_4)^2 - M^2} - \frac{1}{(p_1+p_2)^2 - M^2} \right\} \quad (3)$$

where X is any vector would also make $M_1+M_2+\tilde{M}_3$ gauge invariant. Such terms have often been added in calculations of radiative processes. They must be incorrect however (unless $X = p_1+p_2+p_3+p_4$) since there are no possible diagrams, except for the external radiation diagrams Fig. 1a and 1b (which lead to M_1+M_2) which can generate a $k \cdot X$ in the denominator of the amplitude.[3] A similar example relevant to bremsstrahlung is given in Fig. 2 where one can show that to leading order in k the contribution from the double scattering term Fig. 2c is just that required to make the first two terms gauge invariant to that order in k.[4]

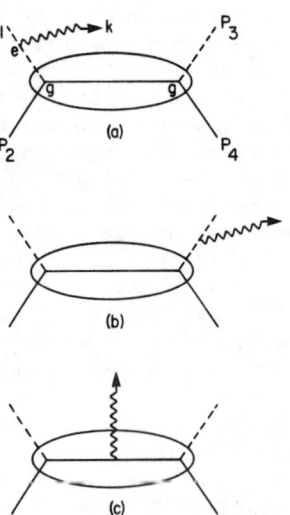
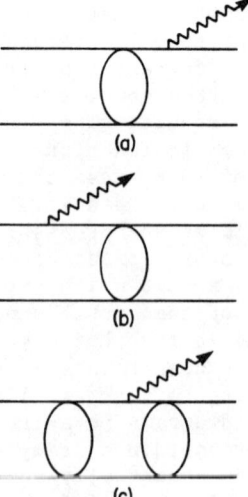

Fig. 1. Bremsstrahlung in a spinless scattering process.

Fig. 2. External radiation (a), (b) and double scattering (c) contributions to bremsstrahlung.

A very important but often ignored consequence of this cancellation enforced by gauge invariance is the fact that the relative magnitudes of the various diagrams are gauge dependent quantities. Thus to compare sizes of non gauge invariant subsets of diagrams, as has often been done for example in calculations purporting to show the importance of meson exchange corrections to radiative processes, is physically meaningless.

Thus to summarize, in this section we have considered some general properties of the electromagnetic interaction. It is "weak" and "known" and so should be a good probe of hadronic systems. It is however at the same time "complicated" and "constrained" by current conservation and gauge invariance and so provides a non-trivial and hence interesting window on these strongly interacting systems.

III. PROTON-PROTON BREMSSTRAHLUNG

To begin our consideration of specific radiative processes let us look first at proton-proton bremsstrahlung (ppγ), one of the simplest of such processes, and one which we clearly must understand if we are to understand any coupling of the photon to a hadronic system. Bremsstrahlung has been advocated as one of the most straightforward ways to probe the off shell aspects of the nucleon-nucleon force. However for a variety of reasons, both theoretical and experimental, not much has been learned yet. A new ppγ experiment[5] is underway at TRIUMF however, which has brought about increased interest in the process and motivated some new theoretical calculations. This experiment will measure the cross section over a much wider kinematic region than ever before and also, for essentially the first time, the analyzing power using polarized protons. This should be very interesting as existing theories predict quite different results for this analyzing power.

There have been two major theoretical approaches to ppγ which have been used in recent years. The first of these is the soft photon approximation[6,7] (SPA) which is based on the Low theorem.[2] The amplitude is expanded in powers of k, and gauge invariance used to fix the first two coefficients in terms of the on shell nucleon-nucleon amplitudes. The approach is simple, and easy to make gauge invariant and relativistic invariant. It however has no off shell content, so one hopes to find places where it fails. One of the puzzles seems to be that it actually works relatively well, even in kinematic situations where one would expect large off shell contributions.[7,8]

The second major approach is a standard non-relativistic potential model approach, based on specific choices for the nucleon-nucleon potential. This approach contains off shell information, but is much harder to correct for gauge invariance or relativistic effects. Such calculations in existance at the time of the most recent medium energy experiment,[8] one at 200 MeV at TRIUMF, did not fit the data as well as the purely on shell SPA.

Motivated by this discrepancy and by the possibility of additional data, particularly for the analyzing power, we have recently carried out a new potential model calculation.[9] The calculation was designed to use one of the modern theoretically motivated potentials such as the Paris[10] or Bonn[11] potential, which had not been done before. It also includes a number of effects such as relativistic spin corrections, Coulomb effects, one pion exchange contribution for higher partial waves, relativistic kinematics, etc. which had not all been combined into one calculation before. The basic approach is to solve the Lippmann-Schwinger equation in momentum space for the half off shell nucleon-nucleon T matrix. This is combined with propagator and electromagnetic vertex functions to generate the amplitude. The important correction is the relativistic spin correction, which consists of terms of higher order in v/c at the electromagnetic vertex.

Some results of these calculations using the Paris potential and an extended version of the Reid soft core potential[12] are shown in Fig. 3 for comparison with the SPA. It is obvious that there are large differences between the on-shell SPA and the potential models, differences which should be measureable in both cross section and analyzing power. What is somewhat surprising however is the similarity of the results from the extended RSC and the Paris potentials. In collaboration with R. Machleidt we have also made calculations using the Bonn potential,[13] with results which are again very similar to those of the others. In retrospect one can understand this in terms of the half off shell function for the various potentials. It turns out that it is essentially the same for all the potentials tried. Thus the similarity of the potential model results simply indicates that the potentials we have tried are very similar off shell as well as on shell.

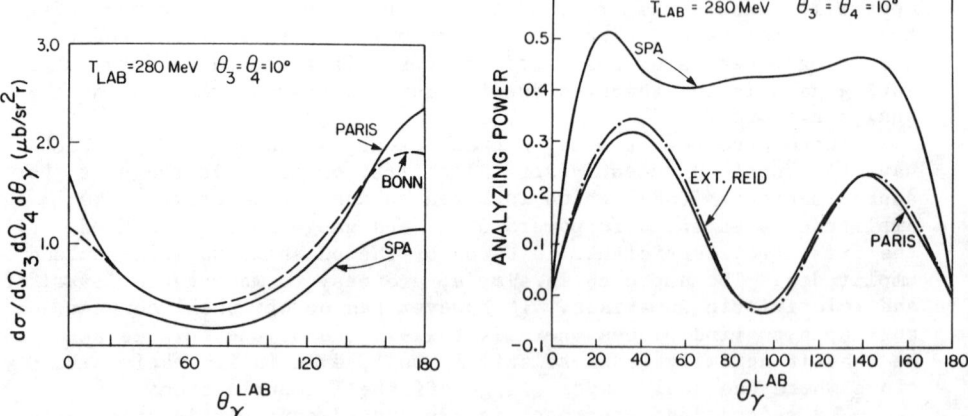

Fig. 3. Cross section and analyzing power for ppγ at incident laboratory proton energy of 280 MeV, with final protons each at 10° from the incident beam, from Ref. 9.

Several things remain to be added to the calculation. Although the amplitude is gauge invariant as it stands there are additional gauge terms coming from nonlocalities in the potential which have never been calculated. Also only the leading term of the double scattering piece has been included so far, although if such terms have the same sign as at lower energies they will increase the cross section[14] and thus increase the discrepancy between the potential results and either SPA or the existing data.

The experimental data for ppγ is rather sparse. The most recent experiment is a TRIUMF experiment at 200 MeV in equal angle geometry.[8] Surprisingly, as can be seen from Fig. 4, it seems to agree better with the SPA than with the potential calculations, all of which are too high, even though calculations indicate that there should be sensitivity to off shell effects in this geometry. A new TRIUMF experiment at 280 MeV is now underway.[5] With both cross section and analyzing power data over a wide kinematic range, it should be able to make a clean distinction between the potential model and SPA calculation.

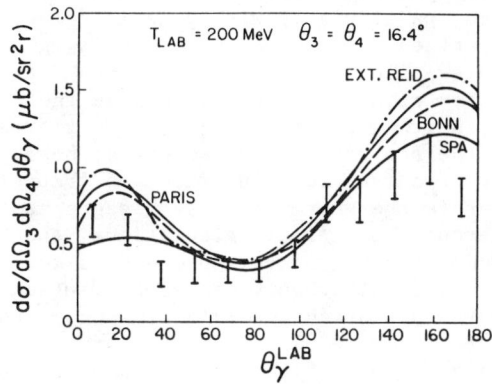

Fig. 4. Comparison of the potential model calculations of Ref. 7 with ppγ data at 200 MeV (Ref. 8).

Given the rather puzzling apparent failure of standard potential models it would be very useful to have additional data at lower energies, say in the 100-200 MeV region where off shell effects are smaller but where some of the corrections to potential calculations are also less important. Such experiments are low rate coincidence experiments and hence almost have to be done at a high duty cycle machine such as IUCF or TRIUMF. Good definition of the target volume and low background are important also. Thus if the luminosity can be made high enough some of the thin target-clean beam aspects of the cooler might be useful.

IV. NEUTRON INITIATED PROCESSES

After ppγ the next most obvious reactions to consider are the closely related neutron processes, neutron-proton bremsstrahlung, np → npγ, and neutron-proton radiative capture np → dγ. These both would require monochromatic external neutron beams.

There have been few calculations[15] and even fewer experiments[16] on neutron-proton bremsstrahlung. It is of course very similar to ppγ. However there are some distinctive features. In particular the leading contribution to the nucleon-nucleon force now involves the exchange of a <u>charged</u> pion so that internal radiation diagrams such as

Fig. 5 will contribute to the bremsstrahlung. For ppγ the simplest such diagrams would involve 2π exchange with an intermediate Δ^{++} rather than simple one pion exchange. Thus one would expect npγ to be described less well than ppγ by standard potential model theories and to be more sensitive to the non-trivial, and thus interesting, contributions of diagrams such as that of Fig. 5. Neutron experiments are more difficult than proton ones, but perhaps partially offsetting this is the fact that the npγ cross section is larger, by as much as an order of magnitude in some kinematic situations, than that for ppγ. This comes about because for npγ the radiation is primarily electric dipole while for ppγ the fact that the nucleons are identical suppresses the dipole radiation and makes the leading term electric quadrupole. Such npγ experiments will obviously be very difficult, but if they could be done, particularly with polarized neutrons, they would be pioneering experiments and might give new insight into the nucleon-nucleon-photon interaction.

ABreaction very closely related to npγ is the radiative capture reaction np → dγ, which would also require an external neutron beam. Unlike either of the bremsstrahlung processes, here there are many experiments, lots of data, and a multitude of theoretical calculations. Unfortunately there are sizeable disagreements among data sets, among different theoretical calculations, and between data and theory, so our understanding of this process is surprisingly shaky. An extensive review of both theory and experiment can be found in the proceedings of a recent workshop held at TRIUMF,[17] so the comments here will be brief.

Clearly np → dγ is one of the most fundamental of the two nucleon radiative interactions. Since it is also a two body interaction it is much easier to handle than ppγ or npγ and as a result there has been much more work on it, both theoretically and experimentally, than on the bremsstrahlung processes. It is an interesting reaction because it provides a testing ground for many of our ideas of how to write down the two nucleon electromagnetic current and to incorporate the various corrections due to meson exchanges and to isobars. One of the first triumphs in this regard was the work of Riska and Brown[18] who showed that meson exchange corrections (MEC), specifically the diagram of Fig. 5 and the pair term of Fig. 6, together with some isobar contributions, were sufficient to explain a long standing discrepancy in the threshold cross section. Later calculations have shown the importance under various circumstances of other meson exchange corrections, of the D state in the deuteron, of off shell aspects of

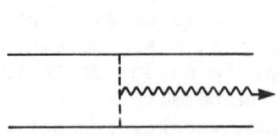

Fig. 5. A term in the neutron-proton electromagnetic current contributing to np → npγ or np → dγ

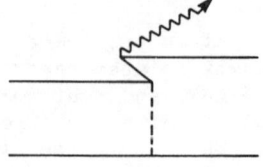

Fig. 6. Pair current contribution to np → dγ.

the nucleon-nucleon interaction, of the Δ(1236) and more recently of relativistic corrections. Not all of these effects are independent, which has been a source of some confusion. For example, as emphasized by Arenhövel,[19] most MEC can be included via gauge invariance using the Siegert theorem, with the result that the residual "MEC" are small. Alternatively all MEC can be calculated explicitly, as for example in Ref. 20 in which case the "MEC" are large. In either case however, it is clear that np → dγ depends on a number of important and interesting effects.

Perhaps because of this multitude of effects, the theoretical situation is rather confused and there are several different approaches which give only qualitatively similar results. The original calculation by Partovi[21] takes the matrix element of a simple one body electromagnetic current between two nucleon wave functions, and includes other corrections via the Siegert theorem. More recent calculations by Hwang, Londergan, and Walker,[22] Rustgi[23] et al., and Friar et al.[24] extend and improve this basic result in a variety of ways. Greben and Woloshyn[25] looked specifically at off shell effects. Laget[20] adopts a completely different approach, based on a sum of an extensive set of diagrams which include a variety of MEC, effects of the Δ, etc. Siegert's theorem is not used, but instead the set of diagrams is made gauge invariant. This approach avoids a partial wave expansion, and so may be more practical at higher energies. Leidemann and Arenhövel[26] use a coupled channel formalism including nucleon-nucleon and nucleon-delta channels analogous approaches which have been used for pp → πd. While the results of the various authors are all qualitatively the same they differ in detail and none provide a real quantitative fit to the data. Clearly a lot of theoretical work is still necessary.

Until very recently the data situation has been equally confused. There are many experiments both as np → dγ and γd → np and the results have not agreed well. For example one of the more modern and comprehensive experiments, that of the Lund group,[27] is as much as 40% higher than most other results in the photon energy region 100 to 200 MeV. There has been some improvement in the last year or so however. A new experiment[28] at IUCF at T_n=185 MeV seems to agree with an equivalent energy experiment[29] at E_γ=100 MeV from Frascati and a new experiment[30] at SIN, although averaging over a large energy bin, seems to be consistent with both. All of these are below the Lund data at this energy. At somewhat higher energies experiments from Bonn[31] and Tokyo[32] seem to agree reasonably well. If we are to make progress in understanding this reaction, highest priority should be given to obtaining a consistent and believable set of cross section data over the medium energy range.

Also in the last year we have seen the first measurements of the analyzing power in the medium energy region.[33] There have been a few calculations of this analyzing power and at the moment there seems to be a clear discrepancy with available calculations at 270 MeV, (as can be seen from Fig. 7), although results at 180 MeV seem okay.

The other persistent discrepancy in np ↔ dγ has been the 0° cross section as a function of energy. The original data of Hughes[34] has now been reasonably well confirmed by new data from Louvain[35] and IUCF.[28] A variety of MEC, deuteron D state effects, relativistic

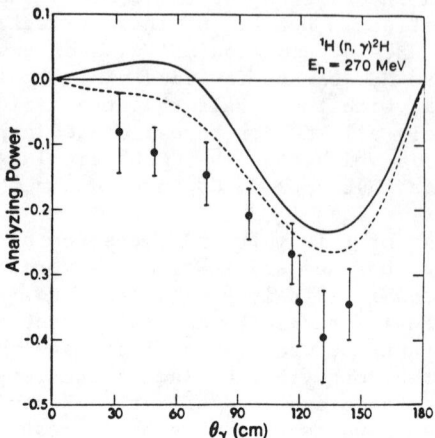

Fig. 7. Comparison, taken from Ref. 33, of the measured analyzing power in np → dγ with theoretical calculations of Partovi[21] (solid line) and of Leidemann and Arenhövel[26] (dashed line).

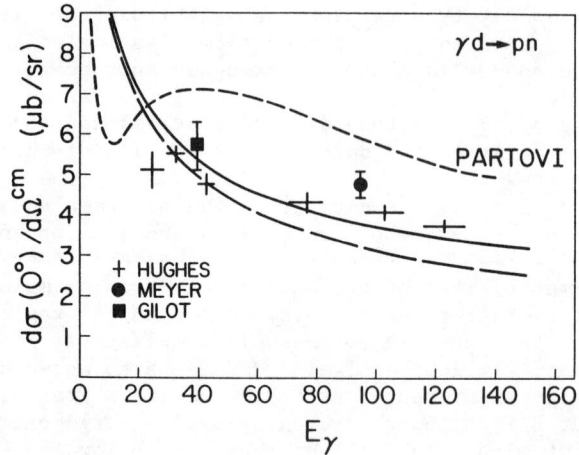

Fig. 8. 0° cross section for np → dγ as calculated in the full soft photon approximation of Ref. 37 (solid line) or an approximation to it of Ref. 36 (dashed line). Data is from Refs. 34, 35, 28.

effects, etc. have been incorporated in various calculations, and usually do produce some improvement over Partovi's original work. One of the better fits to the Hughes data however, as shown in Fig. 8, comes from a simple soft photon approach, which uses gauge invariance to relate the cross section to the amplitude for the non radiative process, just as is done for ppγ.[36,37] In this case, this simply requires the low energy properties of the np → d vertex function. This approach does not fit the angular distributions very well. Its relative success at 0° (and 180°) may simply indicate that it is more important to get relativistic and gauge terms, and hence the leading

MEC, correct to leading orders in k — which is done in the soft photon approach — than to have a good calculation of the various effects to all orders. As regards relativistic effects this is consistent with the results of Ref. 38, 24 who find that the inclusion of relativistic effects is an important improvement at 0°.

What needs to be done next? Clearly good, consistent data is important and one should look carefully to see if the possibility of a high quality tagged neutron beam would make some of these experiments cleaner and easier. It is obviously of interest to continue the measurements of analyzing powers. It is also feasible now to consider double spin experiments, ie. polarized neutrons on polarized protons or a polarized neutron experiment with measurement of outgoing deuteron polarization. Such experiments should be sensitive to subtle interferences of the various amplitudes. There seem to be no calculations; thus theorists should be encouraged to explore such double spin configurations.

V. FEW BODY REACTIONS

We turn now to the next most complicated system, that involving three nucleons, and consider the reaction pd $\rightarrow$ ^{3}Heγ. This will serve as a prototype (and best known example) of a class of few body radiative processes including for example such reactions as nd $\rightarrow$ tγ, pt $\rightarrow$ ^{4}Heγ, n^{3}He $\rightarrow$ ^{4}Heγ, dd $\rightarrow$ ^{4}Heγ, etc. Some of the general comments will apply also to these other reactions.

Why are such processes interesting? In the first place they offer some of the simplest examples of embedding a two nucleon process in a more than two nucleon system, so one might learn something about the techniques required for such embedding which could be useful in looking at similar processes in more complicated nuclear systems. At the same time they are simple enough that there is some hope of obtaining a microscopic understanding of the reaction. Furthermore much is known about the wave functions of such simple systems, particularly for the three nucleon system where there have been extensive Fadeev calculations.

For the specific process pd $\rightarrow$ ^{3}Heγ, low energy calculations suggest that certain spin correlations may give specific information about small components of the ^{3}He wave function.[39] This reaction is also one of the simplest which allows direct comparison with the analogous (p,π) reaction, in this case pd $\rightarrow$ tπ, which has been well studied experimentally. One might hope to learn something extra about both processes by making the comparison. Since many of the ingredients, such as wave functions, proton distortion, and kinematics are essentially the same one might for example, with a full understanding of pd $\rightarrow$ ^{3}Heγ, be able to isolate important features of the pion production mechanism or of pion distortion in pd $\rightarrow$ tπ.

Our theoretical understanding of pd $\rightarrow$ ^{3}Heγ is not particularly good as yet. At low energies there have been detailed and fairly sophisticated calculations, eg. Ref. 40. both of cross section and of analyzing powers. There is some indication that some spin correlations, particularly the analyzing power measured with tensor polarized deuterons, are very sensitive to D state components in the three body wave function.[39] It would be very interesting to extend such considerations to higher energies.

At medium energies the theoretical situation is much worse. Older calculations of Craver et al.[41] and of Finjord[42] are relatively unsuccessful. A new calculation by Laget,[43] based on a series of diagrams does better, and does seem to give qualitative agreement with analyzing power data.

Another approach,[44] which has been relatively successful for the cross section over a range of energies is based on a distorted wave impulse approximation and is motivated by the idea that one should be able to compare pd → ^{3}Heγ and pd → tπ. The model is based on Fig. 9a and is essentially the same as that used earlier for pd → tπ.[45] The cross section is then given by the np → dγ cross section times a kinematic factor times a form factor derived from wave function overlaps. One of the biggest uncertainties is the np → dγ cross section. Some results are shown in Fig. 10 for the Lund[27] np → dγ data which is higher than most and for the average of a set of data which may be low. Results for other energies can be found in Ref. 44.

A very similar calculation, though differing in technical details, has been made by Prats[46] who includes diagrams b and c of Fig. 9 as well as a. His results, which are shown also in Fig. 10, are similar except for the dip in the forward direction which is in

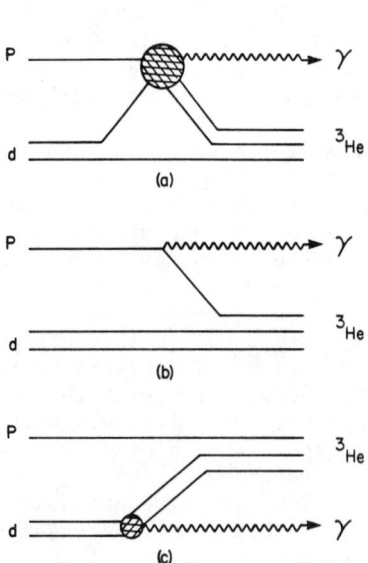

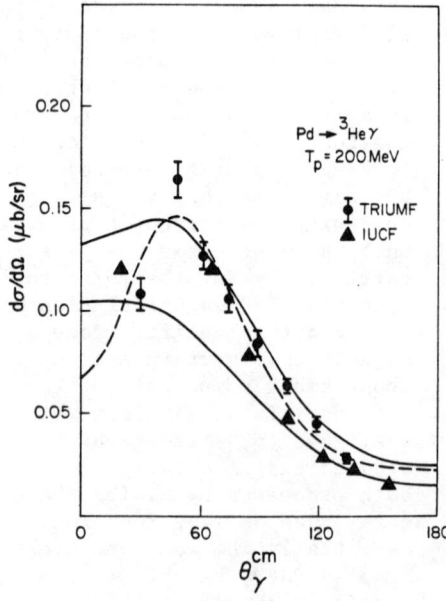

Fig. 9. a) Distorted wave impulse approximation model for pd → ^{3}Heγ used in Ref. 44. b), c) Additional diagrams included also in Ref. 46.

Fig. 10. Results for pd → ^{3}Heγ: Solid curves - DWIA calculation of Ref. 44 using Lund[27] data for np → dγ (upper curve) or all other data (lower curve). Dotted curve — calculation of Ref. 46. Data is from Refs. 49, 50.

fact marginally suggested by the data. In Prats' calculation the dip originates in the interference of diagrams 9a and 9b. However some portion of the proton radiating diagram 9b appears also in a representation of the process np → dγ and hence taking the np → dγ cross section from experiment in Fig. 9a and at the same time including Fig. 9b must involve double counting to some extent. It would be very interesting to have sufficient additional data to see if the dip is really there.

The experimental situation is still not very good, though like for np → dγ, there has been a marked improvement as a result of several recent experiments. There are a number of older measurements[47] in both the γ+^{3}He and the p+d directions which differ among themselves by factors of two or three, and at least one newer experiment[48] designed to resolve the problem, which is also in disagreement with the others. However the three most recent experiments — radiative capture experiments at TRIUMF[49] and IUCF[50] and a photodisintegration experiment at Bates[51] — are in reasonable agreement with each other. Some of this data is shown in Fig. 10 along with the theoretical results.

An exciting recent development is new data on analyzing power. There are data from IUCF[50] at 100, 150, 200 MeV and from TRIUMF[49] at several energies in the range 200-500 MeV and the two agree in the region of overlap. There has also been one as yet unpublished measurement of the tensor analyzing power for this reaction.[52] It would be very interesting to further pursue some of these spin correlation measurements.

To summarize then, it has just been very recently that experiments are beginning to converge on a consistent result and just recently that data on analyzing powers is becoming available. The theorists are behind. There are tantalizing indications from low energy calculations of sensitivities to things like small components of the ^{3}He wave function, but few calculations at medium energies. Additional experiments, particularly of various spin correlations, would undoubtedly stimulate further theoretical work.

Finally we mention very briefly another reaction, dd → ^{4}Heγ, which has been of interest at IUCF where efforts are being made to measure cross section and tensor analyzing powers.[52,53] This reaction has some interesting features: the particles all have isospin zero, so it is hard to find a simple mechanism which involves the delta. The initial state has two identical particles and all particles are bosons, so the spin aspects of the process should be simpler than some to describe. However it is not just a single particle capture, so the electromagnetic current operation may be somewhat complicated and may provide a stringent test of our ability to handle an electromagnetic interaction in a few body system. There however seem to be no published calculations of this process in the medium energy region.

VI. (p,γ) REACTIONS ON "REAL" NUCLEI

So far we have considered only few body radiative reactions which, although complicated, are simple enough so that there may be some hope of understanding the full reaction microscopically. Next we turn to the reaction p + A → γ + (A+1) where A is a "real" nucleus with A > 4. In this case a full understanding of the complete many body process at the microscopic level is probably impossible. Thus the reasons for studying such processes will be somewhat different than for few body processes. On the one hand, we now have the possibility of using the process as a probe of nuclear structure. On the other, we may learn interesting things about how to embed a few nucleon process in a many body system.

Recall that the (p,γ) process, just like (p,π) or (p,d) involves a very high momentum transfer to the nucleus. Figure 11 shows a typical range of momentum and energy transfer accessible at medium energy machines. Thus for example at 200 MeV incident energy, center of mass three-momentum transfers in the 400-700 MeV/c range are accessible. In the simple picture where a single nucleon radiates, the amplitude is proportional to the Fourier transform of the single particle wave function. This idea was used with marginal success to analyze early (γ,p) data to try to extract the high momentum behavior of the wave function.[54] Such a picture is now known to be far too simple, but the very fact of such massive momentum transfers suggests that the reaction will probe regions not easily accessible in other ways.

The reaction may also be very interesting for the possible new insight it may give regarding the (p,π) reaction. The two processes are very similar, with (p,γ) presumably being simpler since no pion distortion is involved. The suggestion that the two processes should be compared was made long ago[55] and such comparisons have been made for very light nuclei.[44] Now that there is a wealth of (p,π) data on a variety of nuclei, similar data for (p,γ) would allow a systematic comparison to be made.

Finally the process may lead to interesting new discoveries in the nuclear structure realm. One of the very exciting results coming from IUCF in recent years was the observation of strong excitation of relatively high lying stretched

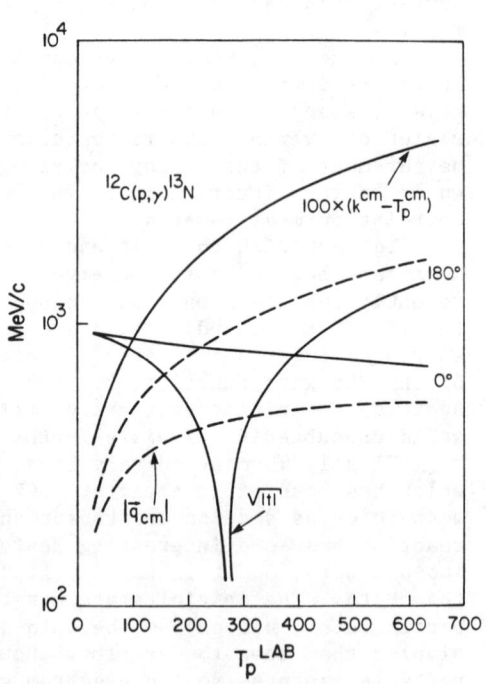

Fig. 11. Energy and momentum transfer range accessible for $\theta_\gamma = 0°$, $180°$ as a function of the incident lab proton energy. Shown are the four momentum transfer $\sqrt{|t|} = |(P_p-k)^2|^{1/2}$, the center of mass three momentum transfer $|\vec{q}| = |\vec{P}_p - \vec{k}|$ and the kinetic energy transfer in the center of mass, $k - T_p$.

configuration states using the (p,γ) reaction at 40-80 MeV proton energy.[56] It would be interesting to see if similar such phenomena exist also at higher energies.

On the theoretical side not much has been done in the medium energy region although there have been a number of calculations at very low energies. The situation is reminiscent of the state of affairs for (p,π) just a few years ago. The several calculations which do exist all give some aspects of the qualitative behavior correctly, but are not quantitatively very good and are at the same time sensitive to the various input ingredients.

Just like for (p,π) there are two major approaches to (p,γ). The direct mechanism is essentially the analog of the single nucleon model (SNM) of (p,π) and the various isobar calculations parallel the two nucleon mechanism (TNM) calculations for (p,π).

In the direct or SNM the photon is radiated from the incoming photon via a single particle charge plus magnetic static coupling. Usually proton distortion is put in as well. Typical calculations of this type are Refs. 57, 58. In the absence of distortion this model gives an amplitude directly proportional to the Fourier transform of the single particle wave function and thus is very sensitive — too sensitive — to high momentum components of the wave functions. For the (γ,p) reaction in the energy region $E_\gamma < 100$ MeV this approach, with distortion, does give qualitative agreement with the available data at a fixed energy.[57] It does not reproduce the energy dependence very well however and there is sensitivity at the level of factors of two or three to the potential chosen for the proton distortion, and probably to other ingredients as well. This SNM also fails to predict the (γ,n) cross section, which experimentally is of the same order as the (γ,p), but which in this model is much smaller than (γ,p) because it can go via magnetic coupling only. There are some indications however that including charge exchange in the distortion might improve the situation.[59]

There have been several different two nucleon models all based on the idea that at medium energies the two nucleon interaction currents, particularly those involving the $\Delta(1236)$, must be important. Gari et al.[60] were one of the first groups to perform such calculations. They found that various non-resonant exchange currents were important, more so than the isobar contributions. Londergan et al.[61] calculate both the SNM and some specific isobar contributions (Fig. 12). They find that the isobar contributions dominate for $E_\gamma > 100$ MeV and give qualitative agreement with the data. More recently Cheung and Keister[62] have performed a similar TNM calculation. Their results, which include the Δ but not all of the one body contributions, reproduce the shape of the cross section, but are much lower in magnitude than the data. They find that the Δ contributions are important, but not as large as found by Londergan, et al.[61] The differences in these three calculations indicate some of the uncertainty which exists in such calculations. Obviously there is a need for more theoretical work.

One of the very interesting discoveries of the past few years has been the importance of relativistic effects in medium energy reactions. A number of groups have shown that particularly spin correlations in p-nucleus reactions are well described in a Dirac formalism. In such an approach the amplitudes are calculated as matrix elements

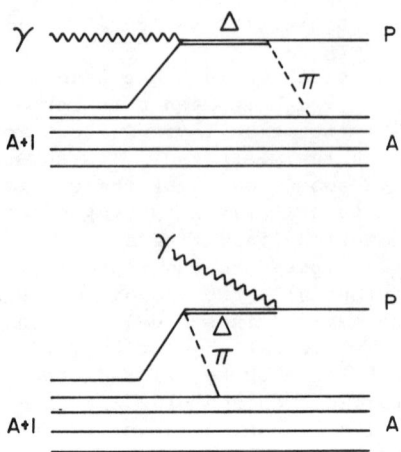

Fig. 12. Some isobar contributions to the (p,γ) and (γ,p) reactions.

of 4×4 Dirac operators taken between 4-component states which are obtained by solving the Dirac equation in the presence of strong scalar and vector potentials. This approach has had a great deal of success for elastic and inelastic proton-nucleus scattering. At least three such calculations of (p,γ) are now underway.[63] Such calculations are particularly aimed at spin variables and so it will be interesting to see what new predictions emerge for the various spin correlations.

On the experimental side, in the medium energy region, the (p,γ) reaction is almost totally unexplored. There is data from both (γ,p) and (p,γ) at lower energies, but the only medium energy data seems to be the IUCF (p,γ) data[56] mentioned above, at the very low energy end of what might be considered "medium energy", and some isolated (γ,p) data at higher energies[54] although there are unpublished experiments from Bates.[64] In the (p,γ) direction one can use polarized protons and measure analyzing powers. This has been done for (p,π) with very interesting, though puzzling results. There seems to be essentially no such data for (p,γ), except on the very light nuclei. The situation is similar to that for (p,π) of 5-10 years ago. Since then we have seen the very nice systematics which have come from the (p,π) program at IUCF, TRIUMF and elsewhere. One could probably anticipate similarly interesting results from a systematic (p,γ) program.

There are two possibilities for new types of experiments which may be feasible as a result of some of the unique features of the machine being planned here. The very good energy resolution, coupled with the possibility of using very thin targets might make it feasible to detect the recoil nucleus directly. Figure 13 shows the maximum and minimum recoil energy obtained in the reaction $^{12}C(p,\gamma)^{13}N$ as a function of the incident proton energy. In the medium energy range for at least at some angles it should be possible to detect such

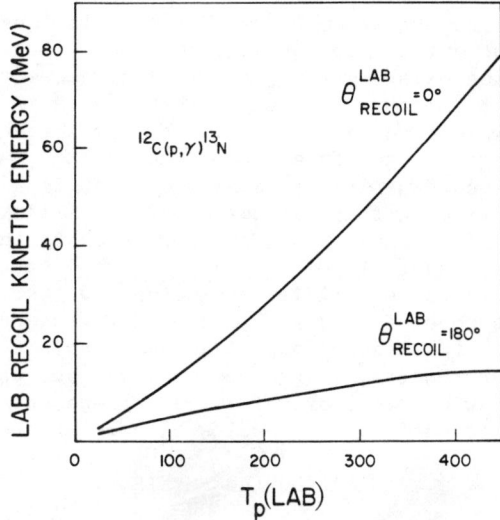

Fig. 13. Maximum and minimum laboratory kinetic energy of the recoil nucleus in $^{12}C(p,\gamma)^{13}N$.

recoils though to separate out some of the low lying excitations would require very good resolution. If the γ were measured in coincidence, one might be able to significantly clean up the experiment.

A second feature, more speculative but very exciting, is the possibility of using polarized atomic beam targets. Such targets would allow polarized proton on polarized target experiments on a variety of targets not now available in a polarized state and without the extraneous material and extra background normally associated with a polarized target. There is no data and no theory for such doubly polarized (p,γ) processes. However, normally the more complicated a spin correlation becomes the more sensitive it is to various interfering amplitudes and so one might expect with such data to provide very sensitive confrontations with the theory.

In summary then, of this section on (p,γ) in real nuclei, it appears that detailed calculations are difficult, probably almost as difficult as have been the very similar calculations for (p,π), even though one does not have the added uncertainty of pion distortion. At present few such calculations have been carried out and those that have give only a very qualitative representation of the data. However there is as yet little data, essentially nothing involving spin correlations, to stimulate such calculations. A systematic study of (p,γ) which would allow detailed comparison with the (p,π) results would undoubtedly stimulate theoretical activity.

VII. SUMMARY AND OUTLOOK

We have surveyed a number of radiative reactions initiated by nucleons. All of these reactions tell us very interesting things about the way a photon couples to a several nucleon system and thus probe aspects of the primary strong interaction. On the theoretical side the common features seem to be that calculations are complex, so that different authors do not always agree. There are still puzzles in fitting existing data, and in many cases the calculations simply have not been done. Obviously then, there is much interesting theoretical work left to be done. On the experimental side, in the medium energy region, when data exists there often have been disagreements among various experiments, and thus there is a need for more, careful work to resolve these disagreements. In many cases there is simply no data. Thus experiments on npγ, on analyzing powers for many of the reactions, and particularly on more complicated or double spin correlations would be essentially exploring new territory. It appears that some of the novel features of a cooled proton beam machine could be useful — the high resolution, clean beam, thin targets which would allow recoil measurements, tagged neutron beam and particularly the possibility of a wide range of polarized targets obtained using atomic beams. We look forward to the results of some of these explorations.

ACKNOWLEDGEMENTS

Support in part from the Natural Sciences and Engineering Research Council of Canada is gratefully acknowledged.

REFERENCES

1. A.J.F. Siegert, Phys. Rev. 52, 787 (1937).
2. F.E. Low, Phys. Rev. 110, 974 (1958).
3. S.L. Adler and Y. Dothan, Phys. Rev. 151, 1267 (1966).
4. See e.g. L. Heller, Phys. Rev. 174, 1580 (1968).
5. Expt. #208, P. Kitching, spokesman, 1984.
6. E.M. Nyman, Phys. Rev. 170, 1628 (1968).
7. H.W. Fearing, Phys. Rev. C22, 1388 (1980) and references cited therein.
8. J.G. Rogers, et al., Phys. Rev. C22, 2512 (1980).
9. R.L. Workman and H.W. Fearing, Proc. of the 10th Int. Conf. on Particles and Nuclei, Heidelberg, 1984, ed. by F. Güttner, et al., Book of Abstracts, Vol. 1, P. C29 and to be published.
10. M. Lacombe, et al., Phys. Rev. C21, 861 (1980).
11. K. Holinde, Phys. Rept. 68, 121 (1981); R. Machleidt, TRIUMF preprint TRI-PP-83-128 (1983).
12. B.D. Day, Phys. Rev. C24, 1269 (1981).
13. H.W. Fearing, R. Machleidt, R.L. Workman, to be published.
14. V.R. Brown, Phys. Rev. 177, 1498 (1969).
15. G.E. Bohannon, Phys. Rev. C17, 865 (1978); M.A. Preston and W. van Dijk, Phys. Rev. C19, 1693 (1979) and references cited therein.
16. J.A. Edgington, et al., Nucl. Phys. A218, 151 (1974) and references cited therein.
17. Proceedings of the Radiative Processes Workshop, TRIUMF, May 1984, ed. by D. Hutcheon and N. Davison, to be published in Canadian Journal of Physics.
18. D.O. Riska and G.E. Brown, Phys. Lett. 38B, 193 (1972).
19. H. Arenhövel, Z. Phys. A. 302, 25 (1981).
20. J.M. Laget, Nucl. Phys. A312, 265 (1978) and Ref. 17.

21. F. Partovi, Annals of Physics $\underline{27}$, 79 (1964).
22. W.-Y.P. Hwang, J.T. Londergan, and G.E. Walker, Annals of Physics $\underline{149}$, 335 (1983) and Ref. 17.
23. M.L. Rustgi, R. Vyas and O.P. Rustgi, Phys. Rev. $\underline{C29}$, 785 (1984), references cited therein, and Ref. 17.
24. J.L. Friar, B.F. Gibson, and G.L. Payne, Phys. Rev. $\underline{C30}$, 441 (1984).
25. J.M. Greben and R.M. Wololshyn, J. Phys. G. $\underline{9}$, 643 (1983).
26. W. Leidemann and H. Arenhövel, Phys. Lett. $\underline{139B}$, 22 (1984) and Ref. 17.
27. P. Dougan et al., Z. Phys. A $\underline{280}$, 341 (1977); $\underline{276}$, 55 (1976).
28. H.O. Meyer, et al., Phys. Rev. Lett. $\underline{52}$, 1759 (1984).
29. G.P. Capitani, et al., Proc. of the 10th Conf. on Particles and Nuclei, Heidelberg, 1984, ed. by F. Güttner, et al., Book of Abstracts, Vol. 1, p. A9 and Ref. 17.
30. G. Nicklas et al., Phys. Lett. $\underline{141B}$, 170 (1984).
31. J. Arends et al., Nucl. Phys. $\underline{A412}$, 509 (1984).
32. K. Baba, et al., Phys. Rev. $\underline{C28}$, 286 (1983).
33. J.M. Cameron, et al., Phys. Lett. $\underline{137B}$, 315 (1984).
34. R.J. Hughes et al., Nucl. Phys. $\underline{A267}$, 329 (1976).
35. J.F. Gilot et al., Phys. Rev. Lett., $\underline{47}$, 304 (1981).
36. J. Govaerts, et al., Nucl. Phys. $\underline{A368}$, 409 (1981).
37. H.W. Fearing, Proceedings of the International Conference on Few Body Problems in Physics, Karlsruhe, ed. by B. Zeitnitz, Vol. II, p. 99 (Elsevier, 1984) and to be published.
38. A. Cambi, B. Mosconi, P. Ricci, J. Phys. G $\underline{10}$, L11 (1984).
39. A. Arriaga and F.D. Santos, Phys. Rev. C $\underline{29}$, 1945 (1984).
40. S. Aufleger and D. Drechsel, Nucl. Phys. $\underline{A364}$, 81 (1981); B.F. Gibson and D.R. Lehman, Phys. Rev. $\underline{C11}$, 29 (1975).
41. B.A. Craver, et al., Nucl. Phys. $\underline{A276}$, 237 (1977).
42. J. Finjord, Nucl. Phys. $\underline{A274}$, 495 (1976).
43. J.M. Laget, to be published.
44. H.W. Fearing, Proceedings of the IX Int. Conf. on the Few Body Problem, Eugene, Oregon 1980, Vol. 1 (contributed papers) p. 39.
45. H.W. Fearing, Phys. Rev. $\underline{C16}$, 313 (1977).
46. F. Prats, Phys. Lett. $\underline{88B}$, 23 (1979).
47. For a review, see e.g. J.M. Cameron, Ref. 17.
48. B.M.K. Nefkens, et al., Phys. Rev. Lett. $\underline{45}$, 168 (1980).
49. R. Abegg et al., Phys. Lett. $\underline{118B}$, 55 (1982) and to be published. See also J.M. Cameron, Ref. 17; A. Stetz, private communication.
50. M.A. Pickar et al., IUCF Annual Report 1983, p. 72; M. Hugi, Ref. 17.
51. D.I. Sober et al., Phys. Rev. $\underline{C28}$, 2234 (1983); W.J. Briscoe et al., Phys. Rev. Lett. $\underline{49}$, 187 (1982).
52. M. Hugi et al., IUCF Annual Report 1983, p. 74.
53. B.H. Silverman, et al., Phys. Rev. $\underline{C29}$, 35 (1984).
54. See e.g. D.J.S. Findlay et al., Phys. Lett. $\underline{74B}$, 305 (1978); D.J.S. Findlay and R.O. Owens, Phys. Rev. Lett. $\underline{37}$, 674 (1976); J.L. Matthews, et al., Phys. Rev. Lett. $\underline{38}$, 8 (1977).
55. J.M. Eisenberg, J.V. Noble, and H.J. Weber, Proc. of the Int. Conf. on Photonuclear Reactions and Applications, ed. by B.L. Berman (Lawrence Livermore Laboratory, 1973), p. 957.

56. S. Blatt et al., Phys. Rev. C30, 423 (1984) and references cited therein.
57. J.T. Londergan and L.D. Ludeking, Phys. Rev. C25, 1722 (1982); S.F. Tsai and J.T. Londergan, Phys. Rev. Lett. 43, 576 (1979).
58. S. Boffi, C. Giusti and F.D. Pacati, Nucl. Phys. A359, 91 (1981).
59. S. Cotanch, Phys. Lett. 76B, 19 (1978).
60. M. Gari and H. Hebach, Phys. Repts 72, 1 (1981) and references therein.
61. J.T. Londergan, G.D. Nixon and G.E. Walker, Phys. Lett. 65B, 427 (1976); J.T. Londergan and G.D. Nixon, Phys. Rev. C19, 998 (1979).
62. C.Y. Cheung and B.D. Keister, TRIUMF preprint TRI-PP-84-97 (1984).
63. J. Sheppard, private communication; A. Picklesimer, private communication; H.W. Fearing, unpublished.
64. See e.g. G.S. Adams et al., Proc. of the 10th Conf. on Particles and Nuclei, Heidelberg, 1984, ed. by F. Güttner, et al., Book of Abstracts, Vol. 1, p. A7, and M. Leitch, et al., quoted in Ref. 62.

DIRECT REACTIONS STUDIES OF THE NUCLEAR CONTINUUM*

Gerhard J. Wagner
Physikalisches Institut der Universität Tübingen
7400 Tübingen
Federal Republic of Germany

ABSTRACT

Progress expected to come from cooled light-ion beams for nuclear continuum studies is discussed. To this aim the status of deep-hole and giant resonance studies is reviewed by way of examples. For investigations of reaction mechanisms the possible measurement of excitation functions and of out-of-plane (p,2p) reactions through recoil nucleus detection is considered.

1. THE ROLE OF ENERGY RESOLUTION IN THE NUCLEAR CONTINUUM

The most conspicuous feature of cooled beams is the excellent energy resolution. It is not immediately clear what the advantage of high resolution would be in the continuum region where one is generally confronted with overlapping levels. Direct reactions such as e.g. the ^{58}Ni($\vec{d}$,^{3}He) reaction [1] at 80 MeV (see fig.1) fortunately are sufficiently selective to allow a spectroscopy of individual levels even above the particle emission thresholds provided the energy resolution is sufficiently good. It is obvious that an improvement of the energy resolution from ≈50 keV (fig.1) to ≈15 keV will substantially increase the potential of spectroscopy near thresholds. Among others this would allow to study the natural line widths of the fragments of nuclear hole states which are related to the amount of 1p-2h admixtures to the 1h components and their escape widths. However, it is equally obvious that with increasing excitation energy the increase of level densities and line widths will always prevent a spectroscopy of individual levels no matter what resolution will be achieved.

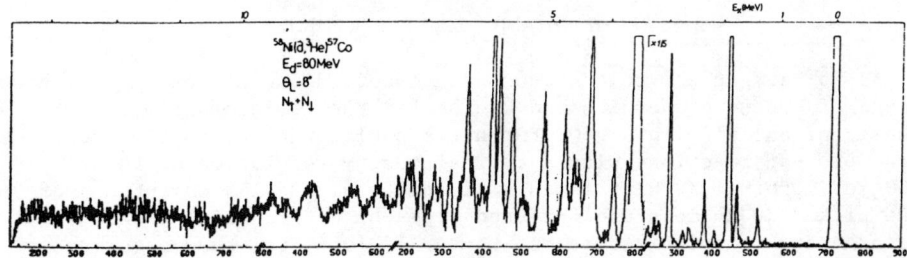

Fig.1. Spectrum of the ^{58}Ni($\vec{d}$,^{3}He)^{57}Co reaction at 80 MeV composed of four momentum bites taken with the QDDM spectrograph of the IUCF. From ref.1.

---

*Supported by the Bundesminister für Forschung und Technologie

One could try to push upwards the excitation energies where information on level widths is inferable from the spectra by adopting statistical methods for the analysis of missing energy spectra. Techniques such as autocorrelation functions have been developped for the extraction of $\langle \Gamma/D \rangle$ from excitation functions. I am aware of two[2] applications to missing energy spectra, namely β-strength functions and (e,e') spectra[3].

An experimental approach to cope with the high level density problem is to increase the selectivity of the reaction mechanism by additional coincidence requirements with specific decay channels.

In concluding these introductory remarks it may be appropriate to emphasize that for the type of reactions to be discussed here, the missing energy resolution will be determined largely by the specifications of the spectrometers (Design goal 15 keV at 200 MeV for the K = 600 spectrometer of the IUCF). A beam resolution of $\simeq$1 keV, if achieved, could be exploited only in excitation functions (see below).

The typical questions in the field of continuum physics are, however, not related to individual levels but to strength function phenomena which represent elementary modes of excitation. Of these I want to discuss first single particle and single hole states (sect.2) and then giant resonances (sect.3) for which a large body of singles and coincidence experiments exists. I shall try to give a very condensed status report meant for those participants who are not active in this field and to present a few examples of current problems. The discussion should make apparent that features of the cooled beams other than the appraised energy resolution will be very decisive indeed for progress in this field. Finally, as typical cooler applications I want to briefly mention the possible use of excitation functions for the study of the "background" from multi-step processes and of out-of-plane (p,2p) or (p,pα) measurements (sect.4).

2. DEEP-HOLE AND SINGLE PARTICLE STATES

As far as nuclear hole states are concerned one may claim fair knowledge of centroid energies and widths for the various occupied subshells of nuclei with A <60. For outer shells the information comes from pick-up reactions with a typical energy resolution of about 100 keV. Orbital angular momentum l and total angular momentum j of the picked-up nucleon are inferred from the differential cross sections and - if polarized beams are available - from the vector-analysing powers, respectively. Typical examples may be found for example in a paper on $1p^{-1}$ strengths in 2s,1d shell nuclei[4]. For separation energies exceeding 20 MeV the absorption of the projectiles and ejectiles so far has largely prevented an investigation of deep-hole states by pick-up reactions. In this energy domain studies of (p,2p) and (e,e'p) reactions have revealed the existence of very short lived inner hole states. The clearest demonstration of nuclear shell structure of light nuclei has come from (e,e'p) work with $\simeq$1 MeV energy resolution[5]. Recently missing energy spectra from (e,e'p) experiments

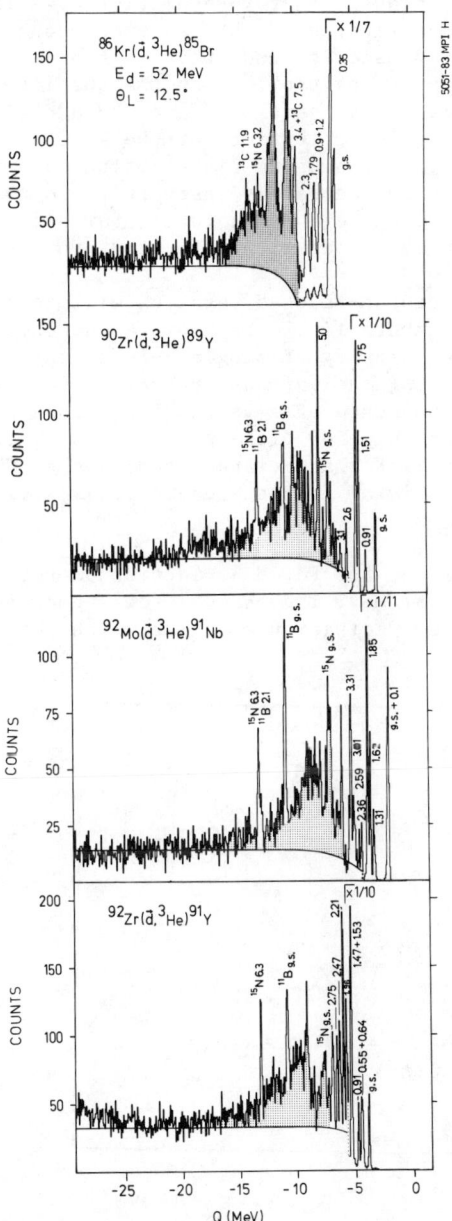

Fig.2. Spectra from (d,^{3}He) reactions on nuclei of the Zr-region taken at $\Theta = 12.5°$. The $1f_{5/2}$, $2p_{1/2}$ and $2p_{3/2}$ hole strengths reside in a few strongly excited levels at low excitation energies. The shaded gross structure on top of an assumed background contains dominantly $1f_{7/2}$ proton hole strength. From ref.7.

at 400 MeV with the fantastic energy resolution of ≈ 150 keV have become available [6]. As the description of the reaction mechanism for (e,e'p) reactions rests on much safer grounds than that of transfer reactions we hope to solve longstanding problems on the absolute strengths of the hole states. While these reactions have the additional advantage of yielding the momentum distributions of the knocked-out protons, it is also true that these momentum distributions are not very characteristic for the l-value and certainly insensitive to the j-value of the protons. It is here where I see a future for high-resolution $(\vec{p},2p)$ reactions and pick-up reactions with polarized protons and deuterons.

This advantage of hadron-induced reaction will be particularly valuable for A >60 nuclei where little is known about deep-lying hole states. The increasing density of single-hole states in heavy nuclei makes a discrimination of the various orbitals more difficult. But it also leads to the presence of several single-hole states within the energy range accessible for pick-up reactions. Therefore, several groups, mostly at Orsay, Karlsruhe and recently the IUCF have started investigations of deep-lying hole states in heavy nuclei by pick-up reactions.

Fig.2 shows spectra of the $(\vec{d},^{3}\text{He})$ reaction on several target nuclei in the ^{90}Zr-region at 52 MeV. They show a few strong states (note the change of scale) at low excitation energies which contain the full

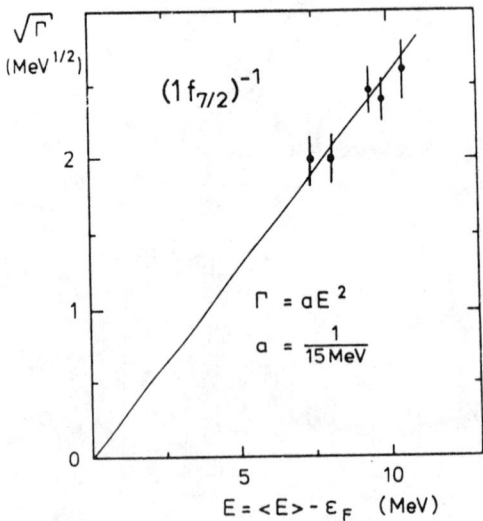

Fig.3. Square-root of the widths of $1f_{7/2}$ proton-hole states of nuclei in the Zr-region vs. distance of their energy from the Fermi surface. From ref.7.

shell model strengths of the $2p_{1/2}$, $2p_{3/2}$ and $1f_{5/2}$ orbitals. With a Q-value of -10 MeV or more one finds a "giant-resonance-like" structure riding on an empirically drawn background. This structure was shown to contain dominantly $1f_{7/2}$ strength. A recent ^{90}Zr (e,e'p) experiment at Amsterdam at a momentum transfer of 190 MeV/c shows a structure with a striking resemblance in many details, however, without the disturbing background present in the (d,^{3}He) reaction.

The systematic trends of energies and widths of the $f_{7/2}$ hole states visible in fig.2 asks for an explanation by nuclear many-body theories. A puzzling result which emerged from these studies so far is the 1f spin-orbit splitting which is about a factor of two larger than predicted by Brueckner-Hartree-Fock theory [8]. The increase of widths of the hole states when going to heavier nuclei is in accordance with the theory of Fermi liquids which predicts a quadratic increase of the widths of single particle and, of single hole states with the distance from the Fermi surface [9]. This is shown in fig.3 for the widths of the $1f_{7/2}$ proton hole in nuclei of the Zr region. This theory qualitatively also explains the observation from recent (α,^{3}He) reaction studies on various Sn isotopes [10] which show a decreasing width with increasing target mass, as obviously for the heavier samples a given single particle state is closer to the Fermi surface. In this context it is important that the (α,^{3}He) reaction populates preferentially high-l orbitals where the centrifugal barrier is high enough to produce a small escape width. It is only for the spreading widths not for the escape widths that a particle-hole symmetry with respect to a Fermi surface should be expected.

A much more detailed picture of the decay mechanism for nuclear hole states emerges from the study of intermediate structures which clearly would profit from a better energy resolution. Fig.4 shows the

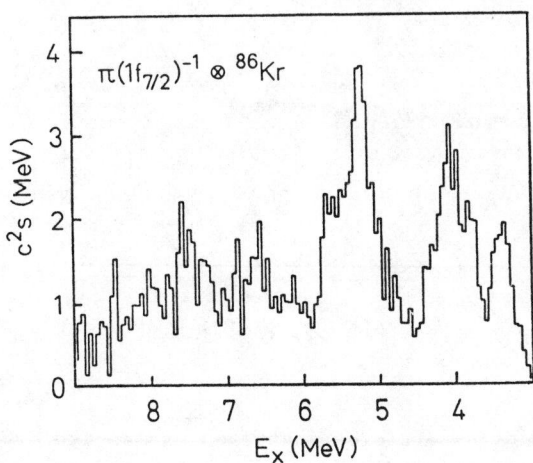

Fig.4. The distribution of $1f_{7/2}$ proton hole strength in ^{85}Br from the analysis of the ^{86}Kr (d,^{3}He) reaction at 52 MeV. From ref.7.

1f$_{7/2}$ proton hole strength in ^{85}Br which shows considerable fine structure. Based on the assumption of hole state decay via phonon-hole coupling which was shown to hold for the 1d$_{5/2}$ strength distributions in ^{89}Y have been performed by two theoretical groups ^{12,13}, which lead to intermediate structures very reminiscent to that of fig.13 both in frequency and width. - The decay mode by particle escape so far has been investigated explicitly by only one coincidence experiment, namely the ^{17}O(d,t) ^{16}O* reaction and the subsequent decay into p+^{15}N and +^{12}C channels^{14}.

3. GIANT RESONANCES AND THEIR DECAY PROPERTIES

Giant resonances which represent another elementary mode of excitation which have received a lot of attention because they represent the fastest known vibrations of a many-body system. They are generally located in the continuum and their short lifetimes of about 10^{-21} s lead to overlapping structures which are difficult to disentangle. Therefore, the selectivity of certain probes to excite only specific classes of giant resonances is of great importance. As an example fig.5 shows a spectrum obtained at the IUCF by bombarding ^{120}Sn with alpha-particles of 151 MeV^{15}. Due to the

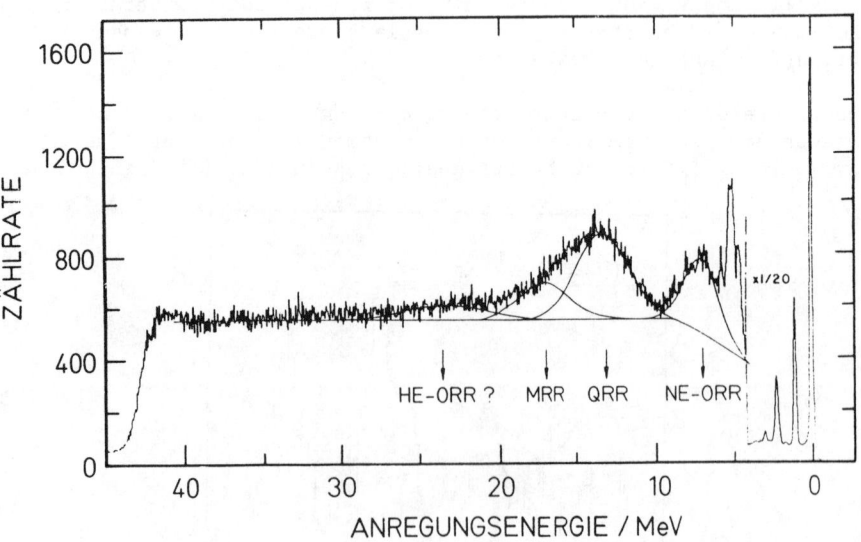

Fig.5. Spectrum of α-particles scattered inelastically of ^{120}Sn at a bombarding energy of 152 MeV and a scattering angle of 13°. Adapted from ref. 15.

symmetry properties of the alpha-particle only isoscalar electric
vibrations are excited. The giant quadrupole resonance (QRR), giant
monopole resonance (MRR), the low energy giant octupole resonance
(NE-ORR) and possibly the high energy giant octupole resonance
(HE-ORR) have been identified on the basis of the angular distributions. For an unambiguous multipole assignment it has turned out
to be important to measure up to the smallest possible scattering
angles. Hence for the planned K = 600 spectrometer it would be very
desirable to reach scattering angles below 6°. Given the weakness
of some of these structures it is crucial to rule out artifacts produced by beam halo and low-energy beam contaminants. In this context
the virtue of the extremely clean cooled beam is quite obvious. This
remark applies equally well to nuclear hole and particle states.

Other probes are needed for magnetic, isovector Gamow-Teller giant
resonances. The energy-dependence of the effective proton-nucleon
interaction [16] allows one to tune on specific operators for the excitation of giant resonances by judicious choice of bombarding energy
and scattering angle. As this technique has been widely applied in
the well-known work on Gamow-Teller resonances at the IUCF [17] it
may suffice to point out that the increase in available energies
which may be expected from the cooler eventually, will enhance the
spectrum of available effective interactions considerably [19]. In
concluding this section on giant resonance studies by singles experiments let me characterize the available information on electric giant
resonances in one of the best studied target nuclei, ^{120}Sn, by fig.6
which contains the resonance energies for isoscalar (left column)
and isovector (right column) giant resonances denoted by their L-
values. The centre column shows the unperturbed 1p-1h energies in
a simple shell model picture.

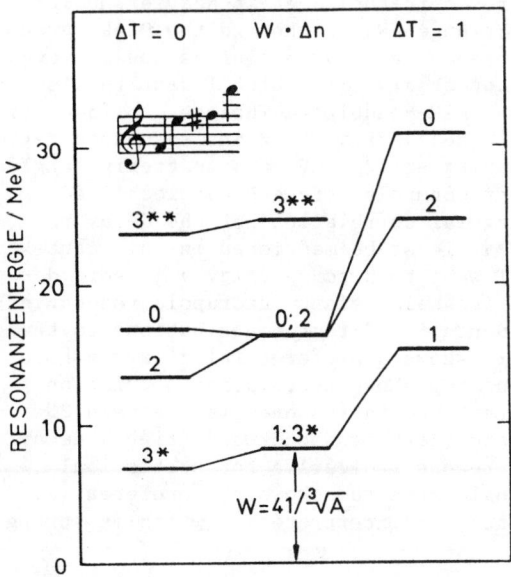

Fig.6. Energies of electric giant resonances in ^{120}Sn denoted by their multipolarity 2^L. Left column: isoscalar, right column: isovector resonances. The center column contains unperturbed 1p-1h energies in an harmonic oscillator shell model. The chord indicates the sound of isoscalar giant resonances transposed by 62 octaves.

What is known about the decay properties of giant resonances? For simplicity one generally discusses two simple and extreme cases of giant resonance decay. First, one has the so-called statistical decay, where the giant resonance which is microscopically described as a coherent sum of one-particle one-hole configurations couples through residual interactions to two-particle two-hole, 3p-3h...configurations until compound states are reached. These are known to emit particles with evaporation-type spectra and with relative branching ratios as obtainable from Hauser-Feshbach calculations. A second and more interesting decay mode is the so-called direct decay, where the 1p-1h configuration emits the particle into the continuum – generally with an energy exceeding that of evaporation products – and leaves the final nucleus in a state with substantial 1h-character. Their decay branches for the population of various final states are intimately connected to the microscopic structure of the giant resonance. The first generation of particle-particle coincidence experiments was performed to establish which of the two decay modes prevails in a given situation. The situation may be briefly summarized as follows[20]: For nuclei with mass A < 40 direct decay prevails, whereas for heavier nuclei the statistical decay dominates. In light nuclei with α-particle structure one finds and understands also a direct alpha-emission in addition to the direct nucleon emission.

These findings pertain to the isoscalar giant quadrupole resonance which is the best studied case so far. Much remains to be done for the future. Neither do we know very much about the decay properties of higher giant resonances very much, nor are experiments and corresponding analysis available which exploit the potential to study the microscopic structure of giant resonances through a detailed investigation of the direct decay properties. Two recent observations indicate that this may probably become a spectroscopic tool even in heavier nuclei. First, in $(\alpha,\alpha'n)$ reactions the Erlangen-Karlsruhe and the Osaka groups observed a 10-20 % direct decay branch in heavy nuclei as was reviewed in ref.21. A second and more recent observation which I want to discuss in some detail was made by a Texas A&M-Heidelberg-Tübingen collaboration when studying the charged particle decay from the giant resonance region excited by inelastic alpha-scattering at 129 MeV at a scattering angle of $0°$ [22]. In this experiment giant monopole strength was localized and identified by studying the angular correlations of the decay products – a technique which should at least be mentioned in this context. The case of ^{58}Ni however, which I want to discuss (fig.7) in some detail, essentially only the known isoscalar giant quadrupole resonance was exciting. Fig.7 contains evidence for direct contributions to the decay of this resonance because it shows a preferential decay to hole states of ^{57}Co. The two spectra on top show the relative population of final states in ^{57}Co for apparent excitation energies between 25 and 28 MeV in ^{58}Ni. Both at forward (left) and backward (right) decay angles one sees evaporationtype protons indicative for statistical decay. In addition the forward angle spectrum shows the preferential population of low-lying hole states, an interpretation which is strongly supported by the ressemblance to the ^{58}Ni(d,^{3}He) spectrum (centre, left). The fact, that these hole states are only populated when the decay detector is close to the recoil detection (top, left) and are practically absent for opposite recoil directions (top, right) suggests that the prevailing process is a direct (α,αp) knockout reaction. The two lower spectra contain the corresponding information now for apparent excitation energies between 16 and 19 MeV, i.e. in the giant resonance region. Because of the lower excitation energy evaporation-type protons are now located underneath the groups leading to nuclear hole states either by direct knockout or direct giant resonance decay. For protons emitted at opposite-to-recoil direction (bottom, right) an origin for knockout processes may be ruled out. Hence they represent strengths from sequential processes via intermediate states in ^{58}Ni. The Hauser-Feshbach calculations which have been performed for various spins of the intermediate states (centre, right) show that low-lying hole states are more strongly excited than expected for statistical decay, while low-energy protons, leading to highly excited states of ^{57}Co, are less abundant than predicted from statistical model calculations. Clearly, detailed and extensive measurements and analysis will be needed in the future if one is to exploit the potential of the direct decay mode to study the microscopic structure of giant resonances.

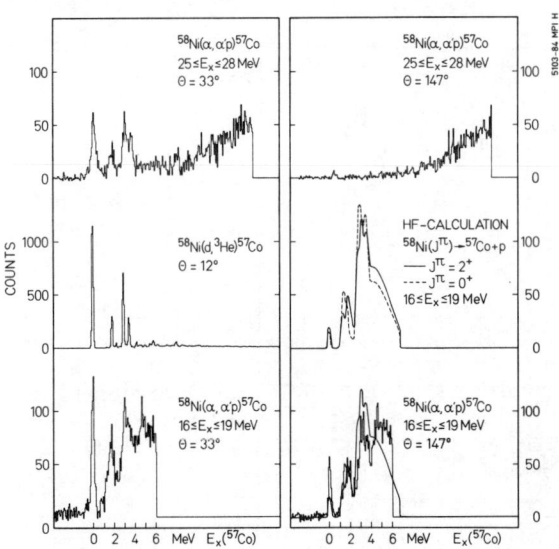

Fig.7. Spectra from a ^{58}Ni (α,α'p) coincidence experiment[23] at 129 MeV with $\Theta_\alpha = 0°$ showing the population of final states in ^{57}Co in comparison with a (d,^{3}He) spectrum at 52 MeV and Hauser-Feshbach calculations assuming various spins for intermediate states in ^{58}Ni. For details see text.

4. TYPICAL COOLER BEAM EXPERIMENTS

Should the cooler technique provide in fact beam energy spreads of $\simeq 1$ keV one could conceive of two possible experiments related to the present questions. The first experiment could provide information on the nature of the so-called background in the continuous spectra. As an example fig.8 shows spectra of α-particles and ^{6}Li-ions, respectively, scattered from ^{28}Si at angles and bombarding energies which favour excitation of the giant quadrupole resonance at an excitation energy of about 20 MeV. Both spectra show this structure, however, sitting on a background of strikingly different shapes. While in α-scattering one has essentially a flat continuum underneath the giant resonance the ^{6}Li spectrum shows a triangular shape characteristic for inelastic heavy ion scattering. In ref. 23 evidence was presented for a one-step nature of the cross section in heavy ion scattering, resulting essentially from the fact that the fragile heavy ions do not survive a second collision in general. The rigid α-particle, however, survives two- and multistep processes which lead to the observed flat continuum. With very little direct evidence at hand as to the reaction times of such multistep processes, it is proposed to measure in a sample case the excitation function for the excitation of narrow energy bins in the inelastic continuum in a search for fluctuations related to the reaction times.

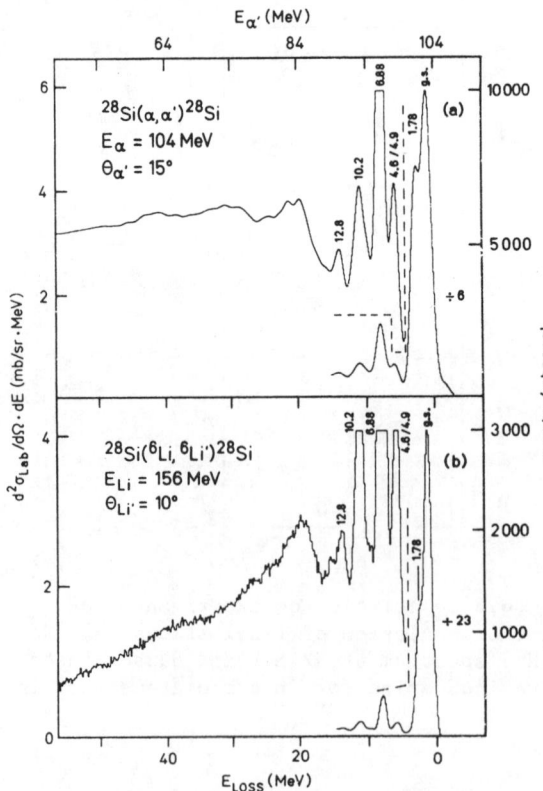

Fig.8. Spectra of inelastically scattered α-particles (top) and ^{6}Li-ions (bottom) from ^{28}Si taken at bombarding energies of 26 MeV/A and at scattering angles favouring excitation of the giant quadrupole resonance.
From ref.23.

High resolution excitation functions at tandem energies typically have been taken for isolated final states, such as the group indicated by (1) in fig.9. The scattering amplitude was assumed to consist of a slowly energy-dependent direct amplitude f_D and a rapidly fluctuating compound nuclear amplitude f_{CN}. For bombarding energies of the order of 100 MeV per nucleon which are being discussed here, f_{CN} presumably will be neglectible, even though in the absence of detailed information about the level densities at high excitation energies in the compound nucleus it might be worthwhile to study an excitation function for such an isolated level as well. The present suggestion, however, refers to a structure of a well-defined narrow energy interval as e.g. (i) in fig.9 where substantial multistep contributions to the amplitude may be expected. Neglecting again the compound nuclear amplitude f_{CN} we view the inelastic scattering amplitude as a sum of a direct amplitude f_D and a multistep amplitude f_{MS}. If reaction times of the order of 10^{-19} s were associated with such amplitudes one should observe fluctuations in the 1 keV range.

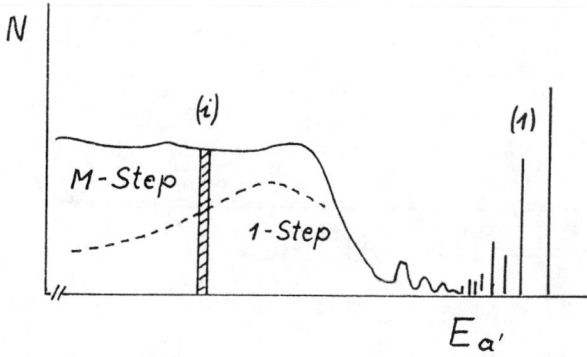

Fig.9. Sketch of the characteristic features of (α,α') spectrum.

The second class of experiments which might become possible with high resolution beams are (p,2p) and (p,pα) quasi-free-scattering experiments, preferentially induced by polarized protons (see sect.2). In conventional (p,2p) experiments one measures the momenta of the scattered and the knocked-out protons, and deduces the recoil momentum of the residual nucleus and the binding energy of the proton. With the cooler one could conceive of an experiment where only one fast proton is detected in addition to the directly observed recoil momentum which is the interesting observable. As the kinematics is again complete the momentum of the second fast proton and consequently the binding energy of the knocked-out proton may be calculated. As we shall see the difficulties of such an experiment are formidable. Nevertheless it should be tried in a few selected cases as it represents for the forseeable future the only way to measure these reactions out-of-plane, because in contrast to the magnetic spectrographs for high energy protons the detectors for recoil products may be easily moved out-of-plane. The need for such reactions both as a test of the reaction mechanism and a source of nuclear structure information, notably for deformed target nuclei has been emphasized since long[24].

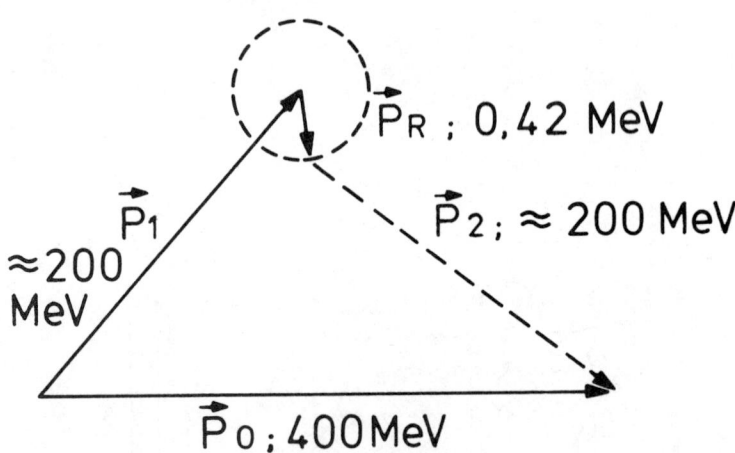

Fig.10. Momenta and corresponding kinetic energies for a (p,2p) reaction on ^{16}O at 400 MeV where one proton and the recoil nucleus are observed.

The difficulty of the experiment consists, of course, in the small recoil energies which makes the experiment feasible only for very light target nuclei and relatively large recoil momenta. Consider a $^{16}O(p,2p)$ reaction at 400 MeV (see fig.10). For a typical shell model momentum of 110 MeV/c the energy of the recoiling ^{15}N nucleus would be 420 keV. This energy would have to be determined with an accuracy of about 1 keV and an angle uncertainty of less than 10 mrad in order to determine the binding energy with an accuracy of better than 0.4 MeV. This example shows that the task would be to measure and identify (!) heavy ions with a kinetic energy of about 1 MeV and a momentum resolution $p/\Delta p \approx 10^3$. At this level one obviously has to worry about atomic excitations and Doppler broadening from the gamma decay of excited final states; for the unfavourable case of the $1p_{3/2}$ hole state in ^{15}N at 6.2 MeV this broadening amounts to about 40 keV. In addition one has to worry about charge state distributions of the outgoing heavy ions in measuring the cross sections. In spite of these difficulties it might be rewarding to attempt such an experiment if one were to start an intensive in-plane (p,2p) programme as a test of the reaction mechanism. Note, that even in the case of the (e,e'p) reaction whose theory rests on much sounder grounds one is not able to reproduce in detail the measured spectral functions[25].

5. CONCLUSIONS

Except for the study of excitation functions and quasifree scattering experiments with detection of the recoil nucleus I have not presented typical cooler applications for the nuclear continuum studies. And I am convinced that these "typical" experiments in view of their difficulties and their call for a 1 keV beam spread will not be experiments of the first generation. It is for a very different reason that I am enthusiastic about the forthcoming facility. I am convinced that the improvement of the overall energy resolution from the typical 150 keV to about 15 keV alone will tremendously boost direct reaction studies also in the nuclear continuum. In addition one must not underestimate the difficulties resulting from halo and low energy tails of many cyclotron beams which have made themselves felt most strongly in small angle experiments. With better-than-tandem beam qualities such problems should disappear. Finally coincidence experiments which constitute a necessity in continuum physics will profit tremendously, first, from the use of ultra-thin targets which will allow a 4 detection also of low-energy charged reaction products and, second, from the duty-factor of unity which will improve the ratio of random-to-true coincidences by about an order of magnitude as compared to conventional cyclotron experiments. With these exciting possibilities ahead one can hope to attack the open problems which I have pointed out in the restricted field of single-particle excitations and giant resonances plus, of course, many other interesting problems which I did not have time and space to cover.

REFERENCES

1. H. Nann, B.M. Spicer, K. Reiner, G.J. Wagner et al., IUCF Experiment E 256, unpublished.
2. P.G. Hanson, Ann. Rev. Nucl. Sci. 29, 69 (1979).
3. S. Müller et al., Phys. Lett. 113B, 362 (1982).
4. G. Mairle et al., Nucl. Phys. A363, 413 (1981).
5. J. Mougey et al., Nucl. Phys. A262, 461 (1976).
6. L. Lapikás, Invited paper, Proc. PANIC, Heidelberg, (1984), to appear.
7. G. Seegert et al., J. de Physique C4, 85 (1984) and A. Pfeiffer, Diplomarbeit, Heidelberg (1984).
8. R.R. Scheerbaum, Nucl. Phys. A 257, 77 (1976).
9. C. Mahaux and H. Ngo, Phys. Lett. 100B, 285 (1981).
10. S. Galès et al. Phys. Lett. 144B, 323 (1984).
11. G.J. Wagner et al., Phys. Lett. 57B, 413 (1975).
12. P.F. Bortingnon et al., J. de Physique C4, 209 (1984) and private communication.
13. V.G. Soloviev, J. de Physique C4, 69 (1984).
14. H. Breuer et al., Phys. Lett. 96B, 35 (1980).
15. F.E. Bertrand et al., Phys. Rev. C 22, 1832 (1980).
16. W.G. Love and M.A. Franey, Phys. Rev. C 24, 1073 (1981).
17. T.N. Taddeucci et al., Phys. Rev. C 25 1094 (1982).
18. R. Klein et al., Phys. Lett. 145B, 25 (1984).
19. G. Berg et al., COSY-Studie, Julich (1984), Report Jül. -Spez.-242.
20. G.J. Wagner, in "Giant Multipole Resonances", ed. F.E. Bertrand, Harwood Academic Publ. (1980).
21. H. Ejiri, J. de Physique C4, 136 (1984).
22. P. Grabmayr et al., Proc. of COPECOS Workshop, Bad Honnef (1984), to appear.
23. W. Nitsche et al., Z. Phys. A300, 109 (1981).
24. D.F. Jackson, Advances in Physics 17, 481 (1968).
25. M. Bernheim et al., Nucl. Phys. A375, 381 (1982).

SESSION D

ENERGY DEPENDENCE STUDIES

ENERGY DEPENDENT NUCLEAR PHENOMENA

T.E.O. Ericson
Theoretical Physics Division, CERN, 1211 Geneva 23, Switzerland

ABSTRACT

Two-body reactions with pion production lead to experimentally favourable kinematics at threshold. The characteristics of charged and neutral production are given as well as the extension to the bound pionic atom states. Some motivations for experiment are presented. At higher energies a "magic energy" occurs with zero momentum transfer. The possibility of using this effect selectively for studies of pion production with excited final nucleus as well as for the selective excitation of nuclear hole states are discussed.

THRESHOLD KINEMATICS

The combination of cooled stored beams of high intensity with the use of very thin targets has important practical consequences in the study of energy-dependent phenomena. In the present discussion, I will concentrate on two of these, both associated with two-body reactions. The first concerns the peculiar kinematics at thresholds in reactions, particularly at thresholds for pion production, and the second concerns the "magical energy" at which two-body reactions can occur with zero momentum transfer.

Consider for example the case of a reaction of the type

$$p + D \to \pi^+ + T \qquad (1)$$

on a stationary target. This is a concrete example of a two-body nuclear reaction with pion production. Exactly at threshold the outgoing triton (and the pion) will go into a pointlike detector at 0°, i.e., the <u>entire</u> 4π cross-section will be registered in a small detector with obvious advantage in count rate. If the binding energy is B this situation occurs at an incident lab. kinetic energy T_1

$$T_1 \simeq \frac{m_{out}}{m_{in}} B, \qquad (2)$$

where the binding can be approximately neglected compared to the nuclear rest mass. In the specific reaction above, the proton lab energy is about 210 MeV.

As a consequence of this kinematic effect: <u>a storage ring is naturally a threshold machine</u>.

Let us look at this phenomenon somewhat more in detail. The charge of the produced pion is very important. Consider the two-body pion production on a nucleus near the threshold energy T_{th}:

Neutral production ("π^0") (see Fig. 1)

For angular momentum $\ell = 0$ which dominates at threshold, the cross-section varies with phase space proportional to momentum k_π:

$$\sigma_{\ell=0} = a_0 (k_\pi/m_\pi). \qquad (3)$$

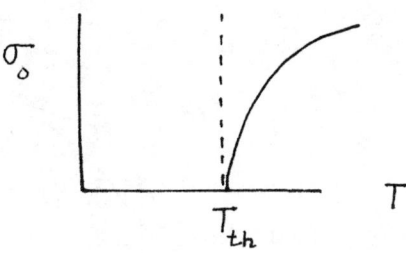

Fig. 1. Neutral π production at threshold.

The cross-section rises quickly to a rather slowly-varying value as $(T-T_{th})^{1/2}$, so that it is rather easy to work relatively near theshold.

Charged production ("$\pi^\pm$") (see Fig. 2)

The threshold cross-section is mainly modified by the Coulomb factor

$$\sigma^\pm \simeq \left(\frac{2\pi\eta}{e^{2\pi\eta}-1}\right) \times \sigma_0 \quad \text{with} \quad \eta = \pm \frac{Z\alpha}{\beta_\pi} \ ; \ \beta_\pi \simeq \frac{k_\pi}{m_\pi}. \qquad (4)$$

Repulsive case ("π^+")

The cross-section is suppressed by the Coulomb penetration factor, but the cross-section has a factor 2π in front, which helps. There is no phase space suppression

$$\sigma^+ \longrightarrow 2\pi Z\alpha \, a_0 \exp\{-2\pi Z\alpha \, m_\pi/k_\pi\}. \qquad (5)$$

Attraction ("π^-")

$$\sigma^- \longrightarrow 2\pi Z\alpha \, a_0 = \text{const.} \qquad (6)$$

The surprising result is a cross-section, which does not vanish at threshold, in spite of lack of phase space. In addition, it is multiplied by $2\pi \simeq 6$, which is quite healthy for count rates.

Energy scale of Coulomb effects

This scale is set by $(2\pi)^2 \times$ (Bohr energy), which, for pions, gives the scale $\simeq 0.14\ Z^2$ MeV.

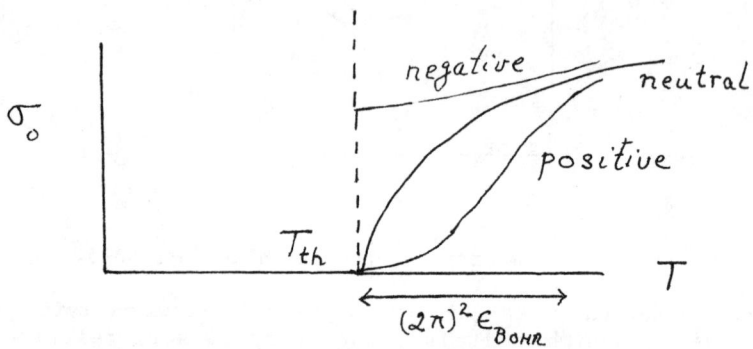

Fig. 2. Threshold behaviour of the positive, negative and neutral cross-sections.

An (unrealistic) numerical example

The NN $\to \pi$D reaction has a cross-section in the absence of Coulomb effects:

$$\sigma_o \simeq 0.27 \left(\frac{k_\pi}{m_\pi}\right) mb. \qquad (7)$$

As a consequence, the nn $\to \pi^-$D cross-section is expected to go to the value $\sigma^- = 12\mu$b at threshold. The energy region over which this effect occurs is 140 keV. Even in this extreme case of $Z\alpha \ll 1$, the cross-section is not small.

PRODUCTION OF π^- BELOW THESHOLD (PIONIC ATOMS) (See Fig. 3)

The $\ell = 0$ cross-section for π^- extends below threshold, forming atomic ns states. According to the correspondence principle, the average cross-section over the dense states of high n close to threshold is the same as the constant threshold cross-section.

The typical energy scale is the binding energy of the 1s state $E_B \simeq 3.6\ Z^2$ keV, i.e., 40 times smaller than the characteristic scale for Coulomb distorsions. Typical values are: Z = 1, 3.6 keV; Z = 5, 90 keV; Z = 10, 360 keV. The bound states of higher n have the characteristic Bohr spectrum with binding energies $\propto n^{-2}$.

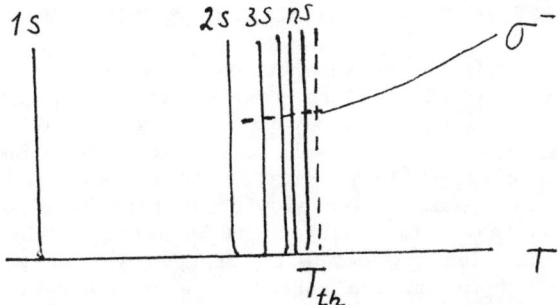

Fig. 3. Occurrence of bound pionic atom states. The dashed extrapolation of σ^- indicates the average cross-section for the atomic states.

CROSS-SECTION FOR EXCITATIONS OF PIONIC ATOMS

The characteristic integrated value of pionic 1s state excitation is obtained from the constant threshold cross-section and binding energy as $\simeq 2\varepsilon_B \sigma^-$. As a numerical example, take the reaction $p + {}^9Be \to \pi^- + {}^{10}C_{gs}$, with an observed production cross-section of $\simeq 30$ nb, and an expected threshold cross-section of over 10 nb. The unknown expected width of the unstable $(\pi^{10}C)$ atom in the 1s is of the order of 3 keV. If a beam of this resolution can be used, the peak cross-section is $\simeq (4\varepsilon_B/\Gamma)\sigma^- \simeq 1\mu b$, which can be studied. Cross-sections of higher n have similar peak values, but Γ_n decreases as n^{-3} and the spacing is smaller.

SIGNATURES OF ATOMIC STATES

In general, there will be very little X-ray emission from the higher ns states, because the main decay process is $\pi^- + A \to$ fast nucleons + debris. Even if the X-rays were observable, one has to face that they are emitted from a moving nucleus (v/c $\simeq 0.02$ to 0.10 in practice), so that experimental set-ups must be able to cope with the Doppler shift.

One way to handle this problem has been proposed by Dr. Meyer. The idea is to use the fact that the pionic state in the atom is simply a resonance in the elastic scattering (e.g., $p + {}^9Be$ elastic scattering) with a strong resonance enhancement right at the resonant position. Another possibility is to exploit the fact that the π^- production cross-sections become constant at threshold for any ℓ. In particular, the neutral cross-section $\sigma^{\pi^o}_{\ell=1} \simeq a_1 (k/\mu)^3$ becomes, for π^- production, $\sigma^{\pi^-}_{\ell=1} \simeq 2\pi a_1 (Z\alpha)^3$ at threshold. Since the absorption cross-section decreases with the centrifugal barrier, the electromagnetic capture is relatively more important in such states. A strategy to exploit this may therefore be to choose an energy just below the threshold for π^- production either for the g.s. or for an excited final nuclear state. In this case, an X-ray cascade will develop and it may be observable.

INTEREST OF DIRECT EXCITATION OF PIONIC ATOMS

Several motivations can be advanced for studies of this kind. In particular, direct excitation of the pionic system e.g., in 1s states of pionic atoms, are unaccessible beyond ^{24}Mg, because of an attenuation of the X-ray intensity by well over two magnitudes. For the heaviest observed cases [^{23}Na, ^{24}Mg], there is a long-standing problem of anomalously small observed widths which have no accepted theoretical explanation. This question can be directly elucidated. In lighter nuclei, the pionic atoms with proton rich unstable nuclei both in the g.s. and in excited states will be accessible. This is quite new information, which will give information on the isospin structure of the interaction. As for pionic atoms built on excited states, we have no present experience, although we of course guess at strong similarities with pionic atoms built on stable nuclei.

In conclusion, it seems realistic to beleive that pionic atoms can be directly studied with a storage ring.

CAN THE THRESHOLD CONSTANCY OF σ^- BE EXPLOITED?

Contrary to the atomic case, there is no difficulty in obtaining a signal when the π^- is physically produced. Let us consider the example of $p + {}^9\text{Be} \to \pi^- + {}^{10}\text{C}$ with σ^- typically about 10 nb, whether to the ground state or to excited states. Exactly at threshold, the ^{10}C projects to 100% into a very small forward counter. Rapidly above the threshold, the ^{10}C defocuses, the count rate will go down, but the condition is repeated for every ^{10}C* excited state until break-up threshold. This suggests that ramping the energy should give a very direct extrapolation of higher energy experiments exactly to threshold, and that spectroscopy of isospin multiplets in very proton-rich nuclei may be possible. The characteristic shape of excitation spectra ε_0, ε_1^*, ε_2^*,... will show cross-sections falling off as $\delta/(\varepsilon-\varepsilon_i^*)$, where all π^- within an energy range δ hit the detector (see Fig. 4).

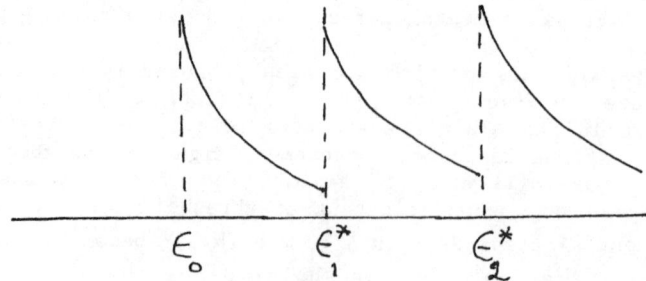

Fig. 4. Rate in a small forward counter on excitating low-lying bound nuclear states.

The same method should work for π° thresholds also, and possibly for multipion thresholds. It is, in addition, quite easy to separate the background from lower thresholds at a new threshold by a very rough energy measurement on the ^{10}C, which has a characteristic energy of $\simeq 14$ MeV. If a lower state is 2.5 MeV below the new threshold, its typical ^{10}C energy corresponds to over 1 MeV above or below that of the new state with rather sharply defined energy. This should give an excellent signal.

A TEST OF CHARGE SYMMETRY USING THRESHOLD KINEMATICS

Charge symmetry is the symmetry in strong interactions corresponding to the replacement of $n \leftrightarrow p$, $\pi^\pm \leftrightarrow \pi^+$ and $\pi^0 \leftrightarrow \pi^0$ (in the cases of concern to us). This symmetry is, of course, broken by Coulomb effects, but is otherwise supposed to hold in nuclear physics. For example, the scattering length $a_{pp} \simeq a_{nn}$ (each about -18 fm), indicating that this symmetry is valid to at least about 0.2%. On a certain level, Coulomb effects can explicitly break charge symmetry as seen in the difference of π^+ and π^- scattering on ^{12}C: near the $n + {}^{11}C$ and $p + {}^{11}B$ thresholds (18.7 MeV and 16.0 MeV), there is a clear difference of cross-sections with a manifest violation due to different n and p asymptotic wave functions.

A particularly direct and interesting test would be the <u>production</u> process

$$D + D \rightarrow \pi^\circ + {}^4He, \qquad (8)$$

which is strictly forbidden by charge symmetry. (Note that the process $D + D \rightarrow \gamma + {}^4He$ and $D + D \rightarrow {}^4He + 2\pi$ are allowed by quantum numbers and not suppressed experimentally.) It is particularly interesting that the mutual polarization of the two deuterons by Coulomb interactions is symmetric and gives no violation of charge symmetry. Present limits are of the order of 10^{-35} cm^2. With a storage ring, and using a forward detection set up, it should be possible to improve this limit by several orders of magnitude in a day with a luminosity of 10^{32}/cm^2 sec. A related, less powerful test is the nonsymmetric reaction $D + {}^4He \rightarrow \pi^0 + {}^6Li$ which can be calibrated on the $\pi^0 + {}^6Li$, $(\pi^- + {}^6Be)$, $\pi^+ + {}^6He$ reactions, once more exploiting forward kinematics.

RECOILLESS PROCESSES[1]

Whenever the outgoing particle is <u>heavier</u> than the incoming one, it is possible to choose the incident lab. kinetic energy such as to have zero momentum transfer for a process with positive Q-value. We call this energy "the magic energy". Approximately,

$$T^{'magic'}_{incident} \simeq B \frac{m_{out}}{(m_{out} - m_{in})}. \qquad (9)$$

It is totally irrelevant whether the Q value corresponds to purely nuclear excitations, meson production at rest or isobars in nuclei; this only affects the specific choice of the incident energy.

As an example, consider (p,d) reactions, which have actually been studied previously at GWI in the near 0° region by Källne and Fagerström[2], using 185 MeV protons. Although the kinematical matching is imperfect ($q \simeq 125$ MeV/c), they observed striking excitations of deep-lying hole states in the 50 MeV region in various elements. (The "magical energy" in this case is about 100 MeV.) The cross-sections for such excitations are practically easy to study (see Fig. 5).

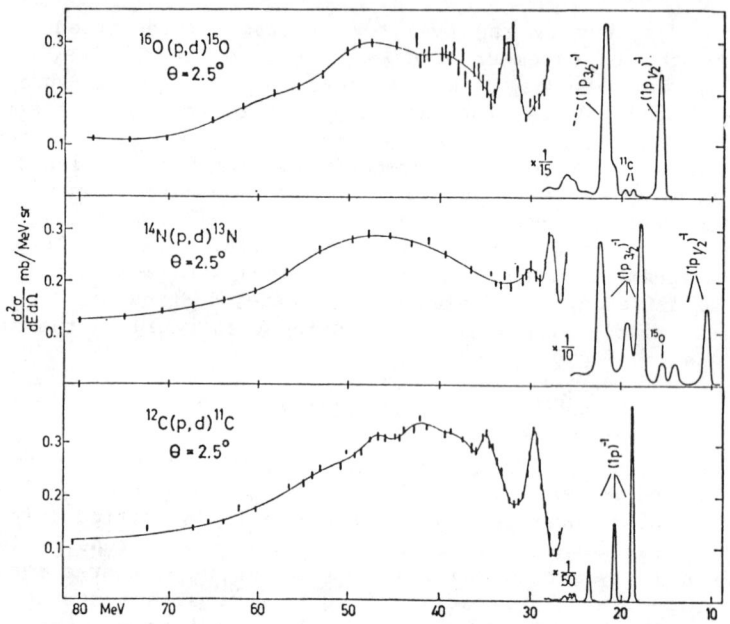

Fig. 5. Separation energy spectra obtained by the (p,d) reaction in ^{12}C, ^{14}N and ^{16}O.

The importance of the $q = 0$ condition lies in the fact that the process matches quasi-free kinematics. It will therefore preferentially pick out the quasi-free states with $q = 0$, which in a shell model are the s-states.

The power of this condition is well known from the production of hypernuclei by the (K,π)-process with $q = 0$: this condition leads to a far simpler spectrum than the one at $q \neq 0$.

The properties of deep-lying nuclear states, whether single particle or multiparticle, are of considerable interest and may be picked out in an enhanced fashion at $q = 0$, which would emphasize $L = 0$ states. One should therefore give considerable thought to a possible programme of such studies using a forward spectrometer. Characteristic processes would be (p,d), (p,t), (p,^{3}He), (p,^{4}He), (d,^{4}He), etc., where the exact choices of combinations in practice are likely to be influenced by practical consideration of how the "magical energy" is best realized at storage ring. It is clear that it may be of additional interest, beyond such classical nuclear physics studies, to "implant" a pion at rest in a nucleus, so as to directly measure its self-energy, or to measure directly Δ nuclear states with very selective kinematical conditions and at near-zero momentum transfer. A specific example is the NN → πD process, which will be discussed in detail by Dr. Kilian.

The condition of q = 0 is also extremely favourable for the production of pionic atoms, since their cross-sections drop quickly for q > $\mu Z\alpha$ which is very small.

The conclusion of this discussion is therefore that careful use of the increased freedom in the choice of kinematical conditions with a storage ring as well as the advantage of very thin target from which the reaction nuclei will emerge in an easily measurable way will lead to a series of new types of nuclear physics experiments.

REFERENCES

1. T.E.O. Ericson and K. Kilian, Contributed paper to the PANIC Conference, Heidelberg, July 1984.
2. B. Fagerström and J. Källne, Phys. Sci. 8 (1973) 14.

TECHNIQUES OF ENERGY CHANGING AND ENERGY MONITORING OF COOLED STORED BEAMS

D. Möhl
PS Division, CERN, CH-1211 Geneva 23
Switzerland

ABSTRACT

It can be desirable - both for machine operation and for machine and particle physics experiments - to change the energy of a beam, circulating in a storage ring, by small and well controlled amounts. This paper compares six different methods of energy changing and five methods of non destructive energy monitoring.

INTRODUCTION

Due to the nature of the Lorentz force, the energy of a circulating charge particle can best (or only) be changed by a longitudinal electric field. The methods to be discussed differ by the way in which this electric field is created.

If the storage ring guide fields are kept constant, the beam changes its radial position with momentum $\Delta x(s) = D(s)(\Delta p/p)$ where the "dispersion function" $D(s)$ [frequently also called $\eta(s)$ or $x_p(s)$ or $\alpha_p(s)$] is a property of the magnetic lattice of the storage ring. It results from the equilibrium between the increased centrifugal force in the arcs as the energy increases and the "restoring" action due to the focusing fields. In general, $D(s)$ varies with the azimuthal position s around the ring.

To keep the radial position unchanged, all guide fields have to be changed by $\Delta B/B = \Delta p/p$.

1. TECHNIQUES OF ENERGY CHANGING

PROGRAMMED RF ACCELERATION

This is the method used to accelerate the beam in synchrotrons and cyclotrons : an RF field is applied with a frequency $f = h\, f_0$ at a harmonic h of the revolution frequency. On each traversal, the particles experience an accelerating kick due to the longitudinal electric field. As discussed in textbooks, the motion can be analysed in terms of a potential well ("bucket") travelling around with the particles.

To capture the coasting beam, the RF field has to be turned on slowly so that particles are trapped into the potential well ("adiabatic capture"). The well must be deep enough to hold the energy spread of the beam. This determines the required minimum RF voltage per turn $U = \int_0^{2\pi R} E_s\, ds$ which is

$$U = \frac{\pi}{2} h \, |\eta| \, \beta^2 \, \gamma \, \left(\frac{\Delta p}{p}\right)^2 U_o \tag{1}$$

where $\eta = 1/\gamma^2_{transition} - 1/\gamma^2 = -\frac{\Delta f_o}{f_o} / \Delta p/p$

is the off energy function of the storage ring, which determines the change of revolution frequency with momentum; $U_o = 938$ MV for protons and $\beta = v/c$; $\gamma^2 = (1 - \beta^2)^{-1}$

For the $\Delta p/p$ of cooled beams, the required voltage is very modest, see the example of Table 1, so that a simple RF system can be used.

Table 1

Example of the RF voltage required to hold a beam

Assumed beam and machine parameters :

RF harmonic number : h = 1
off energy function : η = -1
particle velocity v/c : β = 0.3

Momentum spread $\Delta p/p$	± 10^{-4}	± 10^{-3}	± 10^{-2}
RF voltage to hold the beam U (Volt)	1.3	130	13000

To change the beam energy, the potential is moved by changing the RF frequency. The concurrent momentum change is

$$\Delta p/p = -\eta^{-1} \frac{\Delta f}{f} \tag{2}$$

During energy change, the beam is bunched. However, after the change, the RF can be turned off adiabatically to redistribute particles uniformally around the circumference. If the capture, energy change and debunching are done sufficiently slowly, the blow-up of momentum spread and the residual density modulation can be kept small and the voltage required is only slightly larger than the minimum (1). Sufficiently slowly in this context means a few (2 to 5) synchrotron oscillation periods T_s for each of the 3 operations, where

$$T_s = \beta/f_o \left(\frac{2\pi \gamma U_o}{h \, |\eta| \, U}\right)^{1/2} \tag{3}$$

Usually, T_s is less than 10 ms, so that the procedure can be finished within 100 ms.

PHASE DISPLACEMENT ACCELERATION

This method was used in the CERN ISR to accelerate the beam from about 26 to 31 GeV/c. The procedure is to turn on an RF accelerating field at a frequency just below (or above) a revolution harmonic. In contrast to normal RF acceleration, the potential can have a depth smaller than the beam momentum spread, i.e. the voltage is smaller than (1). Sweeping the frequency slowly to a value above (below) hf_0 particles are pushed in the opposite direction, i.e. towards lower (higher) frequency. By virtue of (2), they are accelerated in a machine working above transition like the ISR and vice versa in a machine below transition.

The voltage is then turned off, the frequency restored to its initial value and the process of sweeping the RF potential well ("the RF bucket") across the range of beam frequencies is repeated (some 100 times in the ISR) until the desired total energy displacement is achieved.

The advantage of this method : the beam stays practically unbunched and very little RF voltage is needed. The price : phase displacement needs more time and tends to introduce more blow-up and distortion of the energy distribution than normal RF acceleration.

In the ISR, this method was used since the voltage (1) to hold a beam of one or several percent $\Delta p/p$ turned out excessive and also since the momentum spread of the bunched beam would have exceeded the acceptance.

BETATRON ACCELERATION

Imagine a torus (or a frame) of ferromagnetic material in which a magnetic flux is created enclosing the beam. From Maxwell's equation, the energy gain per turn of the particles due to a change of this flux is

$$U = \oint E \, ds = - \int \dot{B} \, dF = - \dot{B} \, F \qquad (4)$$

where F is the effective cross-section of the flux ring, i.e. the cross-sectional area of the ring or frame. The arrangement may be viewed as a transformer where the excitation coil is the primary and the beam the secondary winding. The advantage of this method : all particles are uniformly displaced in energy. The disadvantage : a large size magnet is required to give an appreciable energy change. In fact, the total energy change experienced by a particle with f_0 revolutions per second is by virtue of (4)

$$e \Delta U = e \Delta B \, F \, f_0$$

An example : For f_0 = 1 MHz, ΔB = 2 Tesla, to provide an energy change $e\Delta U$ of 2 MeV, one needs a "frame" with a cross-section of F = 1 m^2. For a magnet which takes 1 m of straight section space, this means a volume of $\approx$ 4 m^3 and a weight of $\approx$ 30 tons. For a cooled beam, the same energy change can easily be supplied in a fraction of a second using a very modest RF system, which supplies a few tens of Volts per turn.

ENERGY DISPLACEMENT BY OFFSET OF A MOMENTUM COOLING SYSTEM

When the energy of the cooling electrons is changed slowly or in small steps, the ion beam will follow due to the drag force which acts between the two beams. In the cooling experiments, this has been used to measure the cooling strength. As a rule of thumb, one can assume that the beam can be displaced by an amount equal to its initial energy width within one cooling time. This seems well suited for our present purpose.

Similar considerations also apply to stochastic cooling. Using the filter method, one can change the electric length of the filter line which defines the nominal revolution time and thereby the momentum. For the Palmer method, which detects momentum via the concurrent radial displacement, one can electronically change the "zero position setting" of the difference pick-up.

In the operation of LEAR, the change of beam energy by the stochastic cooling system is clearly observed when the beam is injected with a slightly displaced energy due to a synchronization error between the proton synchrotron (PS) and LEAR.

ENERGY DISPLACEMENT BY RF NOISE

Filtered noise applied on an RF cavity in a band near a harmonic of the revolution frequency causes a diffusion of particles over the corresponding momentum range. This is used in LEAR to transport particles into the extraction resonance. The same method looks interesting to "transport" particles for other applications, for instance to bring a small portion of beam onto an internal target.

Particles perform a random walk over the band covered by noise. The speed $v(\Delta p)$ of this "Brownian motion" is determined by the power level of the noise.

The particle density - which is determined by the inverse of $v(\Delta p)$ - is a smooth function which follows a diffusion type of equation. By displacing the edge of the noisy band into the stack, the number of particles participating in the random walk process can be controlled.

UNSTACKING RF BUCKETS

This is another method by which a small fraction of the beam can be displaced in energy. It is used in the antiproton accumulator (AA) at CERN to extract portions of the stack.

By slowly turning on an RF voltage at a harmonic of a frequency within the range of revolution frequencies of the stack, a fraction of the particles is trapped into the RF potential well ("buckets"). Changing the frequency, the trapped particles are displaced in energy and can be transported out of the stack. This method could also be used to carry small portions of beam onto an internal target. If desired, the small batch can be debunched by slowly turning off the RF.

In this way, a beam of tiny momentum spread can be separated from the main stack. This, together with selective cooling of the small beam, has been considered for special applications in LEAR.

2. MONITORING OF ENERGY CHANGE

MEASUREMENT OF ENERGY FROM RADIAL DISPLACEMENT OF THE BEAM

A position sensitive pick-up electrode can be used to monitor the radial position of the beam centre. For an unbunched beam, one can observe the "Schottky noise" signals due to the fluctuation of the number of particles in a beam sample of a given length. The resolution is limited by the beam size and by the accuracy of the pick-up to typically 1 mm for both Schottky noise and bunched beam pick-ups. It may be possible to gain some factor for cooled beams.

In the presence of guide field errors ($\Delta B/B$), the radial displacement is

$$\Delta x = D \left(\frac{\Delta p}{p} - \frac{\Delta B}{B} \right)$$

The value of the dispersion function $D(s)$, which for small cooling rings is typically in the range of 1 to 10 mm/per mille, is known from lattice calculations to a relative precision of, say, 10^{-2}. The field can be controlled to a precision of 10^{-3} or 10^{-4}. The most critical factor is thus the accuracy by which the beam position can be observed. It limits the method to momentum changes of, say, a per mille which can then be monitored with a relative error of typically 10%. A high precision pick-up is desirable, located at a position where $D(s)$ is large and the beam size due to betatron oscillation small.

CHROMATIC CHANGE OF BETATRON TUNE

The change of betatron tune Q can be measured with a precision of 10^{-2} to 10^{-4} in $\Delta Q/Q$, depending on the effort and on the beam conditions. Once again, the resolution is limited by the beam tune spread and by the measurement device itself. The "chromaticity coefficients" ξ and ξ_B, which determine the Q change are defined by

$$\frac{\Delta Q}{Q} = \xi \frac{\Delta p}{p} - \xi_B \frac{\Delta B}{B}$$

For typical cooling rings with correction sextupoles, these coefficients are tunable in the range of, say, -5 to 1. Their value can be estimated from the storage ring lattice - including the sextupole lenses - to an accuracy $\Delta\xi$ of, say, 0.1. Measurement of energy change from the chormatic tune change is therefore mainly limited by the precision of the Q measurement and by drift and ripple of Q due to power supply imperfections. One can expect a resolution of $\Delta p/p$ of the order of 1 part in 10^3. Large chromaticity $|\xi|$ is helpful in this context.

ENERGY MEASUREMENT FROM REVOLUTION FREQUENCY

Revolution frequency can be measured with high accuracy $\Delta f_0/f_0 = 10^{-5} - 10^{-6}$ by observing the Schottky noise signals of a coasting beam or the bunch signals in the case of a bunched beam. In both cases, a current sensitive (sum) pick-up is to be used - in contrast to the position sensitive (difference) pick-up considered above to detect radial displacement. The revolution frequency is simply $f_0 = \beta c/2\pi R$ and hence

$$\frac{\Delta f}{f} = \frac{\Delta \beta}{\beta} - \frac{\Delta R}{R} = \frac{1}{\gamma^2} \frac{\Delta p}{p} - \frac{1}{\gamma_{tr}^2} (\frac{\Delta p}{p} - \frac{\Delta B}{B}) - \frac{\Delta R_0}{R_0} \quad (5)$$

Estimated relative accuracies are:

γ_{tr} : 10^{-2} (transition energy)

B : 10^{-4} (magnetic guide field)

$2\pi R_0$: 10^{-4} (circumference of central orbit).

Hence, the method looks well suited to measure momentum changes of less than 1 part in 10^4 - perhaps 1 in 10^5. This, especially in machines with a large value of transition energy, as in this case the orbit circumference $2\pi R$, is insensitive to field and momentum errors. In addition, an absolute momentum measurement is feasible to the precision by which the circumference $2\pi R_0$ of the "central orbit" is known. However, both the absolute and the relative measurement become imprecise when the machine works close to transition energy where by virtue of (5) f_0 is insensitive to a change of p. Sufficiently far from transition revolution frequency is a well suited "indicator" of momentum.

FURTHER METHODS OF ENERGY MONITORING

To conclude, we list further methods of energy monitoring and/or calibration:

a) Excite depolarizing RF fields. Measure the frequency of resonant depolarization ("spin k.o."):

$$f_{RF} = (n \pm \nu_{spin}) \cdot f_o$$

where n is an integer, f_o the revolution frequency and

$$\nu_{spin} = (\frac{g-2}{2}) \cdot \gamma$$

the number of spin precessions (per turn) around the magnetic guide field. This method works if the beam is polarized and the gyromagnetic anomaly g-2 of the ions is well known. Used at Novosibirsk for electrons and positrons, it permitted to determine the mass of the J/ψ particle with a precision of 90 eV (A. Skrinsky, Proc. 12 Intern. Acc. Conf., Fermilab 1983, p. 235).

b) Observe known thresholds for reactions on internal target (K. Kilian, CERN EP 84-58 and Proc. of the 4th Workshop on pp collider Phys., Bern, March 1984).

c) Observe energy of the cooling electron beam.

d) Measure time of flight in a straight section using scattering of laser light or correlation of particle signals on pick-up electrodes.

e) Extract beam onto spectrometer.

CONCLUSIONS

Several methods to change the energy of a circulating beam are well established. The choice depends on the application foreseen and the hardware available :

- RF acceleration (normal or phase displacement) seems easiest for a quick and/or large change.

- Use of the cooling system looks natural and well suited for energy sweeping in a cooler ring.

- Diffusion by RF noise or RF unstacking offer themselves to transport beam onto an extraction resonance or an internal target.

Energy can be monitored observing e.g.:

- radial beam displacement;
- betatron frequency;
- revolution frequency;
- "settings" of the cooling system.

Most of these "machine physics methods" are limited - amongst other factors - by the accuracy at which the guide field is known. Highest accuracy is probably possible by deducing momentum from the revolution frequency, especially if the machine has a large value of transition energy and works remote from transition,

Special "particle physics properties" (spin, threshold reactions, ...) look attractive for an absolute calibration.

ON THE POSSIBLE EXISTENCE OF NARROW (N^*N)-RESONANCES

Max G. Huber
Institut für Theoretische Kernphysik, Universität Bonn
Nußallee 14-16, D-5300 Bonn

and

Institut für Kernphysik der KFA Jülich
Postfach 19 13, D-5170 Jülich

and

Hans-Georg Hopf and Manfred Dillig
Institut für Theoretische Physik der Universität
Erlangen-Nürnberg, Glückstraße 6, D-8520 Erlangen
West-Germany

ABSTRACT

Due to the coupling of N^*N configurations with the real pion continuum energy-shifts and modifications of the widths of the corresponding dibaryonic molecules are expected. For configurations, which cannot couple to the absorptive channels this mechanism may lead to a narrowing of specific resonances. Results of model calculations are applied to elastic pion deuteron scattering.

I. INTRODUCTION

It is commonly assumed that the internal quark excitation of a nucleon bound in a nucleus automatically leads to a quark structure with a width being roughly equal to or even larger than the decay width of the free excitation: "Pauli-blocking" tends to narrow resonances whereas the (generally dominating) absorption effects will lead to a broadening. Thus, if a narrow excitation were to be found in a nuclear medium, it would signal automatically an "exotic" excitation, i.e. a transition into a configuration qualitatively different from the conventional modes of motion where individual nucleons (or, to be more precise: individual baryons) are distinctly separated from each other.

For the case of the dibaryon such an "exotic" configuration could be a genuine q^6-structure[1-5].

It is the purpose of this contribution to find out that even in a conventional picture of two individual, conceptually well-separated baryons narrow excitations of the complex system may occur; in fact, there exists a quite natural mechanism which may lead to such structures. The actual existence of narrow resonances, however, depends sensitively on the details of the interaction of the baryons with each other.

II. DIBARYONIC MOLECULES

We first have to define the Hamiltonian of the two-baryon system, which ought to describe both the internal (or quark-) and external degrees of freedom of the two individual baryons. This is done within the general multi-baryon concept, developed first by Klaus Klingenbeck[6] which has been adopted and applied to the two-baryon system by H.-G. Hopf[7].

We start from the two-baryon Hamiltonian

$$H(1,2) = h_{int}(1) + h_{int}(2) + T_{rel} + W(1,2) \qquad (1)$$

where h_{int} describes the baryon resonances

$$h_{int} |N_\nu^*\rangle = \varepsilon_\nu |N_\nu^*\rangle \qquad (2)$$

As explained in detail in[6] h_{int} is manifestly energy-dependent and - above the pion production threshold (i.e. for $\omega > m_\pi$) - non-hermitean; thus the eigenenergies of equations (1) and (2) are generally complex.
The operator $W(1,2)$ denotes the baryon-baryon interaction mediated by the coupling to the meson fields, i.e. the exchange of mesons between the two baryons. As explained in[6,7], the interaction W again is manifestly energy--dependent and for energies $\omega > m_\pi$ even complex.

The eigenmodes, D^*, of the two-baryon system are defined by:

$$H(1,2) |D_\mu^*(1,2)\rangle = E_\mu |D_\mu^*(1,2)\rangle \qquad (3)$$

Those D^*-excitations can be considered as the analogues in the deuteron of the corresponding N^*-excitations.

For the specific case of the (ΔN)-system the eigenmodes of equation (3) have been calculated[7]; they can be used to evaluate the T-matrix for reactions, which are dominated by the internal $\Delta(3,3)$ excitation, such as pion deuteron scattering:

$$T(\underline{k},\underline{k}') = \langle \pi D_i \underline{k}' | L^+ | D_\mu^* \rangle (\omega - E_\mu)^{-1}$$
$$\langle D_\mu^* | L | \pi D, \underline{k} \rangle \qquad (4)$$

Here, L denotes the $\pi N \to \Delta$ transition operator.

In figs. 1 and 2 the ΔN-interaction is shown: Fig. 1 denotes the various contributions to W of equation (1); the corresponding coupling constants and form factors are essentially taken from an analysis of the NN-interaction

in the framework of a boson-exchange interaction[8]. It
turns out, however, that this analysis is not completely
unambiguous: Fig. 2 shows two different ΔN-potentials
obtained for two different sets of potential parameters,
which are essentially equivalent as far as the description of the NN-phase shifts are concerned.

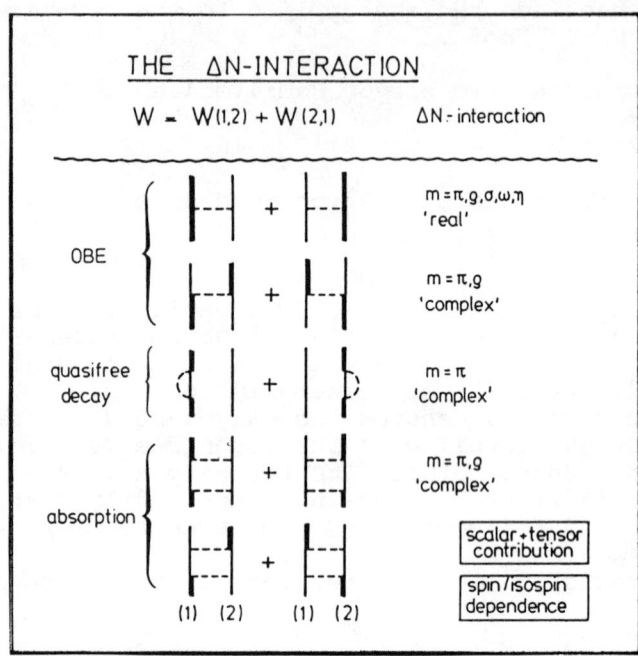

Fig. 1: Schematic picture of the contributions to the
ΔN interaction.

III. RESONANT PION DEUTERON SCATTERING

With the two different ΔN-potentials (a) and (b) of fig. 2
the D^*-spectrum has been calculated and applied to
elastic πD-scattering. The results for the differential
cross-section, iT_{11} and T_{20} are shown in figs. 3 and 4,
repectively, together with the data of [10,11]. In view of
the explorative character of this specific analysis the
non-resonant contributions have been ignored in those
results. Therefore, those calculations are not really
meant to be directly compared to experimental data; (in
passing it is worth being noted, however, that those
results are surprisingly realistic - considering the in-
herent approximations and uncertainties of such a treatment).

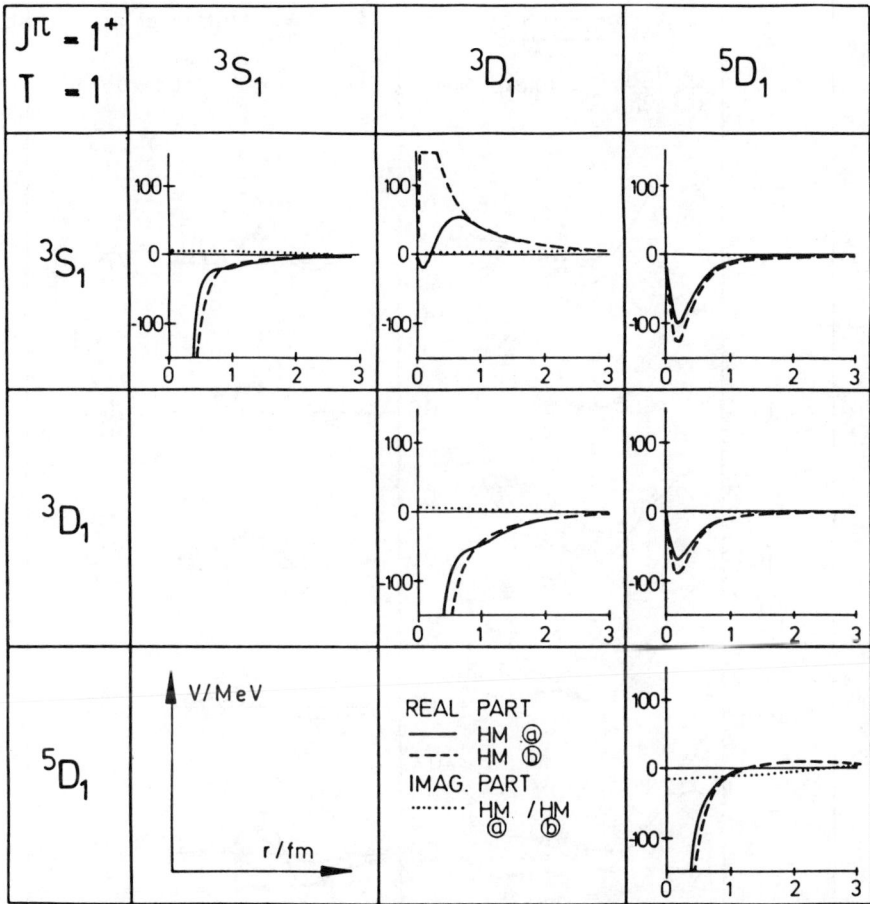

Fig. 2: The ΔN potential for the ($J^\pi = 1^+$, T=1) channel; the two potentials (a) and (b), respectively, refer to different sets of coupling constants and form factors taken from an analysis of the NN-potential.

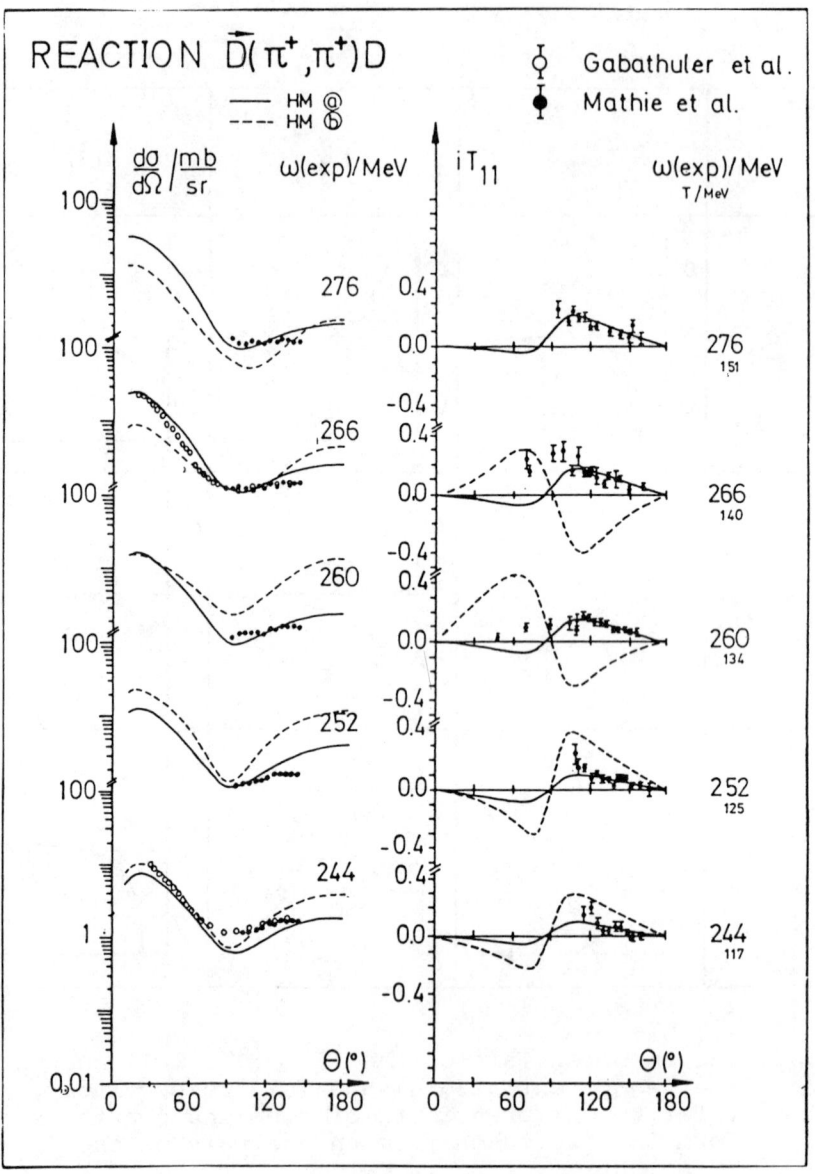

Fig. 3: Comparison of the model predictions for the two potentials (pot. (a), solid line) and (pot. (b), dotted line), respectively, with experimental data for the differential cross section and iT_{11}.

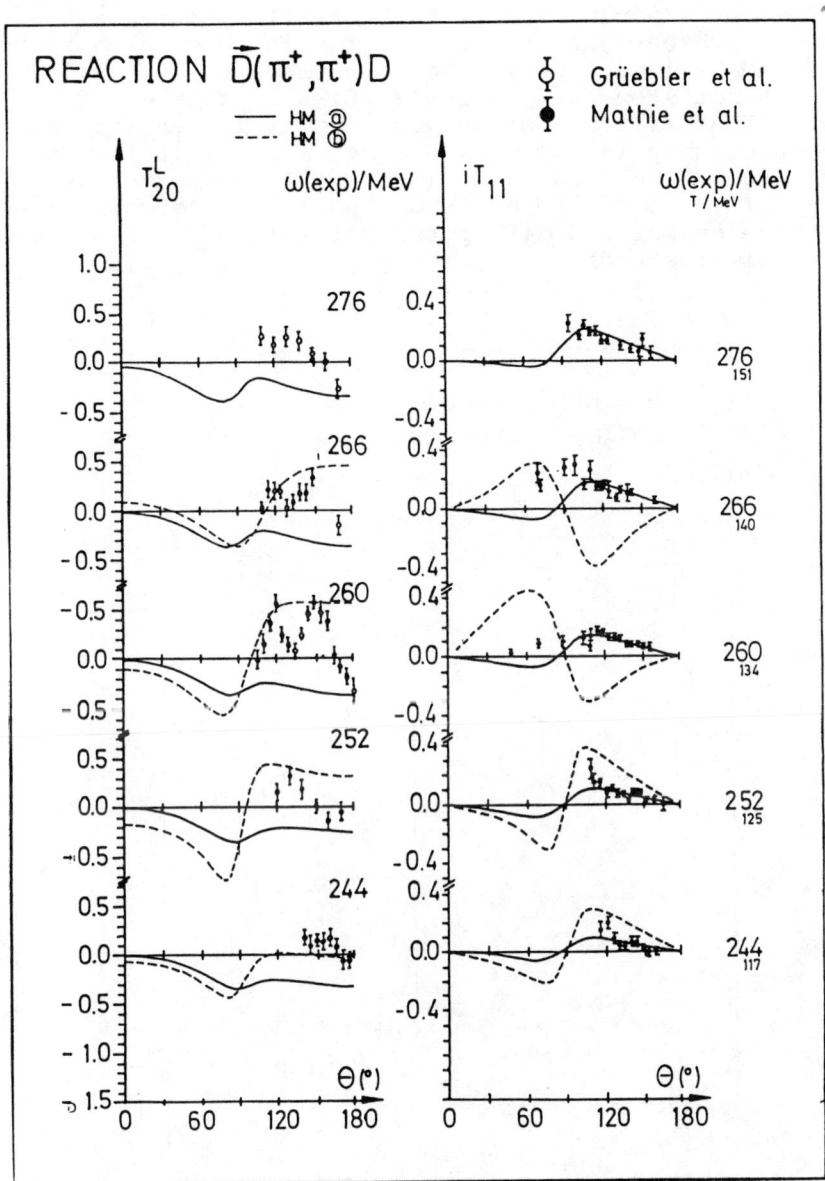

Fig. 4: Same as fig. 3 for iT_{11} and T_{20}.

As can be seen from the comparison of the two theoretical results different ΔN-potentials lead to quite different observables: In particular, the potential set (b) leads to a dramatic behaviour of iT_{11} and T_{20} in the region round $\omega = 260$ MeV (i.e. $T_\pi = 134$ MeV), whereas the potential set (a) does not show any particularities in this energy domain. The differencies between the two potentials show up most clearly in the Argand diagram (fig. 5) for the ΔN-configuration with ($J^\pi = 1^+$, $T = 1$): Whereas the potential (b) leads to a narrow resonance structure, the amplitude calculated with potential (a) does not show any remarkable features.

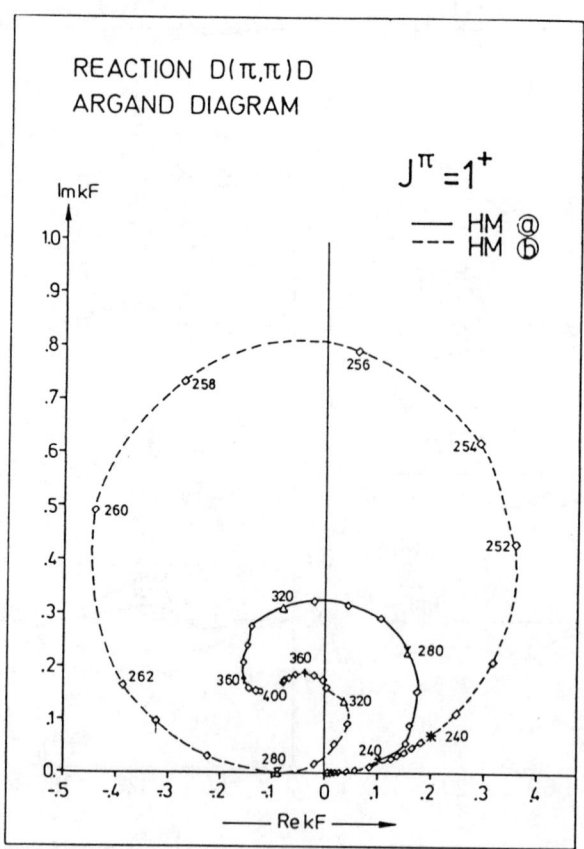

Fig. 5: Argand diagram for elastic πD-scattering for the ($J^\pi = 1^+$, $T = 1$) partial waves; the results are shown for the potentials (a) (solid line) and (b) (dotted line) of fig. 2, respectively.

It is interesting to relate this different behaviour of the
($J^\pi = 1^+$, T = 1) partial wave to the dynamical differencies
of the two potentials shown in fig. 2. This will be done
in a forthcoming paper[12]. Here, we only like to comment
on the analytic and physical origin of the occurrence of
the narrow resonance structure seen in fig. 5.

IV. THE ORIGIN OF NARROW RESONANCES

As mentioned in chapter II both the operators h_{int} and W
are manifestly energy-dependent; they both become non-
-hermitean above the pion-production threshold. It is in
particular the non-hermiticity of W which is responsible
for modification of the imaginary parts of the eigenenergies E_μ (equation (3)). To see this we introduce H_o:

$$H = H_o + W$$

Within a simple two state model it is easy to see that
the free width is modified by the imaginary part of the
matrix element of the coupling potential W; its sign is
clearly determined by the specific form of the interaction
potential (this conclusion also holds for more complicated
systems).

From this argument it is quite obvious that the coupling
of two configurations via the real meson continuum may
have a narrowing effect. If other contributions to the
width, such as absorption processes, are sufficiently
weak it is indeed possible, that specific configurations
may be substantially narrowed by this coupling mechanism.

In the 1^+-case exhibited in fig. 5 it turns out that the
coupling to the two absorption channel is forbidden due
to selection rules; therefore, we have a rather clear
example for the effect to be demonstrated in this contribution.

V. SUMMARY

The coupling of a molecular (N^*N)-configuration to the
real meson continuum may lead to a narrowing of the
resonance of the total system with respect to the free
width. This mechanism is sensitive to rather subtle
details of the N^*N-interaction.

Furthermore: For the occurrence of narrow resonances it is
important that purely absorptive channels are closed -
preferentially by selection rules. This prerequisite can
serve as a guide in the design of specific experiments.

From those results it can be concluded that the search for and the investigation of narrow resonances of the dibaryon system is of utmost importance for a better understanding of the nuclear dynamics above meson production threshold.

ACKNOWLEDGEMENT

The authors enjoyed many stimulating discussions on this problem with Klaus Klingenbeck; they thank C. Elster for providing us with the potential parameters.

REFERENCES

1. K.K.Seth: Experimental Search for Non-Strange Dibaryons, to be published
2. A.Yokasawa, Phys. Reports $\underline{64}$ (1980) 47
3. T.Kamae, Nucl. Phys. $\underline{A374}$ (1982) 25c - 5oc
4. R.Vinh Mau, Nucl. Phys. $\underline{A374}$ (1982) 3c - 23c
5. D.Bugg: Nucl. Phys. $\underline{A374}$ (1982) 95c - 1o6c
6. K.Klingenbeck, Habilitationsschrift Erlangen 1983, to be published
7. H.-G.Hopf, Dissertation Erlangen 1983, to be publ.
   H.-G.Hopf et al, Proceedings Int.Conf. Nucl.Phys., Florence 1983
   H.-G.Hopf et al, Proceedings 1oth Int.Conf. on Particles and Nuclei, Heidelberg 1984
8. K.Holinde, Phys. Rept. $\underline{68}$ (1981) 121
   R. Machleidt, Quarks and Nucl.Structure, ed K.Bleuler Springer Lecture Notes in Physics Vol.197 (1984)
9. C.Elster, private communication
10. K.Gabathuler et al, Nucl.Phys. $\underline{B55}$ (1973) 397
    Nucl.Phys. $\underline{A35o}$ (1980) 253
    E.L.Mathie et al, Phys.Rev. $\underline{C28}$ (1983)2558
    G.R.Smith et al, Phys.Rev. $\underline{C29}$ (1984) 22o6
11. V.König et al, J. Phys. G; Nucl.Phys. $\underline{9}$ (1983)211
    W. Grüebler et al, Phys.Rev.Lett. $\underline{49}$(1982) 444
    J.Ulbricht et al, Phys.Rev.Lett. $\underline{48}$ (1982) 311
    R.J.Holt et al, Phys.Rev.Lett. $\underline{47}$ (1981) 472
12. H.G.Hopf, to be published

SESSION E

PHYSICS WITH POLARIZED BEAMS AND TARGETS

THE NUCLEON-NUCLEON INTERACTION AT MEDIUM ENERGIES

H. Spinka
Argonne National Laboratory, Argonne, Illinois 60439

ABSTRACT

A summary of the medium energy nucleon-nucleon elastic and inelastic scattering data is presented from an experimenter's viewpoint. Motivation for additional measurements is given. Experiments which rely on the unique capabilities of the Indiana Cooler are emphasized.

INTRODUCTION

The nucleon-nucleon (NN) interaction has been studied for many years. Recently there has been great progress in this field due to the wealth of new results from the meson factories at laboratory kinetic energies between about 300 and 800 MeV. As a result, phase shift solutions obtained by various groups have been converging toward unique results.

This paper addresses the question, "Are additional nucleon-nucleon measurements needed in the Indiana Cooler energy range?" In order to answer this question, some general comments on elastic scattering amplitudes are given in Section A and an overview of the existing data base is presented in Section B. This is followed by an extended discussion of pp elastic scattering in Section C. The types of spin observable measurements, tests for basic symmetries or for systematic errors, and the special case of experiments at small angles are all covered. The np elastic scattering situation is presented in Section D, and the nucleon-nucleon inelastic reactions NN → dπ, NNπ and dγ are discussed in Section E. Finally, a summary of these remarks is given in Section F.

A. NN → NN AMPLITUDES

Under the assumptions of parity and time-reversal invariance, there are five complex I=1 (isospin-1 or pp) elastic scattering amplitudes at each angle $\theta = \theta_{c.m.}$ (0 ≤ θ ≤ 90°) and beam kinetic energy T. An overall phase cannot be directly measured except near 0°, where it can be obtained relative to the Coulomb phase. The reason is that all elastic scattering spin observables are bilinear in the amplitudes, that is

$$\text{Spin observable} \sim \text{Re}(\text{Ampl}_2^* \cdot \text{Ampl}_1) \qquad (1)$$
$$\text{or}$$
$$\text{Im}(\text{Ampl}_2^* \cdot \text{Ampl}_1) .$$

There have been several analyses and suggested schemes to determine the I=1 amplitudes in a model independent fashion for beam momenta up to 6 GeV/c.[1-8] These schemes involve measurement at each angle and energy of at least 9 different pp elastic scattering spin parameters.

With the additional assumption of isospin invariance, there are also five amplitudes for the I=0 nucleon-nucleon system. Again, the

I=0 amplitudes at 180° - θ are determined by the amplitudes at θ, just as in the pp case. Some schemes to use np elastic scattering spin observable data to obtain the I=0 amplitudes, assuming the I=1 amplitudes are known, are given in Refs. 2,9.

At energies $T \lesssim 1000$ MeV, the nucleon-nucleon amplitudes are usually determined using a phase shift analysis.[10-15] The amplitudes are decomposed into partial waves, $^{2S+1}L_J$, where $\vec{S} = \vec{s}_1 + \vec{s}_2$ is the sum of the intrinsic nucleon spins, L is the orbital angular momentum, and $\vec{J} = \vec{L} + \vec{S}$ is the total angular momentum. A fit is made to the experimental data in some energy range to obtain the partial wave amplitudes. In order to make the problem tractable, some theoretical inputs and constraints are used. These often differ among the various phase shift analyses, reflecting to some extent the philosophies, styles, or judgement of the physicists doing the analysis.

There are several types of theoretical inputs and constraints which are used. For example, partial waves with J larger than some cutoff are generally determined from meson exchange calculations, which are used to obtain the long-range part of the NN interaction. At a given energy, the cutoff value of J has sometimes grown with time as the total amount of spin parameter data has increased near that energy. Also, at energies below the threshold for π production ($T \lesssim 300$ MeV), the amplitudes are required to be purely elastic. Finally, continuity constraints are applied to the phase shift parameters to require a smooth variation with beam energy. Various phase shift analyses perform these continuity conditions in different ways, and there is some evidence that structure in the amplitudes at $\theta_{c.m.} = 90°$ may have been smoothed too much at higher energies (Ref. 16 and Fig. 1). A particular problem has involved the inelasticities,[17] which are also related to a number of total cross sections (the optical theorem relates total cross sections to elastic scattering amplitudes at 0°).

Another way to look at the problem concerns the connection between the amplitudes and the phase shift parameters.[18] Schematically,

$$\begin{pmatrix}\text{Partial Wave} \\ \text{Amplitude}\end{pmatrix} \sim \begin{pmatrix}\text{Kinematic +} \\ \text{Spin Factors}\end{pmatrix} \cdot \int_{-1}^{1} (\text{Ampl.}) \cdot \begin{pmatrix}\text{Legendre} \\ \text{Polynomial}\end{pmatrix} \cdot d(\cos\theta) \quad (2)$$

The five full amplitudes (Ampl.) can be determined in a model independent way only to an overall angle dependent phase (Phase (θ)), as mentioned before. It is the various theoretical inputs and constraints noted above that permits the evaluation of the overall phase in the phase shift analyses. Furthermore, as described in the next section, a model independent amplitude analysis cannot even be performed for $T \lesssim 400$ MeV because of a lack of sufficient spin parameter data.

It should be noted in Eq. (2) that if J is the total angular momentum for the partial wave, then the order of the Legendre polynomial is J or J ± 1. Therefore, at least J angles must be measured to find the partial wave amplitude with angular momentum J.

Similar considerations also apply to the inelastic reactions such as NN → πd, πNN, etc. In those cases the number of amplitudes is larger than for nucleon-nucleon elastic scattering, making the experimental job of determining the amplitudes more difficult.

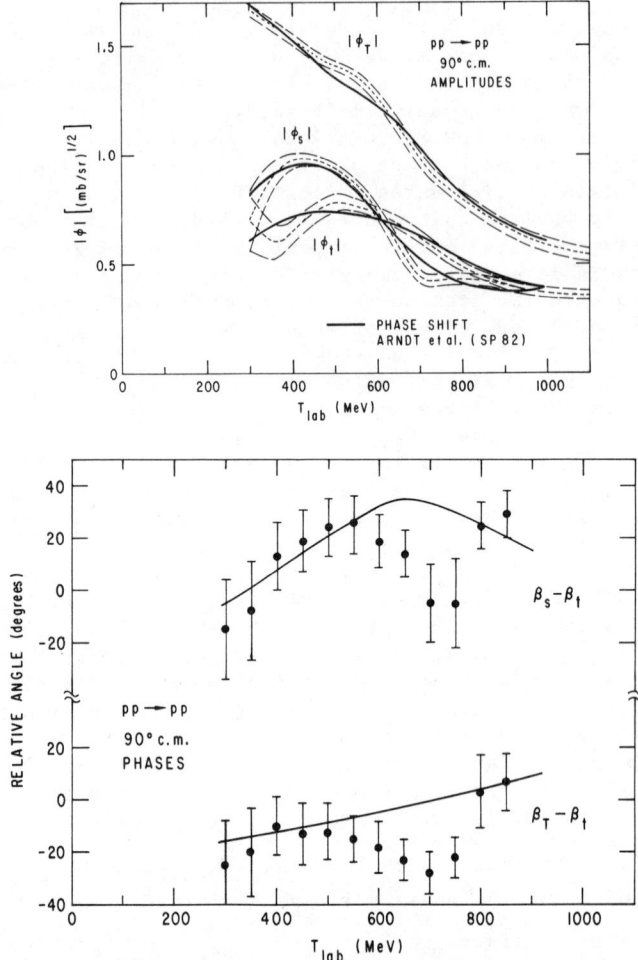

Figure 1. (a) The pp elastic scattering amplitudes at $\theta_{c.m.} = 90°$ as a function of lab kinetic energy. The definition of the amplitudes and the data used in the determination are from Ref. 16. The solid curves are Arndt's predictions (Refs. 10, 19). a) The magnitudes of the amplitudes. (b) The relative phases.

B. EXISTING NN → NN DATA

In order to properly evaluate the quality and quantity of the nucleon-nucleon elastic scattering spin parameter data, a variety of factors must be considered. Some of these factors are the number of points and of different spin parameters, the angles and energies covered, and the quoted errors. The large number of parameters makes it difficult to give a detailed characterization of the data base. Instead, a number of different ways of displaying the data in terms of the parameters above will be presented. In this way, some general features of the data base will be exhibited.

Figure 2 shows one way of presenting the pp and np points measured.[10] Only spin parameters other than differential cross sections and polarization are shown; similar plots of the other two qualities can be found in Ref. 10. The location of the measured data as a function of θ and T is indicated by a box for "old" data (pre-1978) and by an N for "new" data (post-1977). Each symbol plotted may represent more than one spin parameter or point at the same angle and energy.

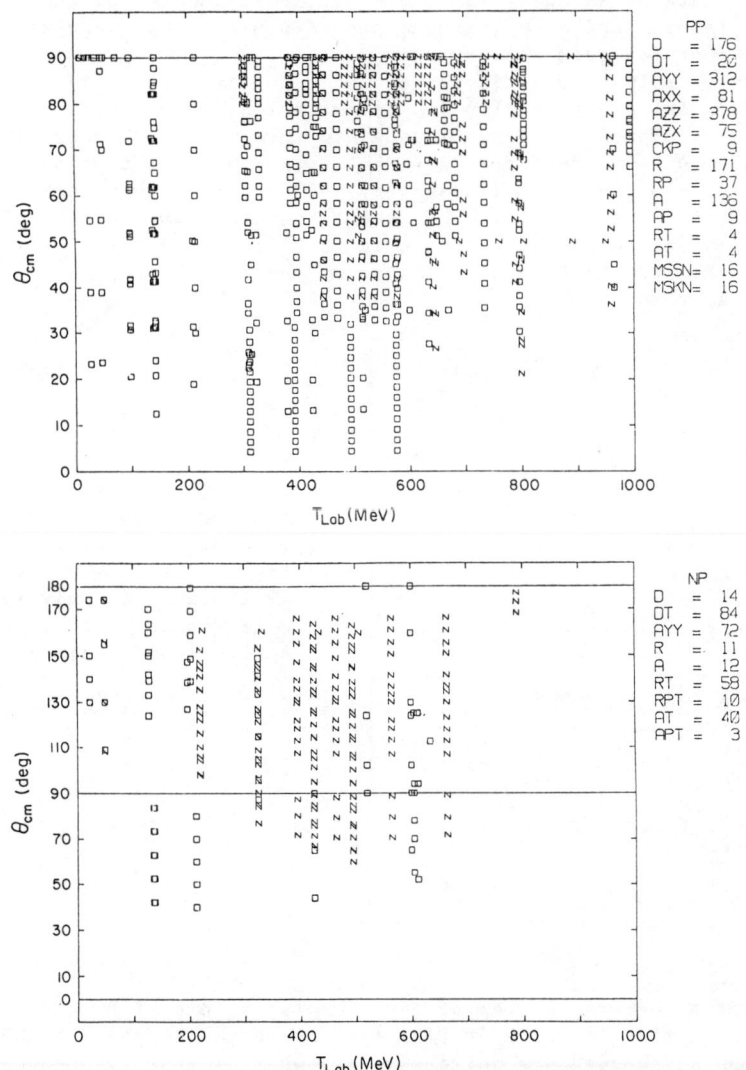

Figure 2. Representation of the nucleon-nucleon elastic scattering data base in terms of $\theta_{c.m.}$ and T_{lab} from Ref. 10. Only spin observables other than cross sections and polarizations are included. (a) pp → pp.   (b) np → np.

Concentrating on the pp results, it is apparent that they are rather sparse between about 150 and 300 MeV. The same is true for the differential cross section and polarization results. Near 140 MeV, the data are from Harvard, Harwell and Orsay. Above about 300 MeV, many laboratories have contributed, including the meson factories. In between, with the exception of some polarization results from Berkeley and TRIUMF, all the other data are from Rochester. These results were published in the early and mid-1960's, with most of the data at an energy of 210 MeV. Some independent checks of these results and some measurements of other spin parameters and other energies seem appropriate.

Using only pp elastic scattering spin parameter data other than differential cross section and polarization, the results from the data base of Ref. 19 were binned into three energy ranges (100-250, 250-400, 400-550 MeV) for Figs. 3-5. In Fig. 3, the pp data were grouped into five year classes (1970-1974, etc.) to illustrate the age of the measurements. The recent results from the meson factories are apparent, especially in the range T = 400-550 MeV. On the other hand, the lack of recent activity for the T = 100-250 MeV class can be clearly seen.

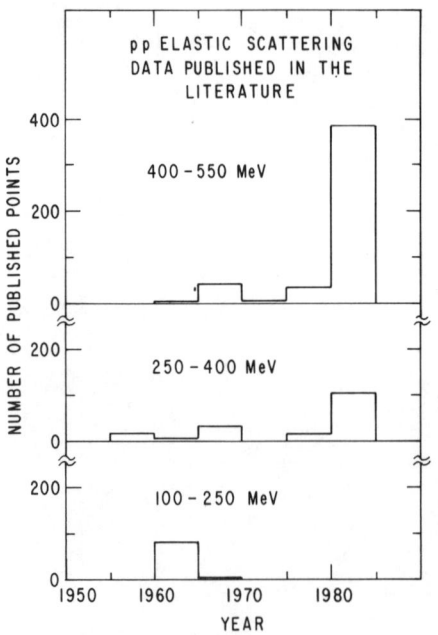

Figure 3. Number of pp elastic scattering points as a function of the year of publication. The data are from Ref. 19 and include only spin parameters other than cross sections and polarizations.

The integrated number of experimental points with errors less than some size is shown in Fig. 4. The shapes of the T = 100-250 and 400-550 MeV curves are quite similar, whereas the data between 250-400 MeV has a larger fraction of points with poorer statistical precision. With the increasing importance of inelastic channels at higher energies, as well as the larger number of partial waves significantly affected by short-range forces, the larger data base at higher energies is certainly justified.

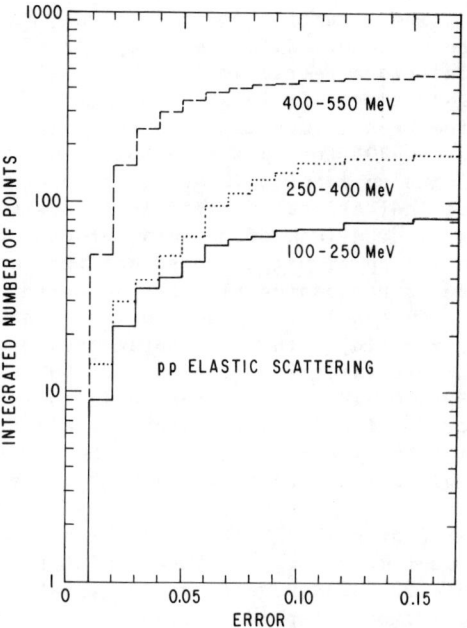

Figure 4. The integrated number of pp elastic scattering points as a function of the size of the quoted statistical errors. The data are from Ref. 19 and include only spin parameters other than cross sections and polarizations.

Finally, in Fig. 5 both the number of points and of different spin parameters for the pp data are shown as a function of angle. A lack of data at small angles is apparent in the range T = 100-250 MeV. For Fig. 5b, the polarization data were included in the count of the different spin parameters. It is clear that there are too few different spin parameters measured to perform a model independent amplitude analysis at any angle (other than 0° and 90°) below 400 MeV.

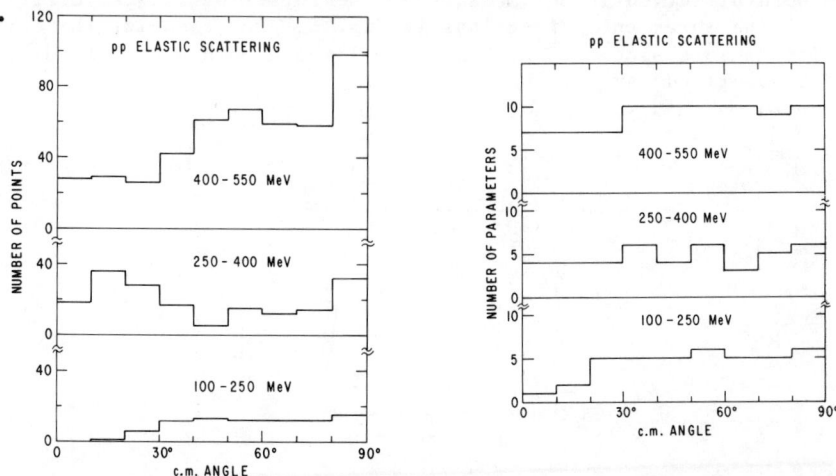

Figures 5. The number of points(a) and different spin observables(b) for pp elastic scattering as a function of c.m. angle. The data are from Ref. 19 and include only spin parameters other than cross sections for(b) and other than polarizations and cross sections for(a).

The situation for the np data base is generally poorer than for the pp case. With the exception of the more extensive np differential cross section data, the total of the other spin parameter measurements is less for np than for pp elastic scattering. This is true even though the relevant angular range is 0-180° for np, but only 0-90° for pp scattering. In addition, the statistical precision of the np results is generally poorer than for pp results, and there are indications of significant systematic errors for some of the np data. The lack of np data between about 150 and 300 MeV is similar to the pp situation noted earlier; this can be seen in Fig. 2b and the corresponding plot for np polarization in Ref. 10.

In conclusion, much of the pp data between 100 and 400 MeV is fairly old, with fewer parameters measured than at higher energies. The statistical precision, on the average, is somewhat poorer between 250-400 MeV, but is comparable to recent data from meson factories for T = 100-250 MeV. There is almost no data between 150 and 300 MeV with the exception of the Rochester results near 210 MeV, and forward angle measurements are lacking at the lower energies. A reasonable experimental case could be made to recheck some of the Rochester data, even though there may be no reason to suspect any problems. Measurements of additional spin parameters, energies and angles in the 150-300 MeV range would probably be useful to more accurately determine the phase shifts (and to test the theoretical inputs and continuity conditions). Finally, the np data base is even less complete than the pp case, and more data are justified.

For the remainder of this paper, the notation (beam, target; forward scattered, recoil) will be used to express the elastic scattering spin observables. The spin directions in the laboratory frame are defined in Fig. 6, and Table 1 gives symbols for the observables. In the table, * indicates the spin direction is known, and 0 means the spin direction is not measured. The subscripts i,j,k refer to one of the three spin directions in Fig. 6. For example, the correlation tensors such as C_{LL}, C_{SS}, etc. involve both a polarized beam and a polarized target.

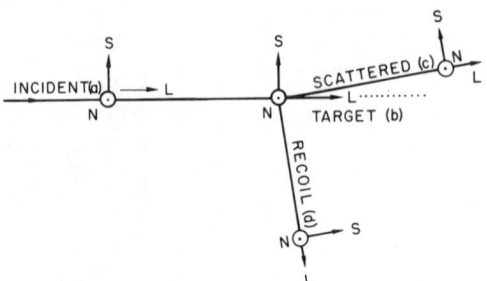

N: NORMAL TO THE SCATTERING PLANE
L: LONGITUDINAL DIRECTION
S = N x L IN THE SCATTERING PLANE

Figure 6. Definition of $\vec{N}$, $\vec{S}$ and $\vec{L}$ spin directions in the lab frame.

Table I. Symbols for observables

Observable	Description	Symbol
(0,0;0,0)	Differential Cross Section	$d\sigma/d\Omega$
(*,0;0,0) or (0,*;0,0)	Polarization	P
(*,*;0,0)	Correlation Tensor	$C_{jk} = A_{jk}$
(*,0;0,*) or (0,*;*,0)	Polarization Transfer Tensor	K_{jk}
(0,*;0,*) or (*,0;*,0)	Depolarization Tensor	D_{jk}
(*,*;0,*)	Triple Spin Tensor	H_{ijk}
(*,*;*,0)	Triple Spin Tensor	J_{ijk}

Assuming the usual conservation laws, there are three independent total cross sections for the nucleon-nucleon system. These three are

$$\sigma^{Tot} = [\sigma^{Tot}(\updownarrow) + \sigma^{Tot}(\updownarrow)]/2$$
$$= [\sigma^{Tot}(\uparrow\downarrow) + \sigma^{Tot}(\uparrow\uparrow)]/2 \quad (3)$$
$$\Delta\sigma_T = \sigma^{Tot}(\uparrow\downarrow) - \sigma^{Tot}(\uparrow\uparrow)$$
$$\Delta\sigma_L = \sigma^{Tot}(\updownarrow) - \sigma^{Tot}(\updownarrow) \quad .$$

The spin averaged total cross section is σ^{Tot}. The quantity $\Delta\sigma_T$ is the difference in total cross sections for beam and target tranversely polarized. The quantity $\Delta\sigma_L$ is the difference in total cross sections for beam and target longitudinally polarized (the arrows denote the spin directions in the laboratory frame).

C. PP ELASTIC MEASUREMENTS

This section will focus on spin parameters other than differential cross sections. Cross sections are also important, but the experimental techniques and problems are different. For example, absolute solid angles, efficiencies and luminosities are needed for differential cross sections, but only relative values for "spin-up" and "spin-down" are required for the other spin observables. Three types of apparatus are employed for these measurements: polarized beams, polarized targets, and polarimeters to measure the spin of an outgoing proton. (This will be referred to as a "carbon" polarimeter, even though under some conditions an analyzer other than carbon may be used. In addition, a magnet may be included to precess L-type to S- or N- type spin.) In order to reduce systematic errors, it is quite important to reverse the beam and target spins frequently. Similarly, if the beam and target are both unpolarized, then the carbon polarimeter should be rotated often for the same reason. (See, for example, the high energy experiment with a gas jet target performed at Fermilab by an Indiana group.[20])

There are several classes of spin parameter measurements, depending on the type of apparatus used. These are given below, along with some comments on each class. The relation of these spin observables to the amplitudes and relations between various sets of amplitudes are given in the comprehensive paper by Bystricky, Lehar and Winternitz.[21]

1) <u>Polarization = P</u>

The polarization parameter can be measured with either the beam or the target or the outgoing proton polarized. In these cases, the spin observable is (N,0;0,0) or (0,N;0,0) or (0,0;N,0) or (0,0;0,N). Comparison between various of the observables can be used to cross-calibrate the polarization or analyzing power of two types of apparatus (polarized target and carbon polarimeter, for example), or to search for systematic effects in the hardware. If the apparatus can be independently calibrated in some way, then a test of time reversal invariance can be performed by comparing (N,0;0,0) and (0,0;N,0).

2) <u>Polarized Target and Carbon Polarimeter</u>

As given in Table 1, the parameters that can be obtained are

$$D_{jk} = (0,j;0,k)$$
$$K_{jk} = (0,j;k,0)$$
(4)

and the observables that are allowed to be different from zero are D_{NN}, K_{NN}, D_{SS}, K_{SS}, D_{SL}, K_{SL}, D_{LS}, K_{LS}, D_{LL} and K_{LL}. Parity conservation requires that only observables with an even number of L and S subscripts are nonzero. Therefore, a measurement of D_{NS}, K_{LN}, etc. can be viewed as a test of parity conservation or of systematic errors in the experiment. There are relations between the D's and K's from time reversal invariance and from the conversion of spin directions in the c.m. frame to the laboratory system

$$\frac{D_{LS} + D_{SL}}{D_{SS} - D_{LL}} = \tan \theta_R$$

$$\frac{K_{LS} + K_{SL}}{K_{SS} - K_{LL}} = -\tan \theta_F .$$
(5)

In Eq. (5), θ_F and θ_R are the laboratory angles of the forward and recoil protons respectively.

3) **Polarized Beam and Carbon Polarimeter**
The spin parameters in this case are

$$D'_{jk} = (j,0;k,0)$$
$$K'_{jk} = (j,0;0,k). \qquad (6)$$

Relations analogous to Eq. (5) can also be derived. The D_{jk} and D'_{jk} are linearly related, and the K_{jk} and K'_{jk} are also linearly related. Hence, if these parameters need to be measured, the choice of D_{jk} or D'_{jk}, etc. can be made on the basis of experimental problems with one technique or another. For example, at high energies, the D_{jk} and K'_{jk} are usually measured because the carbon analyzing power and the scattering cross section in the carbon for the lower energy recoil proton is higher than for the forward proton.

4) **Polarized Beam and Polarized Target**
From Table 1, the spin parameters are $C_{jk} = (j,k;0,0)$ and the nonzero quantities are C_{NN}, C_{SS}, $C_{SL} = C_{LS}$ and C_{LL}. As above, parity conservation requires $C_{NS} = 0 = C_{LN}$ etc. These measurements are good for low count rate situations, since the outgoing proton spin is not measured. A polarized gas target in the Cooler has several advantages over the conventional polarized target technique. One is the lower energy loss for the recoil proton, which permits coincidence measurements to smaller scattering angles. A second advantage is the lack of background nuclei in the target. In the past, good precision small angle polarized target data have only been obtained using a very high resolution magnetic spectrometer (HRS at LAMPF) to separate pp and p + Nucleus elastic scattering. A third advantage is the lack of high magnetic fields at the interaction point, which complicate the tracking and precess the beam spin direction in the case of $C_{SL} = C_{LS}$. A final advantage is the larger angular region without obstruction from polarized target magnet coils; such freedom is presently available only for "frozen spin" targets using dilution refrigerators. Few such experiments have been performed to date using this technique.[22]

5. **Polarized Beam, Polarized Target and Carbon Polarimeter**
The lower statistical precision of these difficult "triple-spin" experiments has limited them to a few special cases. All of the advantages of using polarized gas targets apply here as well, and there is an added advantage that the lack of magnetic field at the interaction point means there is no spin precession of the outgoing protons. In the notation of Table 1,

$$H_{ijk} = (i,j;0,k)$$
$$J_{ijk} = (i,j;k,0), \qquad (7)$$

then there are constraints analogous to Eq. (5),

$$\frac{H_{NSS} - H_{NLL}}{H_{NLS} + H_{NSL}} = -\tan \theta_R$$

$$\frac{J_{NSS} - J_{NLL}}{J_{NLS} + J_{NSL}} = \tan \theta_F .$$
(8)

From the information on the different classes, several patterns emerge. First, there are constraints between various spin observables within each of the classes. Also, the types of spin parameters D_{ij} or D'_{ij}, K_{ij} or K'_{ij}, and C_{ij} each contain four independent quantities. Experiments with fewer spins measured will in general be easier and have higher statistical precision. Finally, there are significant advantages for the use of polarized gas targets over conventional polarized targets.

Before ending the discussion on pp elastic scattering, the special case of 0° results will be mentioned. There are only three independent amplitudes at 0°. Their real parts can be found from small angle elastic scattering measurements in the Coulomb-nuclear interference region (see Fig. 7 and Ref. 23). The imaginary parts of the same amplitudes can be related to the three total cross sections σ^{Tot}, $\Delta\sigma_T$ and $\Delta\sigma_L$ by the optical theorem. Dispersion relation calculations[24] connect the real and imaginary parts. A variety of theoretical input is used to perform these calculations. The small angle polarized target measurements using HRS at LAMPF were performed specifically to test these dispersion relation predictions at 650 and 800 MeV. Data at lower energies do not exist, and such small angle experiments are ideally suited to the Indiana Cooler, where the costs would be significantly less than the present LAMPF experiments.

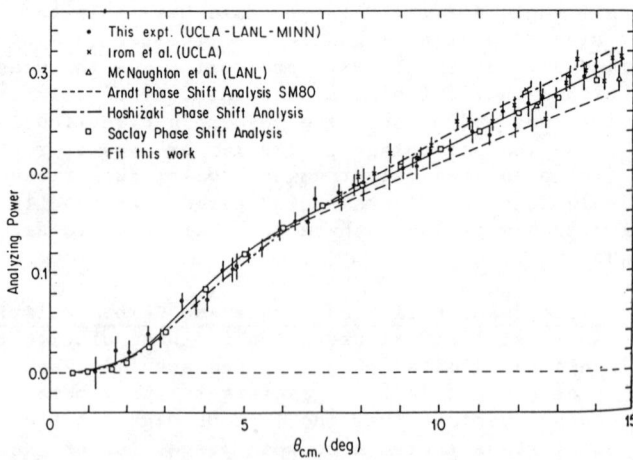

Figure 7. The polarization parameter as a function of c.m. angle for pp elastic scattering at 800 MeV. The data at the smallest angles are from an HRS experiment at LAMPF (Ref. 23).

The $\Delta\sigma_L$ and $\Delta\sigma_T$ measurements were the source of intense investigation at energies above about 400 MeV (Figs. 8, 9 and Refs. 16, 17,

25-31) in the past few years. However, there are few data at lower energies, and the errors are not particularly small. Up to now, various technical problems with the transmission technique, energy losses and nuclear interactions in the target and detectors, etc. have limited the precision of the low energy data. The experimental method in the past has attempted to obtain the amount of transmitted, noninteracting beam for parallel and antiparallel beam and target spins. The technique with a polarized gas target would have to be different, relying instead on detecting all interactions, since detectors cannot be placed in the direct beam. In both cases, only the relative amount of "spin-up" and "spin-down" beam is required, but the absolute target density is important since a total cross section is being measured. It should also be noted that below π-production threshold, $\Delta\sigma_T$ and $\Delta\sigma_L$ must be purely elastic.

$$\Delta\sigma_T = \Delta\sigma_T \text{ (elastic)} = -\int (C_{NN} + C_{SS}) \frac{d\sigma}{d\Omega}\bigg|_{el} d\Omega \quad (9)$$

$$\Delta\sigma_L = \Delta\sigma_L \text{ (elastic)} = -2\int C_{LL} \frac{d\sigma}{d\Omega}\bigg|_{el} d\Omega .$$

These relations can be used as a consistency check on the $\Delta\sigma$, C_{jj} and $d\sigma/d\Omega$ data.

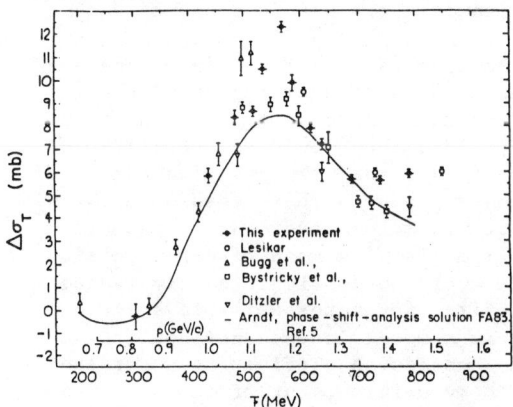

Figure 8. The difference in pp total cross sections for beam and target transversely polarized as a function of lab kinetic energy. The data are from Refs. 17, 25-28.

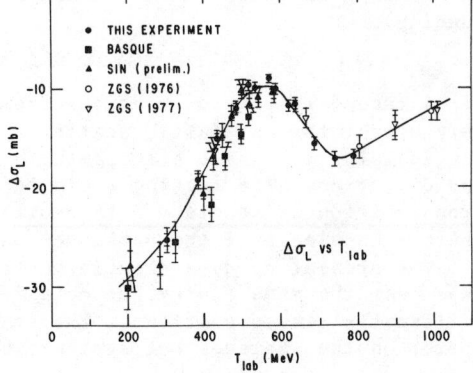

Figure 9. The difference in pp total cross sections for beam and target longitudinally polarized as a function of lab kinetic energy. The data are from Refs. 16, 17, 29, 30.

Given the preceding information on the different classes of experiments and on the amplitudes, the natural question arises as to the number of experiments to perform. For instance, it has been claimed that the pp phase shifts are accurately known up to 650 MeV.[12] On the other hand, past experience at energies above 500 MeV has indicated that when a new type of spin parameter was measured, the data seldom agreed well with the phase shift predictions until sufficient data existed to perform a model independent amplitude analysis.

There are several possible approaches to the problem of how many experiments to perform. Certainly one would be to measure sufficient parameters at several energies in the 100-400 MeV range to do a model independent amplitude analysis. If the statistical precision were good enough and systematic errors could be kept small, then significant tests of the theoretical inputs to the phase shift analyses could be performed. Similarly, the data could be used to search for a time-reversal-invariance violation in the I=1 nucleon-nucleon amplitudes.[32] A major problem is the long-term committment needed to complete the program, such as that exhibited by the U. Geneva group at SIN[7,33]. Insofar as possible, measurements in classes 1-4 should be used, although under certain circumstances, some spin parameters from class 5 might be measured at the same time. For example, the limitation to N-type polarized beam would still permit a model independent amplitude determination if there was fair use of data from class 5.

An alternate approach would be to measure a few spin parameters at a few energies and compare the results to existing phase shift predictions. It would seem prudent to measure some new type of spin parameters in this energy region, since they would be sensitive to different combinations of the amplitudes than the existing data. If there was a significant disagreement, then further measurements could be done; otherwise the effort would be terminated. Still another approach would be to perform experiments with much higher precision than the existing data. The goal would be to significantly reduce the uncertainties on the phase shfits, not necessarily to collect a full set of measurements to perform a model independent amplitude analysis. This approach requires considerable effort to understand and control systematic errors. In my opinion, any of these approaches could be reasonably justified, with the choice dependent on the experimenters and the constraints imposed by the hardware (and funding).

D. NP ELASTIC SCATTERING

A secondary beam of neutrons from the Indiana Cooler could be very useful for np elastic scattering studies. Neutrons could be tagged using the associated particle technique to select the energy and direction. The tagging would be especially useful for total cross section or absolute differential cross section measurements, where a knowledge of the number of incident particles is essential.

The present np data base is generally smaller than the pp data base near the same energy, especially for spin parameters other than differential cross sections. The statistical precision is also poorer on the average, but systematic differences between various sets of measurements of the same spin observable are even more serious.

To illustrate these problems, Fig. 10 contains the np total cross sections at medium energies.$^{34-38}$ The recent LAMPF results^{38} have a different shape and normalization from much of the older data. Another example of serious discrepancies is the backward ($\theta_{c.m.}$ = 180°) charge exchange differential cross section data (Refs. 37, 39-40 and Fig. 11). Precision measurements in the Cooler energy region could help clarify the experimental situation.

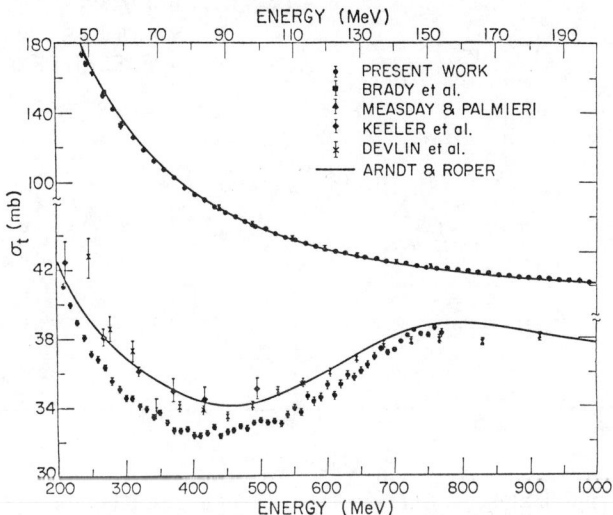

Figure 10. The np total cross section as a function of lab kinetic energy from Refs. 34-38. Note the upper curve corresponds to the energy and cross section scales at the top, etc.

There are numerous np polarization data and these can also be used to demonstrate problems with systematic errors in the np data base. In Fig. 12 is shown the np polarization at two neutron c.m. angles as a function of laboratory kinetic energy.$^{2, 41-47}$ In some cases, when data were not obtained at exactly 90° or 120°, the results of a single set of measurements were interpolated to the proper angles. There is a sizeable variation, outside the quoted statistical errors, evident in some cases. The np elastic scattering situation is significantly worse than the pp case in this regard (for example, see Ref. 48).

It is evident from Fig. 2 that additional spin parameter measurements are needed before the np and pp data bases are comparable. Since the np results are mixtures of I=0 and I=1 amplitudes,

$$\text{Ampl}_{np} = (\text{Ampl}_{I=1} + \text{Ampl}_{I=0})/2, \tag{10}$$

and since systematic errors are present in the np data base, the need for further experiments is strengthened. Note that the lack of np data between 150 and 300 MeV, except near 210 MeV, is reminiscent of the pp situation discussed before. Also, note that $\Delta\sigma_T$ and $\Delta\sigma_L$ for free np interactions do not exist at any energy and there are discrepancies between various dispersion relation predictions.[12,24]

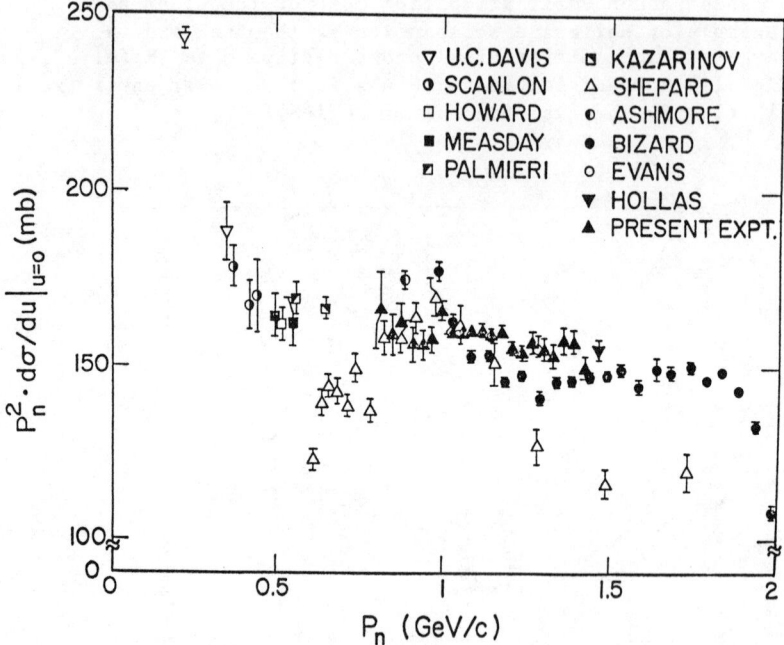

Figure 11. The np backward differential cross section as a function of lab momentum from Ref. 39. Other data are given in Refs. 37, 40.

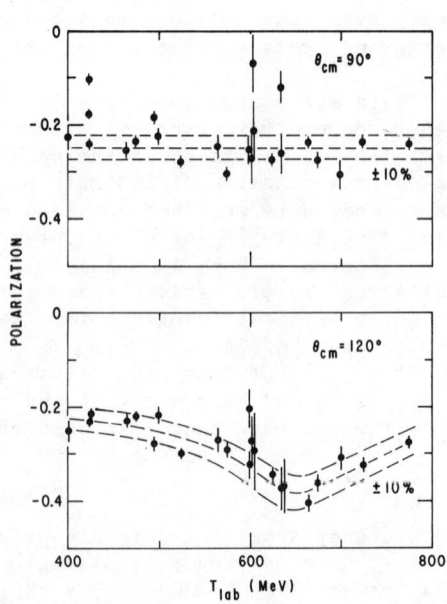

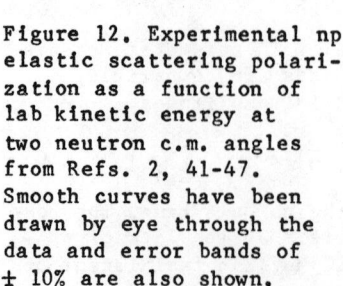

Figure 12. Experimental np elastic scattering polarization as a function of lab kinetic energy at two neutron c.m. angles from Refs. 2, 41-47. Smooth curves have been drawn by eye through the data and error bands of ± 10% are also shown.

Performing sufficient measurements to permit a model independent amplitude analysis may seem a monumental task, but the job may not be as bad as it first seems. Measurements of one spin parameter at both θ = $θ_{c.m.}$ and 180° - θ generally gives two independent bits of information on I=0 amplitudes at θ.[2,9] For example, if the neutron polarized beam intensity is not very high, making extensive measurements with a carbon polarimeter prohibitive, there is a lot of information that can still be obtained on these amplitudes. If np spin observables dσ/dΩ, P, C_{NN}, C_{SS}, C_{LS}, C_{LL} are measured at θ and 180° - θ, and if the I=1 amplitudes are known, then Ref. 9 shows that the I=0 amplitudes can be obtained in a model independent manner with discrete ambiguities. Some additional carbon polarimeter data could resolve the discrete ambiguities and allow checks for systematic errors as well.

E. NN REACTIONS

This section will be split into three parts on the three inelastic reactions NN → πd, NN → NNπ and np → dγ. The experimental and theoretical situation in all three cases is far from satisfactory. Interest in these reactions at energies above 450 MeV has been increasing recently, as large amounts of new nucleon-nucleon elastic scattering data have become available from the meson factories and other accelerators. In one sense, these reactions are the "logical next step" after the elastic scattering studies. The interest in the first two inelastic reactions has also been driven by the presence of resonance-like behavior in some I=1 elastic scattering phase shifts. The problem with inelasticities alluded to before[17] also impacts the understanding of this behavior.

1. NN → πd

In this reaction, many new data have recently become available (for a summary, see Refs. 49, 50). Most of the new results are at energies above about 450 MeV, and they include cross sections, polarizations of the proton and of the deuteron (iT_{11}), and polarized beam - polarized target spin parameters C_{NN}, C_{SS}, C_{LS}, C_{LL}. One reason that polarized target experiments have contributed heavily to the data base is that this is a two-body final state, which is relatively easy to separate from reactions on background nuclei in conventional polarized targets. An example of such data is shown in Fig. 13 from Refs. 51, 52. Various theoretical predictions[53-56] are also shown which are qualitatively similar to the data, but which are quite different quantitatively in some cases.

Phase shift analyses have been attempted up to 800 MeV (proton laboratory kinetic energy) and are given in Refs. 49, 50. There are insufficient data at any energy to perform a model independent amplitude analysis for the six complex amplitudes in this reaction. Below about 450 MeV, only differential cross section and polarization results exist. As a consequence, the phase shift analyses have used a considerable amount of theoretical input, restrictions on the number of partial waves considered, and/or reliance on continuity conditions to arrive at a solution. Additional experiments at lower energies would be beneficial. On the other hand, this reaction can be studied with conventional techniques not requiring the unique characteristics of the Cooler. Furthermore, the most important measurements at this time may involve pions incident on polarized deuterium targets and the use of a carbon polarimeter to determine the spin of an outgoing proton. Such experiments are more appropriate for the meson factories.

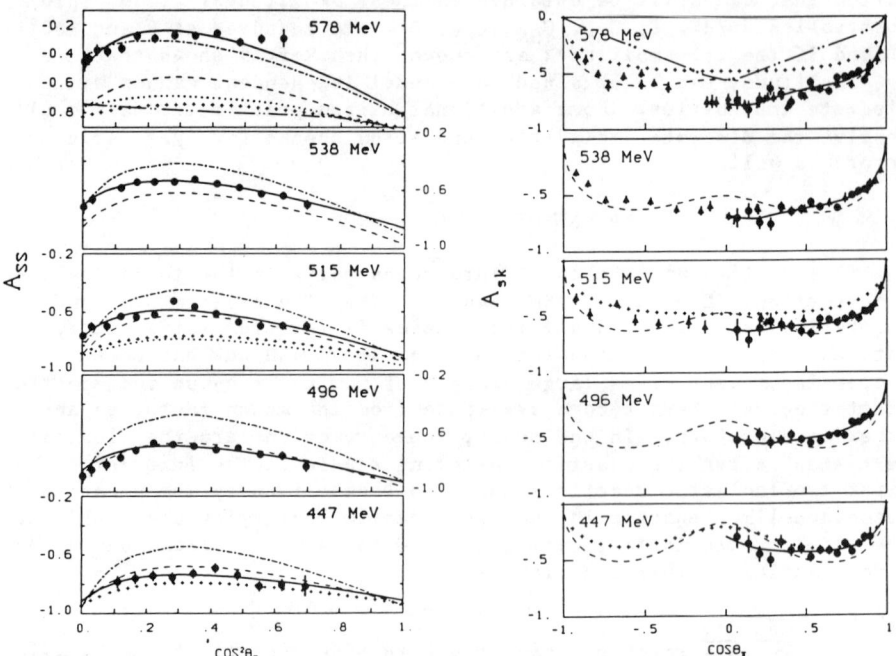

Figure 13. Experimental results on the $pp \to \pi d$ reaction from SIN (Refs. 51,52). The solid lines represent a phenomenological fit, while the other curves correspond to various theoretical predictions. (a) C_{SS} as a function of $\cos^2 \theta_\pi$. (b) C_{SL} as a function of $\cos \theta_\pi$.

One type of measurement on this reaction that may be ideally suited to the Cooler involves a high precision test of charge independence. Using the associated particle techniqiue to produce tagged beams of protons and neutrons at the same time, then the ratio of cross sections for

$$p + p \to d + \pi^+$$
$$n + p \to d + \pi^0$$
(11)

can be measured by detecting the deuteron. Various experimental problems related to deadtimes and other rate effects, solid angles, target thickness, detection efficiencies, absolute beam intensity, etc. could cancel in the ratio.

2. $NN \to NN\pi$

Below 500 MeV, recent experiments at TRIUMF[57] have measured a number of spin observables for the $pp \to pn\pi^+$ reaction. In general, the statistical errors are rather large, and there are few additional data besides cross sections on this channel (see Fig. 14 and Ref. 57). Spin information for pp → ppπ°, np → npπ°, np → ppπ⁻ is basically absent in this energy range. There are many complex amplitudes for each channel, and hence many possible spin parameters that can be measured. However, it does not appear that an adequate number of spin observables will be measured to perform a model independent analysis below 500 MeV in the foreseeable future.

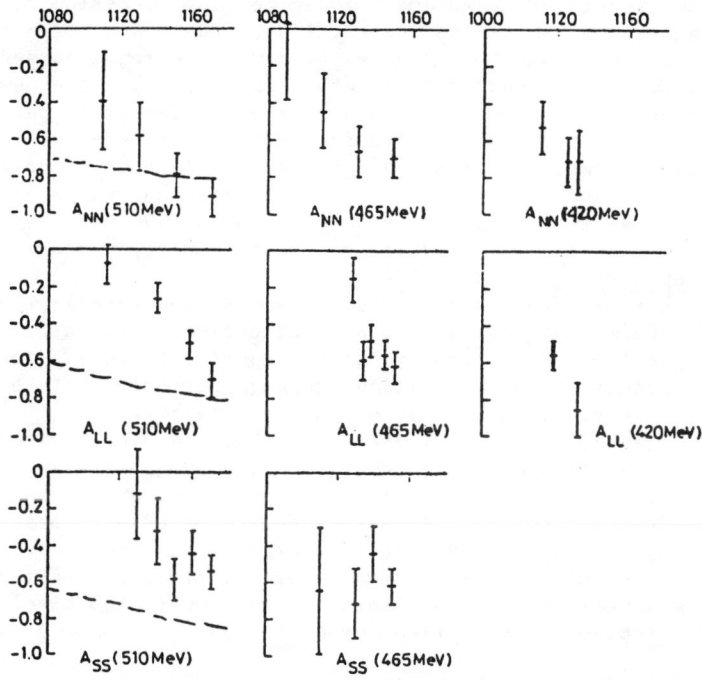

Figure 14. Experimental measurements of C_{NN}, C_{LL} and C_{SS} for the reaction pp → pnπ⁺ at 420, 465 and 510 MeV from TRIUMF (Ref. 57). The data are plotted against the invariant mass of the pπ⁺ system, and the curves are theoretical predictions (Ref. 58).

The standard theoretical model used to predict spin observables in the NN → NNπ reaction is the coupled channels approach of Silbar, Kloet, Dubach and Cass.[58] Some of the predictions at low energy are shown in Fig. 14, where significant discrepancies are observed. In general, the model has had mixed success, doing quite well for some spin observables and some kinematic regions and quite poorly in others.

The Indiana Cooler could make an important contribution to the understanding of which elastic scattering partial waves feed the inelastic channel NN → NNπ. Experiments with a polarized proton beam and a polarized gas target could measure spin observables much more easily than is possible with a conventional polarized target. A larger fraction of phase space could be detected because of the lower energy losses in the target. Magnetic fields that cause the precession of proton spins near conventional polarized targets would also be minimized. The lack of background nuclei in the gas targets would permit simplifications in the detection apparatus; angular measurements alone are adequate to discriminate between events from hydrogen and from nuclei for two-body final states, but they are inadequate for three-body final states. Finally, the lack of polarized target magnet coils near the interaction region allows a more complete coverage of phase space with the gas target.

Another important measurement would be good precision cross section data for the I=0 inelastic reactions. These are often quite difficult because both pp and np experiments are required and systematic errors are frequently dominant (Refs. 59, 60). The use of tagged beams of protons and neutrons at the same time could permit accurate determination of the I=0 inelastic cross sections where many sources of systematic errors would cancel (as in the discussion of the pp → πd and np → πd reactions). There are indications that this inelastic cross section may be very small below 500 MeV, so such measurements may be more appropriate for higher energy coolers.

3. $np \rightarrow d\gamma$

This reaction is quite basic, and should be calculable (in principle) from knowledge of the nucleon-nucleon interaction and electromagnetism. With high precision experimental data, tests of various corrections, such as meson exchange currents, off-shell effects, relativity, etc., should also be possible.

However, in the past there have been many measurements of cross sections and sizeable inconsistencies among different experiments. There have also been lots of calculations which were quite different from each other. Recently, some progress has been made in obtaining agreement between different experiments (see Ref. 61) and this should assist in constraining the theoretical predictions (Fig. 15). Tagged beams of neutrons should be of considerable help in obtaining precise cross section normalizations.

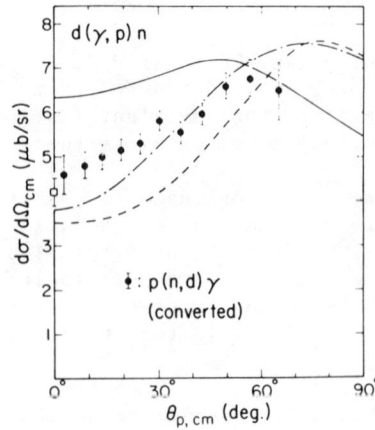

Figure 15. Differential cross section for the reaction $np \rightarrow d\gamma$ at a γ-ray lab energy of 95 MeV from Ref. 61. Three theoretical calculations are also shown.

Few spin observables have been measured in this reaction. Asymmetries for linearly polarized photons[62] or for polarized neutrons[63] (Fig. 16), or protons at higher energies[64] have been obtained. Many additional measurements of these spin parameters and others will probably be required before the np → dγ process is completely understood.[65]

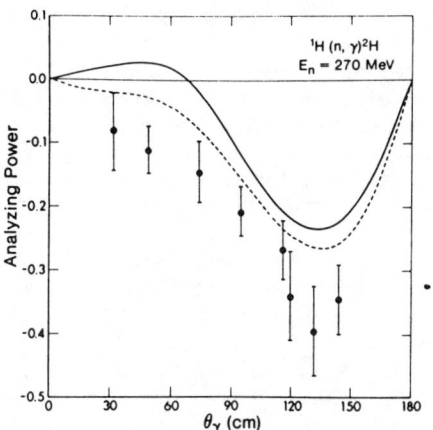

Figure 16. Polarization or analyzing power for the reaction np → dγ using polarized neutrons as a function of the γ-ray angle in the c.m. frame from Ref. 63. The curves are theoretical predictions.

CONCLUSIONS

Various aspects of the nucleon-nucleon interaction were discussed. A summary of the gaps in our knowledge below 500 MeV is given below:

1) Some pp elastic scattering spin parameters should be measured in the energy range from 150 to 300 MeV, especially at small angles. The Cooler and a polarized gas target offer distinct experimental advantages for some of these measurements.

2) There are significant systematic differences in much of the np elastic scattering data base. Measurements with a tagged neutron beam should permit high precision determinations of differential and total cross sections. Other spin parameters should also be measured in the energy range from 150 to at least 300 MeV.

3) The goal of understanding the spin structure of pion production near threshold (up to 500 MeV) has not been achieved. There are few experimental results except differential cross sections and a few spin observables in any of the inelastic channels in this energy range. Measurements of spin parameters in any of these channels, and in particular for pp → NNπ where the Cooler offers special advantages, would be quite useful.

In conclusion, the Cooler has the potential to make unique contributions to the study of the nucleon-nucleon interaction.

ACKNOWLEDGEMENTS

I wish to thank my many colleagues from Argonne, LAMPF, University of Montana, New Mexico State University, Texas A&M University, and Washington State University for useful discussions on these subjects, and especially R. Wagner for a careful reading of the manuscript.

REFERENCES

1. L. Puzikov et al., Nucl. Phys. 3, 436 (1957).
2. B. M. Golovin et al., Sov. Phys. JETP 9, 302 (1959).
3. C. R. Schumacher and H. A. Bethe, Phys. Rev. 121, 1534 (1961).
4. Yu. P. Kumekin et al., Sov. Phys. JETP 19, 36 (1964).
5. W. deBoer and J. Soffer, Nucl. Instrum. Methods 136, 331 (1976).
6. P. W. Johnson et al., Phys. Rev. D15, 1895 (1977); ibid. D16, 2783 (1977).
7. D. Besset et al., Nucl. Instrum. Methods 148, 129 (1978); E. Aprile et al., Phys. Rev. Lett. 46, 1047 (1981).
8. N. Ghahramany et al., Phys. Rev. D28, 1086 (1983).
9. H. Spinka, Phys. Rev. D30, 1461 (1984).
10. R. A. Arndt et al., Phys. Rev. D28, 97 (1983).
11. N. Hoshizaki, Prog. Theor. Phys. 60, 1796 (1978); 61, 129 (1979); K. Hashimoto et al., Prog. Theor. Phys. 64, 1678, 1693 (1980).
12. R. Dubois et al., Nucl. Phys. A377, 554 (1982).
13. J. Bystricky et al., Saclay Report No DPhPE 82-09, 1982 (unpublished).
14. M. I. Dzhgarkara et al., Sov. J. Nucl. Phys. 35, 39 (1982).
15. M. Akemoto et al., Prog. Theor. Phys. 67, 554 (1982).
16. I. P. Auer et al., Phys. Rev. D29, 2435 (1984).
17. J. P. Stanley et al., Nucl. Phys. A403, 525 (1983); D. V. Bugg, Tenth International Conference on Few Body Problems in Phys. (1983), Karlsruhe, ed. B. Zeitnitz.
18. M. L. Goldberger et al., Phys. Rev. 120, 2250 (1960).
19. R. A. Arndt and L. D. Roper, VPI N-N interactive dial-in-computing facility (manual available from Department of Physics, Virginia Polytechnic Institute, Blacksburg, VA 24061). The data base used was dated May, 1984.
20. M. Corcoran et al., Phys. Rev. D22, 2624 (1980); ibid. D24, 3010 (1981).
21. J. Bystricky et al., Jour. de Phys. 39, 1 (1978).
22. W. R. Ditzler et al., Phys. Rev. D29, 2137 (1984).
23. G. Pauletta et al., Phys. Rev. C27, 282 (1983).
24. W. Grein and P. Kroll, Nucl. Phys. B137, 173 (1978); A377, 505 (1982).
25. Ed. K. Biegert et al., Phys. Lett. 73B, 235 (1978).
26. W. R. Ditzler et al., Phys. Rev. D27, 680 (1983).
27. J. Bystricky et al., Ninth International Conference on High Energy Physics and Nuclear Structure, Versailles, 1981 (unpublished).
28. W. P. Madigan et al., Los Alamos National Laboratory Report LA-10070-PR (1984), p. 15.
29. I. P. Auer et al., Phys. Lett. 67B, 113 (1977); ibid. 70B, 475 (1977); Phys. Rev. Lett. 41, 354 (1978).

30. E. Aprile et al., High Energy Physics with Polarized Beams and Polarized Targets, Proceedings of International Symposium, Lausanne, Switzerland, 1980, edited by C. Joseph and J. Soffer (Birkhauser, Basel, 1981), p. 516; C. Lechanoine-Leluc (private communication).
31. J. Bystricky et al., Phys. Lett. 142B, 130 (1984).
32. E. Aprile et al., Phys. Rev. Lett. 47, 1360 (1981).
33. D. Besset et al., Nucl. Phys. A345, 435 (1980); Phys. Rev. D21, 580 (1980); E. Aprile et al., Phys. Rev. D28, 21 (1983).
34. T. J. Devlin et al., Phys. Rev. D8, 136 (1973).
35. D. F. Measday and J. N. Palmieri, Nucl. Phys. 85, 142 (1966).
36. F. P. Brady et al., Phys. Rev. Lett. 25, 1628 (1970).
37. R. K. Keeler et al., Nucl. Phys. A377, 529 (1982).
38. P. W. Lisowski et al., Phys. Rev. Lett. 49, 255 (1982).
39. B. E. Bonner et al., Phys. Rev. Lett. 41, 1200 (1978), and references contained therein.
40. W. Hurster et al., Phys. Lett. 90B, 367 (1980).
41. D. Cheng et al., Phys. Rev. 163, 1470 (1967).
42. Yu. M. Kazarinov, Rev. Mod. Phys. 39, 509 (1967).
43. S. C. Wright et al., Phys. Rev. 175, 1704 (1968).
44. R. Zulkarneev et al., Phys. Lett. 61B, 164 (1976).
45. A. S. Clough et al., Phys. Rev. C21, 988, 998 (1980).
46. M. Sakuda et al., Phys. Rev. D25, 2004 (1982).
47. Data from Bagaturia, Bhatia, Bilenkaya, Leung and Newsom as quoted in Arndt (Ref. 19).
48. M. W. McNaughton and E. P. Chamberlin, Phys. Rev. C24, 1778 (1981).
49. H. Kamo and W. Watari, Prog. Theor. Phys. 62, 1035 (1979); H. Kamo et al., ibid., 64, 2144 (1980); N. Hiroshige et al. ibid., 68, 2074 (1982).
50. D. V. Bugg, J. Phys. G10, 47, 717 (1984).
51. E. Aprile et al., Nucl. Phys. A415, 365, 391 (1984).
52. J. Hoftiezer et al., Nucl. Phys. A412, 273 (1984).
53. B. Blankleider and I. R. Afnan, Phys. Rev. C24, 1572 (1981).
54. J. A. Niskanen, Phys. Lett. 79B, 190 (1978).
55. A. S. Rinat and Y. Starkand, Nucl. Phys. A397, 381 (1983).
56. W. Grein et al., SIN preprint PR-83-08 (1983).
57. R. Shypit et al., Phys. Lett. 124B, 314 (1983); D. V. Bugg, Proceedings of the Tenth International Conference on Few Body Problems in Physics, Karlsruhe, ed. B. Zeitnitz (1983).
58. W. M. Kloet and R. R. Silbar, Nucl. Phys. A338, 281, 317 (1980); J. Dubach et al., Phys. Lett. 106B, 29 (1981); J. Phys. G8, 475 (1982).
59. W. Grein et al., Phys. Lett. 96B, 176 (1980).
60. B. J. VerWest and R. A. Arndt, Phys. Rev. C25, 1979 (1982).
61. H. O. Meyer et al., Phys. Rev. Lett. 52, 1759 (1984).
62. V. G. Gorbenko et al., Nucl. Phys. A381, 330 (1982).
63. J. M. Cameron et al., Phys. Lett. 137B, 315 (1984).
64. T. Kamae et al., Phys. Rev. Lett. 38, 468, 471 (1977); Nucl. Phys. B139, 394 (1978); K. Ogawa et al., Nucl. Phys. A340, 451 (1980).
65. Additional discussion on the np $\rightarrow$ dγ reaction is given in the talk by H. Fearing at this workshop.

FEW-BODY REACTIONS AT MEDIUM ENERGIES

M. Bleszynski
Physics Department, UCLA, Los Angeles, California 90024

ABSTRACT

Several current interesting aspects of medium energy proton reactions with light nuclei are reviewed. Special attention is given to the relativistic effects in proton-deuteron elastic scattering. Few examples of possible experiments in the range of energies 100-500 MeV are discussed.

The study of spin observables in the elastic scattering of protons from few-nucleon systems: ^{2}H, ^{3}He, ^{3}H and ^{4}He constitutes a crucial test of our understanding of the multiple scatttering effects in hadron collisions with nuclei at intermediate energies. Due to the small number of nucleon constituents complete multiple scattering calculations are feasible in which the full spin dependence of the nucleon-nucleon (NN) interaction and complete nuclear wave functions can be taken into account. The succesive terms in the multiple scattering expansion can be evaluated separately without approximations (like the optical model) which are necessary to make the treatment of proton scattering from heavy nuclei technically feasible. The uncertainties in the existing phase shift analyses of NN amplitudes are believed to be very small in the range of energies up to 500 MeV and at the same time spin components of the underlying NN interaction are relatively very large. In such situation comparison of spin observables with the predictions of reaction models provides a very strong and compelling test.

The spin structure of the NN amplitude leads to the dependence of the proton-nucleus amplitude on the combinations of formfactors of matter and spin distribution which, in general do not enter into amplitudes describing electron-nucleus scattering. This important feature makes the investigation of proton reactions not only a complementary but rather a unique method of extracting information on nuclear structure effects. For example measurements of polarization observables in scattering of protons from ^{3}He and ^{3}H can provide potentially a rich source of such information. In contrast to the elastic scattering of electrons from these three-nucleon systems which involves contribution from only 2 (charge and magnetic) formfactors, proton reactions are sensitive to 6 different formfactors of matter and spin distribution[1]. Extraction of the components of the colision matrix depending on these formfactors can be achieved by making suitable polarization experiments with polarized proton beams and possibly polarized targets.

Another reason for studying proton reactions with few nucleon systems is the possibility of finding effects which cannot be explained in terms of a superposition of the underlying two-nucleon interactions but, might require inclusion of quark degrees of freedom or three-body forces. It is however essential to verify first to what extent reactions with few nucleon systems can be described in terms of standard models in which the transition amplitude for a given process is constructed from the two-body NN interactions. In this context both experimental and theoretical studies of the simplest nuclear reaction: proton-deuteron (pd) elastic scattering are of the fundamental importance.

A very interesting problem which should be investigated in the context of scattering of protons from light nuclei is the relevance of the relativistic effects of the similar nature as those which have been found to play an important role in spin observables in scattering of medium energy (~1 GeV) protons from heavier spin zero nuclei[2,3]. The approach of refs 2 and 3 is based on the relativistic impulse approximation Dirac optical potential, composed of scalar and vector (and typically very small tensor) components. The resulting predictions for spin observables are in much better agreement with experimental data than those obtained using standard nonrelativistic approaches based on the Schrodinger equation. The optical potential is constructed by folding the NN amplitudes with the nuclear wave function. A two dimensional Pauli matrix representation of the amplitude is used in conventional nonrelativistic analyses but a Dirac optical potential requires the use of a four dimensional Dirac representation. In ref. 1 the NN amplitude has been assumed to be of the form

$$F = \sum_{n=1}^{5} \hat{O}_n F_n$$

where $\hat{O}_n$, n=1,5 are the operators acting in the four dimensional Dirac spaces of the two nucleons and have been chosen as scalar, vector tensor, pseudovector and axial terms. The on-shell (i.e. on mass shell) part of the above amplitude is uniquely determined by being consistent with a phase shift analysis of NN scattering data. The amplitude for the off-shell scattering is not unique and follows from the form chosen for the matrix representation and from the assumed dependence on kinematical invariants (momentum transfer, energy, and possibly off-shell parameters)[4]. (In fact, the most complete representation of the relativistic amplitude should include 32 (instead of 5) terms[5]. Most of these terms can not be determined from the experimental data and must follow from a theoretical model describing NN scattering). Thus, both Dirac and Schrodinger models are sensitive to a parametrization of the amplitude to the extent that the off-shell states contribute to multiple scattering terms[4]. Such sensitivity is especially important in the Dirac approach which includes large contributions from negative energy waves which are always off-shell in addition to the off-shell positive energy contributions which are present in both treatments. The contributions of the negative energies account for much of the difference between Schrodinger and Dirac models and are largely responsible for the improved agreement of the latter. It is therefore imperative that the negative energy parts of the NN amplitude be determined in an accurate and physically sensible manner.

Since it is possible to do a fairly complete multiple scattering calculation for pd scattering, this process should provide a good test of models for the NN amplitude and could be possibly used to obtain constrains on the ambiguities connected with the choice of the representation of the relativistic NN amplitude. Also, because the deuteron has spin one, there are many independent spin observables which are sensitive to various spin components of the NN amplitude. For example, contributions from the pseudevector and axial components of the NN amplitude which do not contribute to the optical potential describing scattering of protons from spherically symmetric spin zero nuclei, are very important in some spin observables in pd scattering.

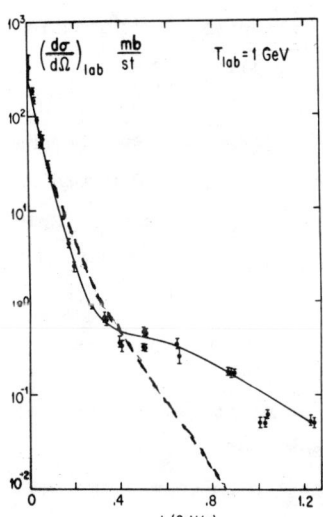
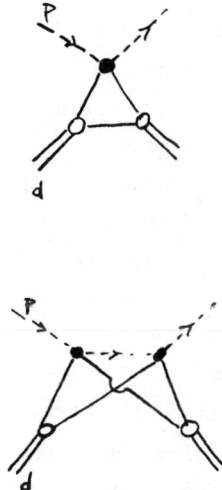

Fig 1. The differential cross section for p-d elastic scattering at 1 GeV. Data are from ref 6. The dashed curve represents the single collision approximation and the solid curve was obtained with single and double collision terms.

Fig 2. Multiple scattering diagrams representing the the single and double scattering processes.

section for the elastic scattering of 1 GeV protons[6] followed by the succesful description of this process made in the framework of the Glauber theory. The p-d collision amplitude was assumed to be composed of three terms. The first two terms represent the case in which the projectile interacts with one of the two nucleons in the deuteron (single scattering terms), while the third term represents the process in which both the target nucleons have been hit succesively by the incident proton (double scattering process). These single and double collision terms may be represented with the multiple scattering diagrams displayed in Fig 2. The change in slope in $d\sigma/dt$ seen in Fig. 1 has been attributed to the contribution from the double collision term.

The p-d reaction stimulated considerable interest as a potential source of information on the dynamics of the double collision process.

Recently, with the improving experimental technique it has become possible to measure various spin observables in p-d reactions at medium energies. The fact that deuterium has spin 1 leads to 23 independent spin observables. Measurements of these observables allow one to study in a selective way contributions from different ingredients of the multiple scattering theory.

The most complete theoretical decription of p-d elastic scattering at medium energies iis given in ref 7. The model of ref. 7 is a generalization of the Glauber theory which includes finite energy, recoil, several kinematical corrections as well as the complete treatment of spin dependence of the projectile-target nucleon interaction and of the deuteron wave function.

From the general symmetry arguments it follows that the collision matrix for the elastic proton-deuteron scattering is composed of 12 independent complex amplitudes and can be decomposed suitably in terms of products of the projectile spin $\vec{\sigma}$, the deuteron spin $\vec{J}$ and quadrupole moment operators $Q_{ik} = 1/2(J_iJ_k+J_kJ_i) -2/3\delta_{ik}$ as follows:

$$F = F_0^0 + F_y^0 J_y + F_{xx}^0 Q_{xx} + F_{yy}^0 Q_{yy} + F_x^x o_x J_x + F_{xy}^x o_x Q_{xy} +$$
$$+ F_0^y o_y + F_y^y o_y J_y + F_{xx}^y Q_{xx} o_y + F_{yy}^y Q_{yy} o_y + F_z^z J_z o_z + F_{yz}^z Q_{yz} o_z \ .$$

Here the x-axis is parallel to the momentum transfer and the y-axis is normal to the scattering plane. The coefficients F_0^0, F_{xx}^0, ... enable us to calculate the observables which can be written in general as

$$C(\alpha,i;\beta,k) = \mathrm{Tr}\{ F\ o_\alpha\ O_i\ F^+ o_\beta\ O_k\ \}/\ \mathrm{Tr}\{F\ F^+\},$$

where $O_i = 1$, J_i, Q_{mn}, $m,n = x,y,z$. The current status of polarization experiments in pd elastic scattering at medium energies is illustrated in Table 1 where we have listed the observables already measured (denoted with *) as well as the planned experiments at LAMPF and Saturne (denoted with X). The letters L and N appearing as two first arguments denote, in accordance with the Madison convention, the directions parallel to the initial beam momentum and normal to the scattering plane. The letter S appearing as the first argument corresponds to the direction forming the right-handed coordinate frame (N,L,S). The letters N,L and S appearing as the third or fourth arguments of the C observables correspond to the right handed coordinate frame with L taken along the scattered beam momentum. We note that some of the observables listed in Table 1 are not linearly independent. The general completness condition for the spin observables in the case of elastic scattering of spin 1/2 and spin 1 particles is discussed in ref. 8.

Observable	\multicolumn{5}{c}{Energy in MeV}				
	500	600	650	800	1000
C(0,SS;0,0)		*		*	*
C(0,NN;0,0)		*		*	*
C(0,N ;0,0)	*		*	*	
C(N,0 ;N,0)	*		*	*	
C(S,0 ;S,0)	*		*	*	
C(L,0 ;L,0)	*		*	*	
C(S,0 ;L,0)	*		*	*	
C(L,0 ;S,0)	*		*	*	
C(L,SS;0,0)				X	
C(S,SS;0,0)				X	
C(S,L ;0,0)				X	
C(S,S ;0,0)				X	
C(L,L ;0,0)				X	
C(N,SS;0,0)				X	
C(N,NN;0,0)				X	
C(N,N ;0,0)				X	
C(S,S ;0,0)				X	
C(L,L ;0,0)				X	
C(S,N ;S,0)				X	
C(N,N ;N,0)				X	
C(L,N ;L,0)				X	
C(S,N ;L,0)				X	
C(L,N ;S,0)				X	
C(S,NN;S,0)				X	
C(N,NN;N,0)				X	
C(L,NN;L,0)				X	
C(S,NN;L,0)				X	
C(L,NN;S,0)				X	

Table 1. Completed (*) and planned or currently analyzed (X) measurements of spin observables in elastic scattering of protons from deuterium at medium energies[9-12]

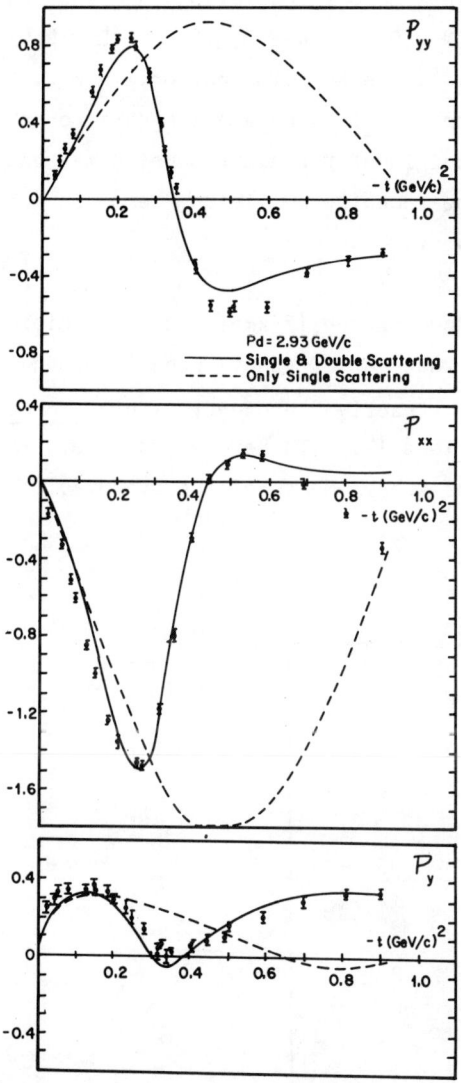

Fig. 3. The measured quantities $\mathcal{P}_{yy}$, $\mathcal{P}_{xx}$ and $\mathcal{P}_y$ as functions of $-t$. The solid lines correspond to the full calculation, the dashed lines were obtained by neglecting the double scattering contribution in the p–d scattering matrix.

Each of the 12 amplitudes F_k^i describing the proton-deuteron scattering depends in general on different ingredients of the reaction theory, like different spin components of the NN amplitude or different deuteron formfactors. Thus, for example the observables $C(0,xx;0,0)$ and $C(0,yy,0,0)$ depend mainly on the D-wave component of the deuteron wave function. Also by taking suitable linear combinations of the observables $C(0,\alpha;0,\beta)$ it is possible to construct quantities which depend selectively on the double spin flip components of the NN amplitude[14]. The contributions from the single and double collision terms enter with different strenghts to each of the amplitudes F_k^i. As an illustration of this point, in Fig. 3 we display the theory predictions for the observables $P_{xx}=C(0,SS;0,0)$, $P_{yy}=C(0,NN;0,0)$ and $P_y= C(0,N;0,0)$ for 800 MeV incident proton energy, obtained using the complete p-d collision amplitude (solid lines) and the single collision approximation (dashed lines).

Although several of the observables already measured, like those displayed in Fig 3, are quite well described in the framework of the nonrelativistic multiple scattering theory[7], in general, agreement with the data and the existing experimental results is far from being satisfactory.

Largest discrepancies between the theory predictions and the data have been found in the observables $C(N,0;0,0)$, $C(S,0;S,0)$, $C(L,0;L,0)$, $C(L,0;S,0)$ and $C(S,0;L,0)$ at 800 MeV[13,14]. As an illustration, in Fig. 4 we show the comparison between the theory predictions and the data for the observable $C(N,0;0,0)$. A possible origin of the discrepancies may come from the inadequate treatment of the double collision term in the nonrelativistic multiple scattering theory.

From the inspection of the structure of the 12 amplitudes describing p-d elastic scattering[7] it follows that 4 of them: the amplitudes F^0_{yy}, F^y_{yy}, F^x_{xy}, and F^z_{yz} have a very attractive property — they <u>do not contain the single scattering contributions</u> [15]. Therefore it would be very useful to perform a complete set of measurements of the observables

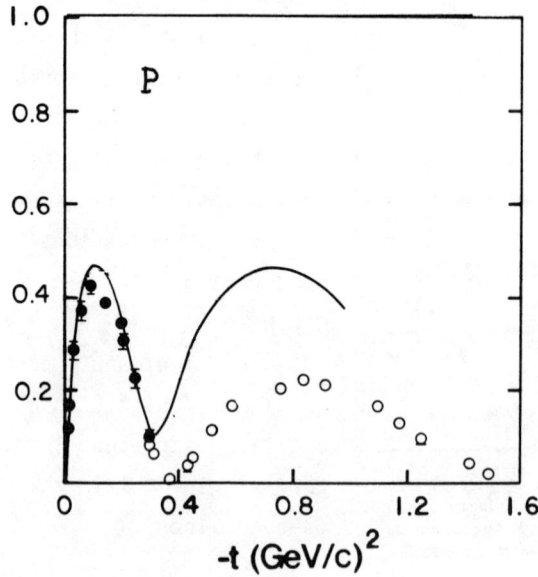

Fig. 4. The proton analyzing power in p-d elastic scattering at 800 MeV data from Ref. 10. Theory predictions have been obtained in the framework of the approach of Ref. 7.

$C(\alpha,i;\beta,k)$ which could be used to determine the amplitudes involving exclusively double collision contributions.

Such complete set of measurements would constitute presumably the most crucial test of the multiple scattering theory in the presence of strong spin dependence of the interaction and, in particular, could be used to assess the importance of the relativistic effects in the double collision term which, as we discuss below, may be very large.

In order to contrast relativistic and nonrelativistic treatments of proton-deuteron elastic scattering, we constructed a simple model[16] which can incorporate either a two dimensional Pauli matrix or four dimensional Dirac matrix parametrization of the nucleon-nucleon scattering amplitude. The model includes the S and D-wave components of the deuteron wave function but neglects interactions with the spins of the target nucleons, and, is intended to provide an estimate of the magnitude of the differences between relativistic and nonrelativistic treatments.

With the normalization conventions of ref. 7, both the relativistic and the nonrelativistic proton-deuteron elastic collision matrix in the Breit frame are given as a sum of single and double scattering terms,

$$F(\vec{Q}) = F_s(\vec{Q}) + F_d(\vec{Q}),$$

where

$$F_s(\vec{Q}) = \frac{M}{m} H(\tfrac{1}{2}\vec{Q}) \left[f_1(\vec{Q}) + f_2(\vec{Q}) \right]$$

$$F_d(\vec{Q}) = \frac{M}{4m^3} \int d^3p\, H(\vec{p}) \left[f_1(\tfrac{1}{2}\vec{Q}+\vec{p}) G(\vec{l}+\vec{p}) f_2(\tfrac{1}{2}\vec{Q}-\vec{p}) + f_2(\tfrac{1}{2}\vec{Q}+\vec{p}) G(\vec{l}+\vec{p}) f_1(\tfrac{1}{2}\vec{Q}-\vec{p}) \right].$$

Here $\vec{Q}$ is the overall momentum transfer, $\vec{l}+\tfrac{1}{2}\vec{Q}$ and $\vec{l}-\tfrac{1}{2}\vec{Q}$ are the initial and final momenta of the incident proton, $k^2 = l^2 + \tfrac{1}{4}Q^2$ is the square of the magnitude of these momenta, and m and M are the proton and the deuteron masses respectively. With $H(\vec{p})$ we denote the deuteron form factor matrix, f_1 and f_2 are the proton-proton and proton-neutron scattering amplitudes, and $G(\vec{p})$ is the projectile propagator. We note that in addition to being a matrix in the spin space of the deuteron, $F(\vec{Q})$ is an operator in the two-dimensional Pauli spin space of the projectile in our nonrelativistic analysis and is an operator acting between the four-dimensional projectile Dirac spinors in the relativistic treatment.

The operator $H(\vec{p})$ acting in the deuteron spin space can be expressed in terms of the spherical H_0 and quadrupole H_2 form factors:

$$H(\vec{p}) = \int d^3r \, \Psi^\dagger(r) \, \Psi(r) \, e^{i\vec{p}\cdot\vec{r}}$$

where $\Psi(\vec{r})$ is the nonrelativistic deuteron wave function obtained using the Reid soft-core potential[17]. $H(\vec{p})$ is taken to be the same for both relativistic and nonrelativistic calculations.

Our neglect of the projectile interaction with spins of the target nucleons leaves the nonrelativistic NN amplitude in the form

$$f(q) = f_{ce}(q) + f_{so}(q) \, \vec{\sigma}\cdot\hat{n}. \qquad (1)$$

Here $\vec{\sigma}$ refers to the projectile spin and $\hat{n}$ is a unit vector normal to the scattering plane. The functions f_{ce} and f_{so} are assumed to be functions of momentum transfer only and are taken from the recent compilation of phase shifts[18] at a fixed "optimal" energy[7]. The relativistic NN amplitude is constructed by replacing the nonrelativistic amplitude with the Dirac matrix form

$$f(q) = f_S(q) + f_V(q)\gamma_0 \qquad (2)$$

where the coefficients f_S and f_V are determined by the on-shell relation in the Breit frame

$$\bar{u}(\vec{k}_0 - \tfrac{1}{2}\vec{q}, s_f) \, [f_S(q) + f_V(q)\gamma_0] \, u(\vec{k}_0 + \tfrac{1}{2}\vec{q}, s_i) = \chi(s_f)^\dagger \, [f_{ce}(q) + f_{so}(q)\vec{\sigma}\cdot\hat{n}] \, \chi(s_i) \, .$$

Here $\vec{k}_0$ is the Breit frame average momentum determined by the "optimal" energy, $\chi(s)$ is a Pauli spinor, and $u(\vec{p}, s)$ is a positive energy Dirac spinor[19].

In the nonrelativistic treatment, we employ the free wave Schrodinger propagator

$$G(\vec{p}) = \frac{2m}{k^2 - p^2 + i\varepsilon} \tag{3}$$

and for the relativistic analysis, we use the Dirac propagator

$$G(\vec{p}) = \frac{\slashed{p} + m}{k^2 - p^2 + i\varepsilon}$$

$$= \frac{2m}{k^2 - p^2 + i\varepsilon} \left[\frac{E(k) + E(p)}{2E(p)} \Lambda^+(\vec{p}) + \frac{E(k) - E(p)}{2E(p)} \Lambda^-(\vec{p}) \right] \tag{4}$$

Here $p = |\vec{p}|$ is the magnitude of the intermediate momentum, k is defined by the energy of the incident proton, $E(k) = \sqrt{k^2 + m^2}$, and $\slashed{p} = E(k)\gamma^0 - \vec{p}\cdot\vec{\gamma}$. The last line of Eq.(4) expresses the relativistic propagator in terms of positive and negative free wave projection operators, $\Lambda^+(\vec{p})$ and $\Lambda^-(\vec{p})$.[18]

In either the relativistic or nonrelativistic case, the calculated proton deuteron amplitude can be expressed in terms of the projectile spin $\vec{\sigma}$, and the deuteron tensor polarization operators Q_{xx} and Q_{yy}[7]:

$$F = F_0^0 + F_{xx}^0 Q_{xx} + F_{yy}^0 Q_{yy} + \left[F_0^y + F_{xx}^y Q_{xx} + F_{yy}^y Q_{yy} \right] \sigma_y \; .$$

Here the x-axis is parallel to the momentum transfer and the y-axis is normal to the scattering plane. The coefficients F_0^0, F_{xx}^0, ... enable us to calculate the leading terms for the proton polarization and deuteron tensor polarization observables. In Fig. 5 we display our results at 500 and 800 MeV for the differential cross section $d\sigma/dt$, the induced proton polarization $P = C(N,0;0,0)$ the proton Wolfenstein parameters $D_{SS} = C(S,0;S,0)$ and $D_{LS} = C(L,0;S,0)$, and the induced deuteron tensor polarizations $P_{xx} = C(0,XX;0,0)$ and $P_{yy} = C(0,YY;0,0)$. There are large differences between our nonrelativistic calculations (solid lines) and the relativistic calculations (dashed lines) for many of of these observables.

It should be stressed that the contributions from the single scattering term and on-shell part of the double scattering term are

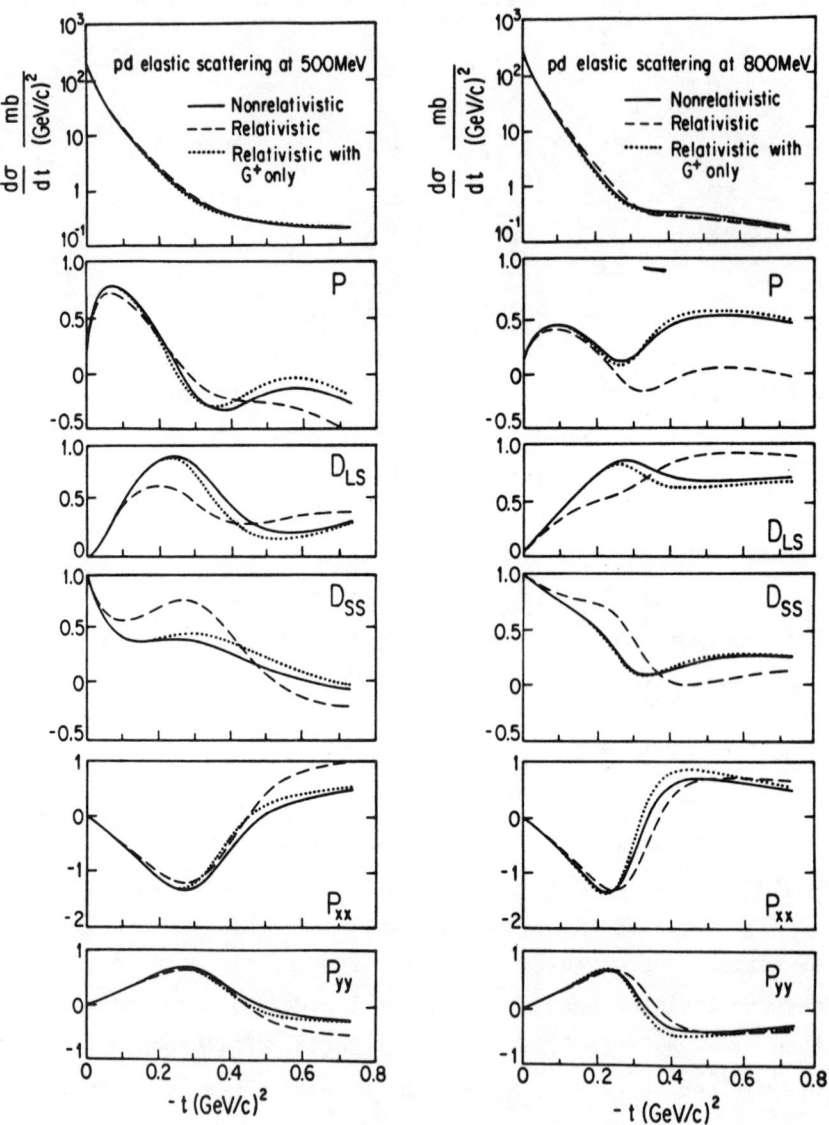

Fig. 5. Comparison between the relativistic (dashed lines) and nonrelativistic (solid lines) calculations for the spin observables in elastic pd scattering.

identical in our nonrelativistic and relativistic models because the on-shell parts of the propagators (3) and (4) as well as the on-shell matrix elements of the amplitudes (1) and (2) are equal. The only difference is in the off-shell part of the double scattering term which is large only in the relativistic treatment. The dotted curves were obtained by keeping only the positive energy part of the propagator in the relativistic calculation.

In order to illustrate the importance of negative energy states we have also done a second relativistic calculation (the dotted curves in Fig. 5) in which we have kept only the positive energy intermediate states in the double collision term, i.e. we have calculated F_d using only the positive energy part of the relativistic propagator :

$$G(\vec{p}) = \frac{2m}{k^2-p^2+i\varepsilon} \frac{E(k)+E(p)}{2E(p)} \Lambda^+(\vec{p}) .$$

The relatively small differences between the solid and dotted curves indicate that most of the difference between the nonrelativistic and relativistic treatment can be attributed to contributions from negative energy states. The remaining difference between this relativistic calculation without negative energy states and the nonrelativistic analysis follows from the different treatment of off-shell positive energy states. (It is interesting to note that the contributions from the negative energy part of the Dirac propagator can be interpreted, in the framework of nonrelativistic multiple scattering theory, as a contribution from the three-body force[20], of the analogous origin as that considered in electromagnetic interaction of electrons[21].)

We emphasize that the size of the negative energy contributions follows from our particular choice for the form of the relativistic amplitude given by Eq. (2). A different choice which still has the same on-shell behavior can give very different results.[4] The sensitivity to the contributions of negative energy states suggests that the data on pd spin observables might be combined with relativistic calculations to obtain information about the negative energy part of the NN amplitude.

Another important aspects of proton-deuteron elastic scattering worth mentioning is its relation to the studies of particle (e.g. antiproton or kaon)- neutron interaction. Measurements of spin observables in p-d elastic scattering for energies up to ~ 500 MeV where the NN amplitudes are known best and are strongly spin dependent provide a nontrivial and strong test of the method used to obtain information on particle-neutron amplitudes from particle-d scattering data. Below we discuss an example of a test which can be made in the range of energies below 500 MeV.

Recent measurement of the total p-d cross sections in pure states of spin[22] have had as an objective extraction of the total crosss sections $\Delta\sigma_L$ and $\Delta\sigma_T$ for pn scattering from the data on $\Delta\sigma_L$ and $\Delta\sigma_T$ for pd scattering. The extraction procedure is based on the formulas similar to those following from the Glauber theory, which can be written as

$$\Delta\sigma_T^{pd} = \Delta\sigma_T^{pp} + \Delta\sigma_T^{pn} - \delta_T$$

$$\Delta\sigma_L^{pd} = \Delta\sigma_L^{pp} + \Delta\sigma_L^{pn} - \delta_L$$

where δ_T and δ_L are the "defect terms" -contributions from the double scattering term which, in the framework of the nonrelativistic multiple scattering theory, constitute approximately 10-15 % corrections to the single collision terms[23]. Currently we carry out the calculations of the quantities δ_T and δ_L in the relativistic framework in which we include all five spin components of the NN amplitude. Our preliminary results indicate that these quantities are strongly affected by relativistic corrections. Consequently, the findings on the values $\Delta\sigma_T^{pn}$ and $\Delta\sigma_L^{pn}$ obtained from the data on the corresponding pd cross sections may need significant revision. In this situation it might be very interesting to measure the quantities $\Delta\sigma_T^{pd}$ and $\Delta\sigma_L^{pd}$ (as well as total cross sections with an unpolarized beam and target) for energies below 500 MeV, extract the values of the corresponding pn cross sections, and compare them with the values resulting from the phase shift analysis. Such comparisons would be very valuable because below 500 meV the ambiguities in the phase shift analyses of NN amplitudes are supposedly the smallest and at the same time the spin components of the NN amplitude related to the quantities $\Delta\sigma_T^{NN}$ and $\Delta\sigma_L^{NN}$ are quite large. The results of these comparisons would serve as a very strong and sensitive test of the method used to extract the values of $\Delta\sigma_T^{pn}$ and $\Delta\sigma_L^{pn}$ from the pd data at higher energies where the resonant structures in $\Delta\sigma_T^{pp}$ and $\Delta\sigma_L^{pp}$ were found, and where the possible presence of similar structures in $\Delta\sigma_T^{pn}$ and $\Delta\sigma_L^{pn}$ has been searched for.

As another example we consider scattering of protons from ^{4}He. In order to verify the quantitive importance of the relativistic effects in this reaction we have carried out preliminary multiple scattering calculations similar to that p-d calculation described above. In case of ^{4}He, the amplitude for the elastic scattering has been assumed to be composed of the single, double, triple, and quadruple collision terms

$$F(\vec{Q}) = F_1(\vec{Q}) + F_2(\vec{Q}) + F_3(\vec{Q}) + F_4(\vec{Q})$$

which have been calculated in analogous manner as those for p-d scattering, by using the Schrodinger and the Dirac propagators respectively. The amplitudes F_j, $j=1,4$ can be written schematically as follows:

$$F_1 = A \frac{M}{m} \langle\Psi|h_1(\vec{Q})|\Psi\rangle,$$

$$F_2 = A(A-1) \frac{M}{2m^2} \int d^3p \, \langle\Psi|h_2(\vec{p}-\vec{k}_f)G(p)h_1(\vec{k}_i-\vec{p})|\Psi\rangle,$$

$$F_3 = A(A-1)(A-2)\frac{M}{4m^3}\int d^3p_1 \, d^3p_2 \langle\Psi|h_3(\vec{p}_2-\vec{k}_f)G(p_2)h_2(\vec{p}_1-\vec{p}_2)G(p_1)h_3(\vec{k}_i-\vec{p}_1)|\Psi\rangle,$$

$$F_4 = A(A-1)(A-2)(A-3) \frac{M}{8m^3} \int d^3p_1 \, d^3p_2 \, d^3p_3$$

$$\langle\Psi|h_3(\vec{p}_3-\vec{k}_f)G(p_3)h_3(\vec{p}_2-\vec{p}_3)G(p_2)h_2(\vec{p}_1-\vec{p}_2)G(p_1)h_3(\vec{k}_i-\vec{p}_1)|\Psi\rangle.$$

In the above formulas $h_n(\vec{q}) = e^{i\vec{q}\cdot\vec{r}_n} f_n(\vec{q})$, f_n n=1,4 are the isospin averaged amplitudes for scattering off the n-th nucleon which were taken in the forms given by eqs. (4) and (5) in the respective nonrelativistic and relativistic calculations, A=4 is the target mass number and M is the ^{4}He mass. The ^{4}He wave function Ψ was parametrized as the product of the single particle s-wave functions with the harmonic oscillator parameter b=1.37 fm.

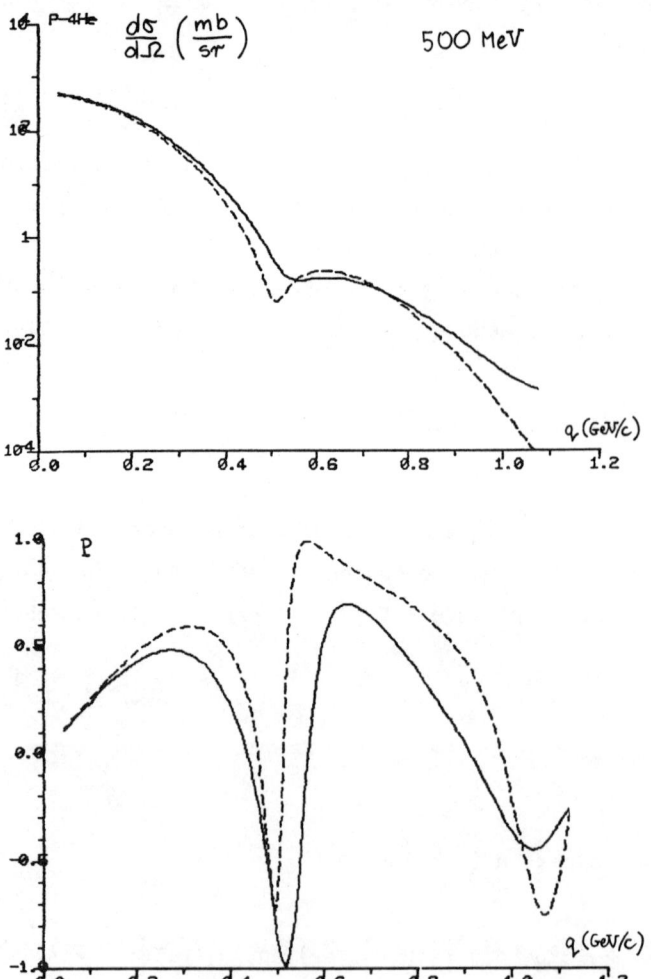

Fig. 6. Differential cross section and analyzing power for p-^{4}He elastic scattering at 500 MeV. Solid and dashed lines represent the relativistic and nonrelativistic calculations respectively.

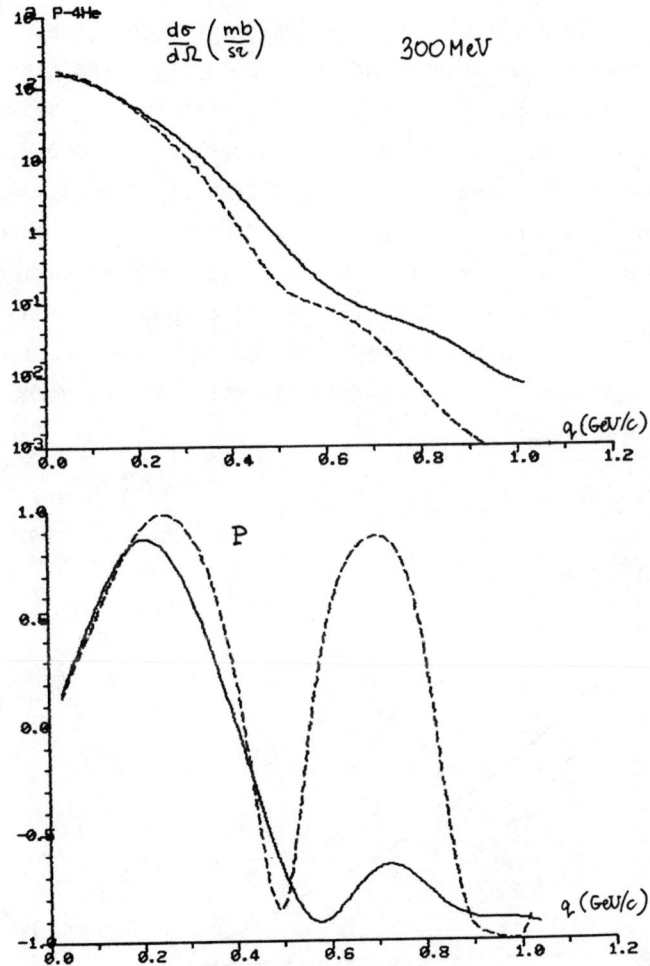

Fig. 7. Differential cross section and analyzing power for p-^{4}He elastic scattering at 300 MeV. The solid and the dashed lines represent the relativistic and the nonrelativistic calculation, respectively.

The results displayed in Figs 6 and 7 indicate that the relativistic effects play a very important role in both the cross sections and the analyzing powers. In Fig. 8 we also display the quantitiy $\alpha_{^4\text{He}}=\text{Re}[F(0)]/\text{Im}[F(0)]$ plotted as a function of the incident proton energy. It is seen that this quantity is quite sensitive to the relativistic corrections. The largest difference between the nonrelativistic and relativistic calculation occurs around 300 MeV. These comparisons suggest that measurememts of differential cross sections in elastic scattering of protons from ^{4}He in the Coulomb-nuclear interference region in the range of energies 100-500 MeV would serve as an interesting test of the presence of the relativistic effects in nuclear reactions at medium energies. We emphasize that by experimental determination of $\alpha_{^4\text{He}}$ one could verify the importance of the relativistic effects without making polarization measurements.

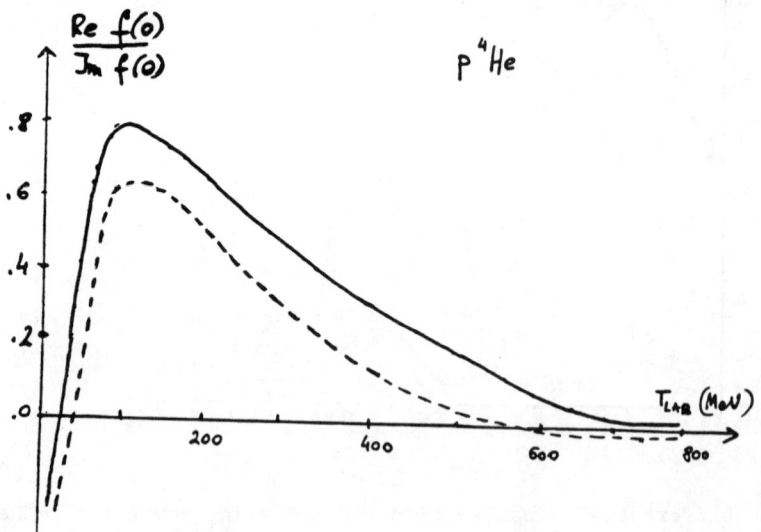

Fig. 8. $\alpha = \text{Re}[F(0)]/\text{Im}[F(0)]$ for p-^{4}He scattering plotted as a function of the incident proton energy. The solid and the dashed line represent the relativistic and the nonrelativistic calculation, respectively.

To summarize our discussion, proton reactions with few nucleon systems present a potentially rich source of information on many aspects of nuclear reactions.

We have considered here three interesting examples of possible experiments, very well suited for the incident proton energy range of 100-500 MeV:

a) A Complete measurement of spin observables in p-d elastic scattering which could be used to determine transition amplitudes depending exclusively on double collision term.

b) Measurements of total cross sections in pure states of spin in p-d elastic scattering,

c) Measurement of the energy dependence of the ratio of the real to the imaginary part of the scattering amplitude for the elastic scattering of protons from ^{4}He.

Such experiments should provide very valuable information on one of the currently most interesting aspects of nuclear reactions at medium energies - the importance of relativistic effects, and, might be used to obtain constraints on ambiguities present in relativistic models connected with the choice of the relativistic representation of the NN amplitude.

ACKNOWLEDGEMENT

The theoretical calculations discussed in this paper have been done in collaboration with D. Adams and E. Bleszynski.

REFERENCES

1. A. Azizi, E. Bleszynski and M. Bleszynski to be published.

2. J. Shepard, J. A. McNeil, and S. J. Wallace, Phys. Rev. Lett. $\underline{50}$ (1983) 1443

3. B. C. Clark, S. Hama, R. L. Mercer, L. Ray, and B. D. Serot, Phys. Rev. Lett. $\underline{50}$ (1983) 1644

4. D. L. Adams and M. Bleszynski Phys. Lett. $\underline{136B}$ (1984) 10

5. H. Fearing, G. Goldstein and M. Moravcsik, Phys. Rev $\underline{29D}$ (1984) 2612

6. G. W. Bennet et al.,Phys. Rev. Lett. $\underline{19}$ (1967) 387

7. G. Alberi, M. Bleszynski, and T. Jaroszewicz, Ann. Phys. $\underline{142}$ (1982) 299

8. M. Moravcsik, H. M. Sinky, and G. Goldstein, preprint 1984

9. M. Bleszynski et al., Phys. Lett. $\underline{87B}$ (1979) 198

10. M. Bleszynski Phys. Lett., $\underline{106B}$ (1981) 42

11. T. H. Sun et al., Phys. Rev C, in press, A. Rahbar et al., to be published, G. Weston et al., to be published

12. R. Winkelman et al.,Phys. Rev $\underline{21C}$ (1980) 2535

13. G. Igo, Nucl. Pys. $\underline{A374}$ (1982) 253

14. G. Igo and M. Bleszynski, Saturne proposal nr 38 (1983)

15. This property of the amplitudes F^0_{yy}, F^y_{yy}, F^x_{xy}, and F^z_{yz} is common to both nonrelativistic and the relativistic calculations.

16. D. L. Adams and M. Bleszynski Phys. Lett. $\underline{B}$, in press

17. R. V. Reid, Ann. Phys. $\underline{50}$ (1978) 411.

18. R. A. Arndt and L. D. Roper, Phys. Rev D $\underline{25}$ (1982) 2011

19. J. D. Bjorken and S. Drell. Relativistic Quantum Mechanics (McGraw-Hill, New York, 1964)

20. J. Friar, "Three-Body Forces", in AIP Conference Proceedings $\underline{97}$, (1983) 378

21. H. Primakoff and T. Holstein, Phys. Rev. $\underline{55}$,(1939) 1218

22. I. P. Auer et al,Phys. Rev. Lett. $\underline{46}$ (1981) 1177

23. G. Alberi, M. Bleszynski, and T. Jaroszewicz,and S. Santos, Phys. Rev. $\underline{D20}$ (1979),2437

STORING AND COOLING OF POLARIZED IONS

E. Steffens
Max-Planck Institut für Kernphysik, D-6900 Heidelberg,
Federal Republic of Germany,
and visiting scientist at PS Division, CERN, CH-1211 Geneva,
Switzerland

ABSTRACT

Storing and synchrotron acceleration of medium energy ions for nuclear physics is reviewed. Apart from the standard vertical spin orientation, other orientations at the target location become possible by using a Siberian snake. In particular, longitudinal vector polarization and second rank tensor polarization with arbitrary orientation of the symmetry axis can be produced. It is shown that electron cooling does not cause depolarization of the circulating ions.

INTRODUCTION

Due to the strong spin dependence found in medium energy proton-proton and proton-nucleus collisions, there is a growing interest for polarized beams in this energy domain. Storing and electron cooling of polarized ions have been proposed so far for the cooling ring projects at IUCF[1], Uppsala (Celsius)[2] and Jülich (COSY)[3]. This paper reviews briefly the problem to accelerate polarized ions and the solutions found in high energy machines. On this basis, the special requirements for nuclear physics experiments in a cooler ring are discussed.

As for unpolarized circulating ions, the goal for polarized stored particles like protons and deuterons will be to perform high resolution studies with thin internal targets. In addition, polarized circulating ions may open up the possibilities to produce intense beams of polarized secondary particles like neutrons with excellent energy resolution. In all cases, permanent cooling is necessary.

What are the requirements imposed by the experiment? We assume that a polarized beam is injected into the ring and that the polarization is high and all the necessary polarization states can be produced by the source. For the various observables, different spin orientations at the target are necessary:

(i) <u>Vertical spin</u>. This is the standard orientation in a storage ring. It is used for measurement of iT_{11} (vector analyzing power) and C_{nn} (spin correlation coefficient with transverse spin).

(ii) <u>Longitudinal spin</u>. This orientation is required for the measurement of T_{20} (tensor analyzing power for $s > 1$, e.g. deuterons) and C_{ll} (spin correlation coefficient with longitudinal spin).

(iii) Other orientations. An orientation of 45° to the beam in the scattering plane is needed to measure T_{21} (tensor analyzing power for $s \geq 1$).

In the following, we will try to answer the question to which extent we can hope that the above requirement can be met.

SPIN RESONANCES IN CIRCULAR MACHINES

We restrict ourselves to heavy particles like protons, where no self-polarizing and self-cooling mechanisms, as in the case of electrons, are present at the energies under discussion. A number of excellent papers on the acceleration of polarized ions in synchrotrons exist[4,5,6,7,8] and will be referred to in the following.

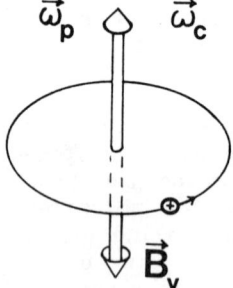

Fig. 1: Orientation of the precession frequency $\vec{\omega}_p$ for a positive ion with $G > 0$.

Let us start to consider the effect of the vertical guiding field on the spin motion. We write (see Figure 1):

$$\frac{d\vec{S}}{dt} = \frac{e}{\gamma mc}[1 + \gamma G]\vec{S} \times \vec{B} = \vec{\omega}_p \times \vec{S}$$

$$\vec{\omega}_p = -\frac{e}{\gamma mc}[1 + \gamma G]\vec{B} = \omega_c [1 + \gamma G]\hat{y} \qquad (1)$$

with $\quad \omega_c = \frac{eB_v}{\gamma mc}$

Here $\vec{S}$ is the spin vector in the rest frame, $\vec{B}$ is the magnetic field strength in the laboratory frame and $G = (g-2)/2$ the gyromagnetic anomaly (see Table 1). For a vertical orientation of $\vec{S}$ the spin is stationary; otherwise, the horizontal component precesses with angular frequency $\vec{\omega}_p$, which can be expressed as a multiple of the revolution frequency ω_c. For example, protons of 108 MeV ($\gamma G = 2$) perform 3 spin rotations per turn, protons of 631 MeV ($\gamma G = 3$) 4 rotations, etc.

Ion	Gyromagnetic anomaly G	Number of spin rev. per turn ($\omega_p - \omega_c$)	Lowest imperfection res.	
			n	T/GeV
p	1.793	few (≥ 2)	2	0.108
			3	0.631
d	-0.143	fewer	-1	11.25
paramagnetic ion $\mu \approx \mu_{Bohr}$ $m \approx m_p$	$\approx 10^3$	$\geq 10^3$ ($\rightarrow$ depol.)	-	-
e, μ	1.16×10^{-3}	≈ 0	1	0.440

Table 1: Some parameters relevant for storage of polarized ions

In a real machine, there are radial and longitudinal components of the magnetic field from the focussing and stray fields, and from magnet imperfections. We refer the generalized precession frequency ($\vec{\omega}_p$) relative to the frame rotating with ω_c [6]:

$$(\vec{\omega})_{rot} = -\frac{eB_v}{\gamma mc} \left[(1+G)\frac{B_{||}}{B_v} + \gamma G \frac{\vec{B}_r}{B_v} + \gamma G \hat{y} \right] \quad (2)$$

The main sources of these fast-varying fields are field errors leading to vertical closed orbit distortions ("wave number" n=integer) or vertical betatron oscillations ("wave number" of the corresponding field distortion: $kP \pm Q_v$; P = machine periodicity, Q_v = vertical betatron tune).

Under most conditions, the spin precession caused by horizontal field components averages to zero and the vertical spin direction would still be quasi-stationary. Only if the wave number γG of the vertical precession comes close to one of the horizontal wave numbers, these small horizontal precession angles add up coherently and a sizeable spin precession arises [4]. The conditions for these so-called depolarizing resonances are the following (for more general conditions, see ref.7):

(i) imperfection resonances: $\gamma G = n$ (3a)
(ii) intrinsic resonances : $\gamma G = kP \pm Q_v$ (3b)

Imperfection resonances are fixed in energy, whereas the position of intrinsic resonances depends on the betatron tune of the machine. The location of the lowest imperfection resonances for some typical ion species are given in Table 1.

STORAGE AND ACCELERATION

If a polarized beam with vertical spin is injected at a certain energy γ_0 sufficiently distant from a resonance and stored, one would not expect any depolarization.

To my knowledge, only one experimental study has been done so far on this problem. At the ZGS, using an extended 21 s flat-top at 3.25 GeV/c, an upper limit for the depolarization rate of 0.025% per second was found[9]. This value corresponds to a 1/e polarization life time of more than one hour, which is much more than the cycling times envisaged for proton storage rings [3,10]. Therefore, even if one admits the influence of other neighbouring resonances, the depolarization of the stored beam due to spin resonances should be negligible, at least under the condition of the ZGS experiment. But depolarization rates in a proton storage ring with strong cooling by a cooler and simultaneous strong heating by an internal target have not been studied yet. It was pointed out in ref. 11 that depolarization away from resonances requires the <u>simultaneous</u> presence of scattering and damping processes. This led to the conjecture[12] that polarized proton beams simultaneously heated and cooled would be subject to depolarization, by a process similar to that of quantum fluctuations and radiation damping in electron storage rings. This would suggest a polarization lifetime of the order of the beam lifetime <u>without</u> cooling, but requires a more specific calculation for this situation.

If the particle energy in the cooler synchrotron is varied, the left hand side of eq.(3) changes and the resonance condition may be fulfilled. In their pioneering paper[4], Froissart and Stora obtained the following formula for the polarization P_f after having crossed an isolated resonance:

$$P_f/P_i = 2 \exp[-\pi \varepsilon^2/2\alpha] - 1 \qquad (4)$$

with

P_i = initial polarization
ε = strength of the resonance
α = parameter which describes the speed of the resonance crossing (e.g. $\alpha = G\gamma/\omega_c$ for an intrinsic resonance).

The relevant parameter is $\varepsilon^2/2\alpha$. The two limiting cases are:
(i) $\varepsilon^2/2\alpha \ll 1$. This means a weak resonance and/or fast crossing and we obtain: $P_f \approx P_i$
(ii) $\varepsilon^2/2\alpha \gg 1$. Here we have a strong resonance and/or slow crossing speed, which results in $P_f \approx -P_i$, that means a complete spin flip will take place.

As mentioned before, imperfection resonances (3a) are caused by radial field components B_r which in turn are related to the vertical closed orbit distortions y. Under certain statistical assumptions, the strength of an imperfection resonance $\gamma G = n$ can be expressed in terms of the rms value of y [6]:

$$|\varepsilon| \simeq n\, y_{rms}/\pi <\beta> \qquad (5)$$

where $<\beta>$ is the average β function. In Figure 2, the ratio P_f/P_i for crossing of a resonance with $n = 2$ is plotted as function of the "crossing speed" $d\gamma/dt$ with y_{rms} as parameter. $<\beta> = 10$ m has been assumed.

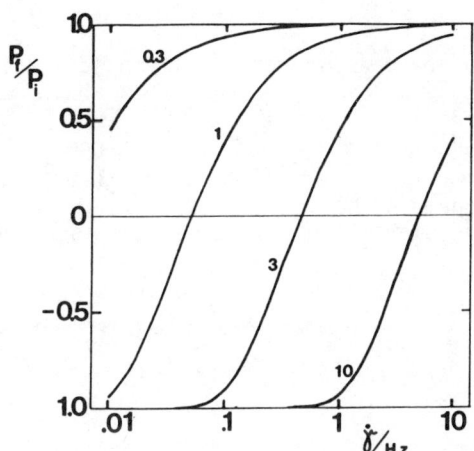

Fig. 2 : The relative polarization $P_f P_i$ after crossing an imperfection resonance with $\gamma G = 2$ as function of the "crossing speed" $\dot\gamma$. The vertical closed orbit distortion y_{rms} (in mm) was taken as a parameter ($<\beta> = 10$ m, $\omega_c = 10^7$ Hz).

The two limiting cases of eq.(4) are reproduced, that is $P_f \simeq P_i$ for small y_{rms} and fast crossing, and $P_f \simeq -P_i$ for large y_{rms} and slow crossing. Only in these cases the magnitude of P is conserved as required; the intermediate region has to be avoided.

A better account for the resonance strength of a given lattice can be obtained by using the computer program "DEPOL" written by E.D. Courant[13]. In addition, the simple picture of resonance crossing is strongly modified by the presence of synchrotron oscillations, which cause multiple resonance crossing[4,9].

For a small storage ring with synchrotron acceleration, where the number of resonances is small, it is not necessary to device elaborate techniques like pulsed quadrupoles or Siberian snakes, which are used or proposed for the big accelerators[14,8]. As for Saturne II, it seems possible to cross the weak resonances quickly without change and the strong ones slowly to perform a proper spin flip[15]. Therefore, a flexible system of ramping the magnets of different velocity is required. Using the correction dipoles, the strength of an imperfection resonance can be modified by correcting or amplifying closed orbit distortions in order to optimize the conditions for resonance crossing[15].

Finally, we want to emphasize that for deuterons, there are no depolarizing resonances in the relevant energy range (see Table 1), as the gyromagnetic anomaly is about 12 times smaller than for protons.

STORAGE WITH NON-VERTICAL SPIN DIRECTION

Up to now, we have only considered vertical spin orientation. As mentioned earlier, another spin direction important for the experiment is the longitudinal one.

The most simple idea to produce longitudinal polarization[15] is sketched in Figure 3. The beam with "spin up" is bent vertically by

$$\alpha_b = 90°/\gamma G \qquad (6)$$

and becomes longitudinally polarized. Further downstream, the beam is bent by $-2\alpha_b$ and α_b, which restores the original momentum and spin direction. Unfortunately, the necessary bending angle is large for low and medium energy protons and energy-dependent ($\alpha_b = 50.2°$ for $\gamma = 1$; $\alpha_b = 32.8°$ for 500 MeV). Therefore, this scheme cannot be used as a universal method, in particular if heavy equipment like a spectrometer is needed for the experiment.

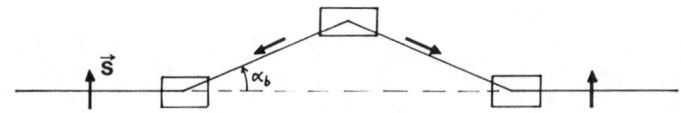

Fig. 3: Simple scheme to obtain longitudinal polarization[15].

Let us know consider a situation where the spin is longitudinal in one straight section of the ring. As we have seen in section 2, this corresponds to a spin resonance. In general, spin resonances can be cured by a Siberian snake[7,15,16], which rotates the spin around the beam axis by 180°. Consider a ring with two parallel straights and a snake in one of them as shown in Figure 4 for three different spin orientations in the straight opposite to the snake[16].

Only for the longitudinal orientation, the spin vector is stable (lower figure). Therefore, such an arrangement can be used to perform experiments with longitudinal vector polarization[15]. The beam has to be injected with longitudinal spin into the straight section or at any other position on the circumference with the local (energy-dependent) spin direction.

For spin-1/2-particles like protons, the vertical and longitudinal spin orientations are sufficient. For particles with spin > 1/2 like deuterons (s = 1), other orientations are required to measure in particular observables related to the (2nd rank) tensor polarization t_{20} of the beam. The quantity t_{20} is invariant if we invert the spin axis. Going back to Fig. 4, we see that for the other two cases, the spin axis is inverted after each turn, which does not change tensor polarization.

We conclude that, by using a Siberian snake, particles with stable longitudinal vector polarization at the target and with arbitrary (2nd rank) tensor polarization can be stored.

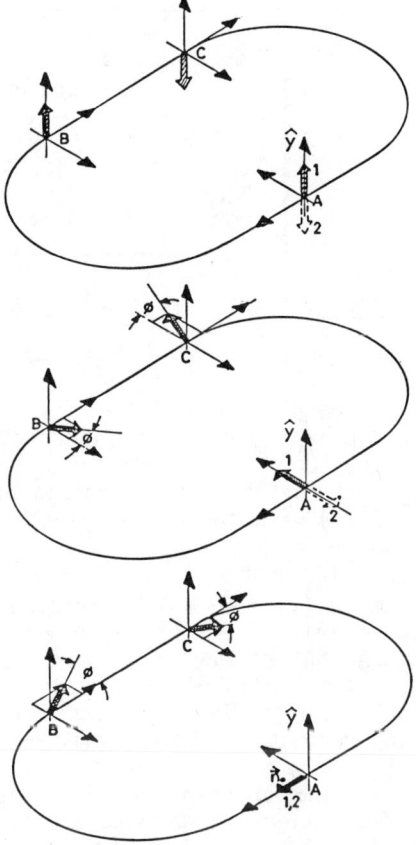

Fig. 4: Principle of the Siberian snake (taken from ref. 16), showing the evolution of spin vector components in one revolution. Section A is field-free, section BC contain the 180° spin rotator.

At low and medium energies, 180° spin rotation can be produced by a solenoid. The spin precession angle is given by

$$\theta_{//} = \theta_o \frac{B_{//}}{T m} \frac{L}{\beta \gamma} \frac{1}{} \qquad (7)$$

where $B_{//}$ is the axial field, L the length and θ_o(prot.) = 51.2°, θ_o(deut.) = 7.9°. In Table 2, the B L values for $\theta_{//}$ = 180° are given. Superconducting solenoids are required. Their perturbation of the orbital motion has to be compensated for by additional elements. Such an insertion has been designed for the e^+e^- storage ring VEPP-4 at Novosibirsk[17]. It consists of two solenoids with a total maximum field integral of 21 Tm, one normal and four skew quadrupoles.

T/MeV	$B_\| L$ / Tesla m	
	protons	deuterons
30	0.89	4.11
100	1.67	7.51
300	3.03	13.40
1000	6.37	26.42

<u>Table 2</u>: Integrated solenoid field strength $B_\| L$ for $\Theta_\| = 180°$

ELECTRON COOLING

Two effects might be caused by the electron cooler:
a) the spin motion is distorted by the longitudinal field $B_\|$ of the cooler solenoid;
b) part of the polarization is lost by hyperfine interaction (hfi) with the cooler electrons.

a)   If we take $B_\| L = 0.2$ Tm for the cooler solenoid, we get from eq.(7) for 100 MeV protons $\theta = 21.6°$ per passage for transverse spin direction. This would make a stable spin motion impossible, but of course it can easily be compensated for by an additional solenoid with opposite field direction, as proposed for the IUCF cooler[1]. Small compensation errors contribute to the imperfection resonances, which have to be avoided or suppressed anyhow.

b)   The question whether there is depolarization due to hfi between electron and nuclear spin is related to the very interesting problem whether one can polarize circulating ions by using polarized electrons[18].
    Let us consider an interesting example, where by charge exchange from an optically-pumped Na-vapor target a proton picks up a polarized electron[19]. In Figure 5, the calculated proton polarization as function of the external magnetic field is shown. At weak field for full coupling, the polarization approaches 50%. In reality, it is somewhat lower because of electron capture into excited levels.

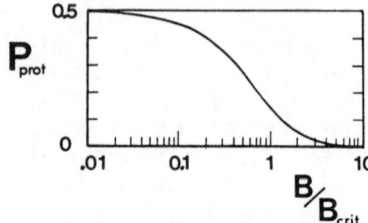

Fig. 5: Proton polarization as function of external field B (in units of B_{crit}= 507 Gs) after hfi of polarized electrons with unpolarized protons (electron ground state assumed).

The important difference to our case is that the system p (or $\bar{p}$) plus cooling electron is not bound. In fact, in the proton-electron system, with a very small probability, such a "radiative capture" takes place and these neutral hydrogen atoms are lost immediately. But for circulating ions, the interaction takes place only during the very short collision times in the order of 10^{-16} s, which is small compared to the hyperfine period of the H atom (7×10^{-10} s). Thus, a coupling with a subsequent polarization transfer cannot take place.

A possible argument could be that although the interaction is weak, due to the high repetition rate in a storage ring, there might be some net effect. Let us estimate in a very crude model the spin-flip probability of a proton or antiproton by a transverse field $B_\perp = 100$ T, which is in the order of the hydrogen 1 s field and which is applied for $t_c = 10^{-16}$ s during each "close collision". For the spin-flip probability, we get:

$$W_{close\ coll.} = \sin^2(B_\perp t_c \mu/h) \qquad (8)$$

which yields $W_{cc} \approx 10^{-13}$ per close collision. If we consider a particle in an electron gas of temperature 10 eV and density $n_e = 3 \cdot 10^8 /cm^3$, we obtain for the rate of close collisions:

$$\dot{N}_{cc} \approx 3 j_x \sigma_{cc} \qquad (9)$$

with
$$j_x = n_e v_x$$

$$\sigma_{cc} = \pi a_0^2 \qquad (a_0 = \text{Bohr radius})$$

We get $\dot{N} \approx 3/s$ and for the spin-flip probability per unit time ($\eta = L_{cooler}/C \approx 3\%$):

$$dW/dt = \eta \dot{N} W \approx 10^{-14}/s \qquad (10)$$

From this extremely low number, we conclude that there is no depolarization of the circulating polarized ions by hfi. On the other hand, it can be stated that polarizing circulating ions using polarized electrons is not feasible. Only in the case of circulating paramagnetic ions (that is ions carrying an unpaired electron), one might take advantage[20] of the large spin-exchange cross-sections. But it seems doubtful whether the polarization of paramagnetic ions (see Table 1) can be conserved in a storage ring for a sufficiently long time.

ACKNOWLEDGEMENTS

I wish to thank B.W. Montague for pointing out the possibility of polarization lifetime limitations for simultaneous cooling and heating of the beam, and D. Möhl for encouragement and guidance in the preparation of this talk.

REFERENCES

1. The IUCF cooler-tripler proposal, IUCF, December 1980.
2. A. Johansson, CELSIUS-Note 83-1, 1983.
3. COSY-Studie, Jül-Spez-242, February 1984.
4. M. Froissart and R. Stora, Nucl. Instr. Meth. 7 297 (1960).
5. J. Khoe et al., Part. Accel. 6, 213 (1975).
6. L.C. Teng, in: High Energy Physics with Polarized beams and targets (Argonne 1978); G.H. Thomas (Ed.), AIP Conference Proceedings 51, p. 248 (1979).
7. R.D. Ruth, in: High Energy Spin Physics (Brookhaven 1982); G.M. bruce (Ed.), AIP Conference Proceedings 95, p. 378 (1983).
8. B.W. Montague, Physics Report 113, 1 (1984).
9. Y. Cho et al., IEEE NS24 1509 (1977).
10. R.E. Pollock, IEEE NS30, 2056 (1983).
11. Ya.S. Derbenev and A.M. Kondratenko, Sov. Phys. JETP 35, p. 230 (1972).
12. B.W. Montague, in: High Energy Polarized Proton Beams (Ann Arbor 1977), A.J. Krisch and A.J. Salthouse (Ed.), p. 46, and private communication (1984).
13. E.D. Courant, in: High Energy Polarized Proton Beams (Ann Arbor 1977), A.D. Krisch and A.J. Salthouse (Ed.), AIP Conference Proceedings 42, p. 94.
14. L.G. Ratner, IEEE NS30 (1983) 2690.
15. Ya.S. Derbenev et al., Part. Accel. 8 (1978) 115.
16. B.W. Montague, in: High Energy Polarized Beams and Targets (Argonne 1978), G.H. Thomas (Ed.), AIP Conf. Proc. 51, p. 129 (1979).
17. Ya.S. Derbenev et al., Proc. of the Int. Acc. Conf., Fermilab p. 410 (1983).
18. e.g. remark by L. Dick, in: High Energy Physics with Polarized Beams and Targets, Proc. (Lausanne 1980), Birkhäuser Verlag, p. 109 (Basel 1981).
19. Y. Mori et al., IEEE NS30 (1983) 2740.
20. E.W. Otten, Workshop on the Physics with Heavy Ion Cooler Rings, Heidelberg (May 1984).

FREE AND STORED ATOMIC BEAMS AS INTERNAL POLARIZED TARGETS

W. Haeberli
University of Wisconsin, Madison, Wisconsin 53706

ABSTRACT

The use of a gas target of spin-polarized hydrogen or deuterium atoms as an internal polarized target for a storage ring is discussed. It is shown that a gas target of polarized atomic hydrogen or deuterium has significant advantages over a conventional polarized target. Production of polarized atomic beams is reviewed and the limitations on target density are discussed. It is concluded that, based on present technology, an atomic-beam device can be built which offers a target thickness of polarized atoms of about $10^{12} cm^{-2}$ to the circulating beam in routine operation. Work is in progress to increase the available density by producing an atomic beam of reduced average velocity and reduced velocity spread. It is shown that the target density can be increased by injecting the atomic beam into a vessel which has openings for entry and exit of the circulating beam. A volume density of nearly $10^{13} cm^{-3}$ may be expected, corresponding to a target thickness of $10^{14} cm^{-2}$ if one allows 10 cm target length. The resulting luminosity of 10^{30} $s^{-1} cm^{-2}$ is adequate for many experiments since the relatively open geometry allows detectors covering a large solid angle.

1. INTRODUCTION

Excellent targets of polarized protons have been developed for use in nuclear reaction experiments at intermediate and high energies, where target thicknesses of the order g/cm^2 are appropriate. The much smaller target thickness required for use as an internal polarized target in a storage ring suggests the use of a gaseous target of spin-polarized hydrogen atoms. Here, we will in particular consider targets based on the production of polarized atoms by the atomic-beam method, i.e. separation of spin states by deflection in inhomogeneous magnetic fields. Such a target has fundamental advantages over the solid targets now in use: (1) A polarized atomic beam provides a chemically and isotopically pure target of highly polarized protons or deuterons. (2) The atomic beam can be prepared in such a way that the target requires only a weak (<1Gauss) magnetic guide field. This eliminates the problems of beam deflection and spin precession associated with the strong magnetic fields used in conventional targets. It also eliminates the restrictions on detector placement and choice of target spin orientation imposed by the large magnets required by conventional targets. (3) The target polarization can be reversed or reoriented within a fraction of a second, as opposed to hours; (4) for deuterium targets, large deuteron alignment (tensor polarization) can be obtained, as well as large vector polarization. Experiments requiring large deuteron alignment have not been possible because of target limitations. (5) gas targets are insensitive to radiation damage.

In spite of these advantages, the use of polarized gas targets for nuclear reaction experiments has been precluded so far by the very small available target thickness. However, gas targets should be considered for use in a storage ring, because the ring requires thin targets, and the small available target thickness is compensated partly by the large circulating beam intensity.

We will specifically consider two types of polarized internal targets: a polarized atomic beam of hydrogen or deuterium atoms which intercepts the circulating beam (Sect. 2); and a storage bottle containing a gas of atomic spin-polarized hydrogen or deuterium as a possible means to increase the available target thickness (Sect. 3). The storage bottle, of course, requires openings for unobstructed entry and exit of the circulating beam, and a third opening for continuous injection of a polarized atomic beam into the bottle. The density will be the result of a dynamic equilibirum of atoms flowing in and out of the bottle. In the simplest terms, a storage bottle has the advantage that each polarized atom entering the bottle passes through the circulating beam a number of times before it is pumped away, in contrast to the free atomic beam where each atom passes through the circulating beam but once. The use of a storage bottle was first proposed and tested by the group at Wisconsin[1], while a first test with a free atomic beam as a polarized target was performed at Stanford.[2]

Below, we inspect the fundamental and the practical aspects of obtaining useful target densities. Our intention here is to concentrate the discussion on goals which can be reached with good probability within a development and construction period of some three years. Thus we neglect other schemes which are promising but untested, such as spin exchange between optically pumped alkali atoms and hydrogen atoms. It is certainly possible that a new method may rapidly replace the methods considered below, once feasibility has been demonstrated.

The present discussion is limited to internal targets of hydrogen and deuterium, since these targets hold particular interest for nuclear physics. There are, however, other polarized targets which could more readily be produced with sufficient target thickness. These include beam or vapor targets of Li and Na which can be polarized by optical pumping, as well as targets of ^{3}He.

2. AN ATOMIC BEAM AS A POLARIZED TARGET

2.1 Principles.

The atomic beam method produces beams of polarized atoms by deflection in an inhomogeneous magnetic field (Stern-Gerlach). This method has been developed over the last 25 years for the production of polarized protons and deuterons by ionization of the atomic beam. Much of this work has been summarized in review papers and conference proceedings concerning polarized-ion sources.[3-5]

Fig 1. shows in a schematic way the deflection of hydrogen atoms in the field of a six-pole magnet. The magnetic field B is zero on axis and increases in proportion to r^2 until it reaches a value B_m at the pole tip radius r_m. The Breit-Rabi diagram (Fig. 2a) shows that the energy of atoms with electron spin projection $m_j = +1/2$ (states 1 and 2) increases roughly in proportion to B (provided B is larger than the "critical field" B_c = 507 Gauss), while the energy of $m_j = -1/2$ atoms (states 3 and 4) decreases. Thus atoms in states 1 and 2 feel an axially-symmetric attractive harmonic-oscillator potential inside the magnet, while the atoms in the other two states are driven outward and removed by pumps.

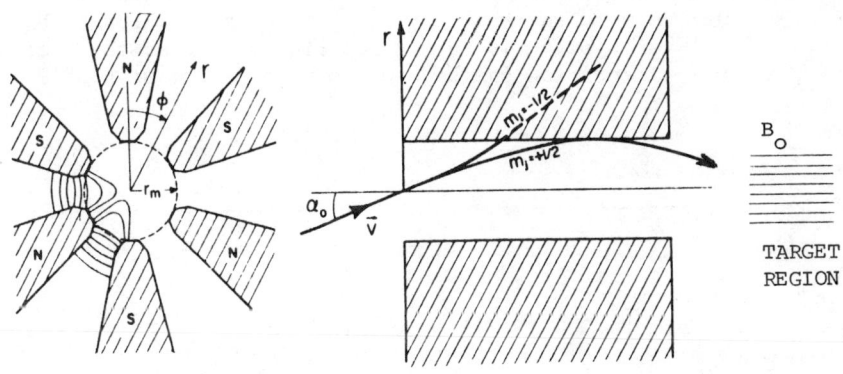

Fig. 1. Schematic of six-pole magnet and spin-separation of atomic beam.

After the atoms exit from the spin-separation magnet, they pass into a region of homogeneous magnetic field. The velocity of a thermal atom is slow enough that the spin direction follows the complicated magnetic field direction between the sixpole magnet and the dipole field adiabatically, so that the atoms are now polarized in electron spin with respect to the uniform field B_o.

Eventually, we are interested in the nuclear polarization of the atoms in the external field B_o. In a strong external field ($B_o \gg B_c$), nuclear polarization P = $\underline{+1}$ is obtained by exposing the atomic beam to RF transitions between hyperfine states. These transitions adiabatically exchange populations between pairs of states. A so-called week-field transition (1 ↔ 3) results in equal population of states 2 and 3, producing proton polarization P = -1, while a strong field transition (2 ↔ 4) yields equal population of states 1 and 4, and thus P = +1. Thus reversal of polarization is accomplished by switching RF transitions on and off, without changing the magnetic field direction. Transitions of this type are commonly employed in polarized-ion sources. Transition probabilities ≥90% are readily obtained.

For application to the cooler-ring, we will assume that the magnetic field B_o in the target region will be made very weak ($B_o \ll B_c$) so that the circulating beam is not perturbed no matter what spatial orientation is chosen for the target polarization. Fig. 2b shows that in a weak field, P = +1 for state 1 (pure spin state), while for state 2 (mixed state) the time-averaged polarization is P = 0 (proton and electron spin precess about one another). Thus the atomic beam from the spin separation magnet will only yield P = 1/2 in a weak field. Larger weak-field polarization (P = ±1) can be obtained only at the sacrifice of half the atoms in the beam, since one needs to reject the atoms in the mixed state. In practice, this is accomplished by following the spin separation magnet with a 2↔4 transition unit and a second spin separation magnet, so that only state 1 passes through both magnets. Such a system is shown in Fig. 3. After the second magnet, a 1↔3 transition can be used to reverse the sign of P.

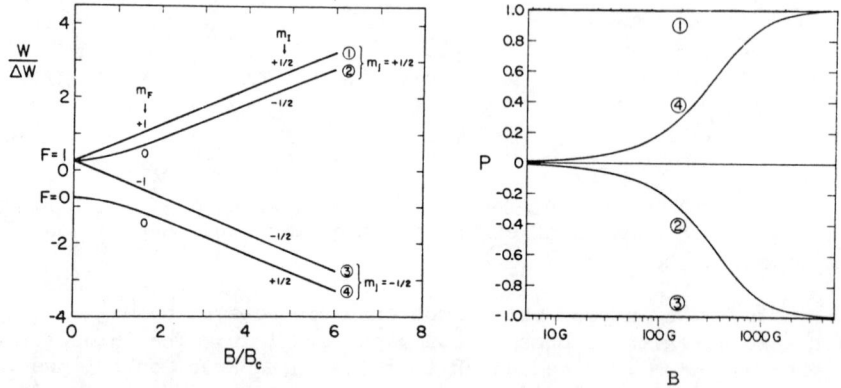

Fig. 2. a) Energy level diagram of the hydrogen atom in a magnetic field. Subscripts I and j refer to nuclear and electron spin, respectively. b) Proton polarization of the four hyperfine states in an external magnetic field B. The figures are from Ref. 3.

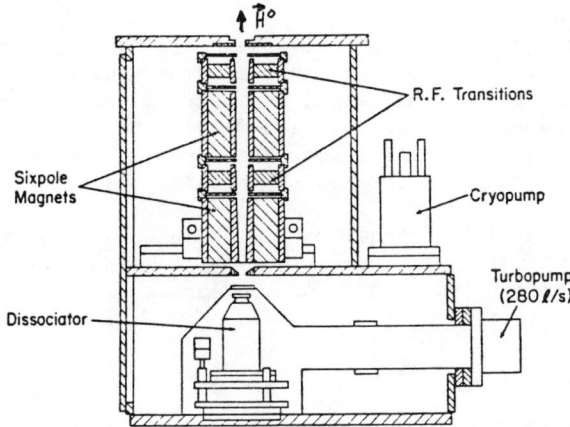

Fig. 3. Atomic-beam source used at Bonn (Ref. 7). The position of the dissociator is adjustable. Two cryopumps of 9000 l/s for H_2 are used to evacuate the second atomic-beam stage and the magnet chamber. The sixpole magnets are 16 cm long.

2.2 Factors Determining Target Density

The atomic beam is produced by hydrogen atoms emerging from an opening (e.g. 2mm diameter) in a vessel in which H_2 gas is dissociated by an RF discharge. The differential intensity of atoms in the forward direction reaches a maximum at some pressure (3 Torr) in the discharge tube. At this point, gain from further increase in pressure is offset by losses from gas scattering in the exit nozzle and the region immediately outside the nozzle, and by increasing volume recombination. The observed maximum forward intensity is of the order $(dN/d\Omega) = 10^{19}$ atoms/sec sr.

The second important factor is the maximum angle α_o accepted by the spin-separation magnet (Fig. 1), and the corresponding solid angle $\Delta\Omega$. The angle α_o is determined by the elementary energy consideration of the radial motion, i.e. by the condition that the kinetic energy at r=o associated with the radial velocity equals the magnetic potential energy at the pole tip radius. Since α_o depends on the velocity v of the atom entering the magnet, we are interested in the weighted mean acceptance solid angle $\langle\Delta\Omega\rangle$ averaged over the velocity distribution. For a six-pole magnet[3]

$$\langle\Delta\Omega\rangle = 2.1\ \mu_B B_m/kT \quad (2.1)$$

The atomic beam intensity N can now be expressed as

$$N = (dN/d\Omega)\langle\Delta\Omega\rangle, \quad (2.2)$$

where μ_B is the Bohr magneton. For $B_m = 0.8$ T and T = 400 K, one obtains $\alpha_o \approx 2°$ and $\langle\Delta\Omega\rangle \approx 4$ msr for atoms entering the six-pole magnet at r=o. In practice, however, the source of atoms (nozzle of the dissociator) is several cm from the magnet entrance to provide space for pumping, so that most atoms enter the magnet at a radius r several mm off axis. This reduces the effective solid angle, roughly in proportion to $(r_m^2 - r^2)$. The obvious solution is to increase the pole-tip radius r_m, but this has the undesirable feature that the beam diameter in the target region increases. The loss in solid angle is not a severe problem for atomic beams at room temperature where α_o is small, but the problem becomes increasingly difficult as α_o is increased by cooling the beam to low temperatures.

The density n of atoms in the target region depends not only on the atomic beam intensity N, but also on the atomic beam diameter in the target region, $2r_T$, and on the reciprocal velocity of the atoms in the beam, $1/v$:

$$n = \langle N/v \rangle / \pi r_T^2, \qquad (2.3)$$

where for simplicity we assumed that the target density is uniform over the diameter $2r_T$. If the circulating beam intercepts the atomic beam at an angle θ, the target thickness t becomes

$$t = 2\langle N/v \rangle / \pi r_T \sin\theta. \qquad (2.4)$$

Because of the small diameter of the cooled beam, the assumption was made that the circulating beam is small compared to the atomic beam ($r_T \simeq 4$mm). It should be noted that the gain in target thickness with decreasing angle θ is limited by the fact that for hardware reasons the target area must be further from the exit of the atomic-beam source as θ is decreased, so that part of the gain is offset by the fact that eventually r_T increases because atoms of different velocities exit at different angles from the spin-separation magnet. In analogy to optics, one says that the beam transport through the magnet has large "chromatic aberrations." Reduction of chromatic aberrations obviously is important because proper beam transport reduces r_T and correspondingly increases t. Aberrations are reduced somewhat by tapering the pole pieces of the six-pole magnet such that the magnet aperture is smallest near the dissociator. Further improvements are obtained by using a second six-pole magnet a small distance beyond the first one ("compressor magnet") as proposed in Ref. 6. In some recent sources, additional flexibility in the magnetic field configuration is obtained by subdividing the spin-separation magnet into two to four short, individually adjustable sections.

It is now generally recognized that for optimum performance, the magnet system needs to be designed to match the velocity distribution from the dissociator, which is not necessarily Maxwellian. The fact that magnet design and beam velocity distribution are interdependent is a considerable impediment in experimenting with beams cooled to different temperatures, since there is no way to rapidly change the length and the diameter of the magnets.

A potentially important way to increase target thickness is to reduce chromatic aberrations by narrowing the velocity spread of atoms from the dissociator below that of a Maxwell distribution. Narrowed velocity distributions are observed when the flow of gas out of the nozzle becomes supersonic.

2.3 Observed Beam Density and Target Thickness

We first characterize the performance of conventional atomic-beam sources of the type in use at many laboratories for the production of polarized ion beams. Typically, the atomic beam originates from a water-cooled dissociator using RF power of a few hundred Watts. The average velocity of the atomic beam is $v \simeq 3 \times 10^3$ m/sec, corresponding to a beam temperature $T \simeq 400$K. Typically, a tapered six-pole magnet of some 30 cm length is used, and sometimes a compressor magnet is added. Typically, the measured atomic-beam intensity is $N \simeq 2 \times 10^{16}$ atoms/sec over a 0.5 cm^2 area some 30 cm from the end of the spin-separation magnet. Thus the target density is $n \simeq 1.3 \times 10^{11}$ cm^{-3}, and the target thickness for a circulating beam at right angle to the atomic beam

$$t = 1 \times 10^{11} \text{cm}^{-2} \qquad (2.5)$$

with a target length $\simeq 1$ cm. Improved target thickness may be obtained by using an oblique angle between the atomic beam and the circulating beam. Taking into account that the atomic-beam diameter will increase (see Sect. 2.2), a factor five increase might be achieved by optimum choice of θ, giving

$$t = 5 \times 10^{11} \text{cm}^{-2} \qquad (2.6)$$

with a target length of 5 to 10cm. The above numbers should be reduced by a factor two if one intends to produce a pure spin state.

Improvements over the preformance figures quoted above have been demonstrated by a number of development projects over the last few years. The main thrust has been to lower the average velocity of the atomic beam, by cooling the beam as it emerges from the dissociator. Lower beam temperature has two beneficial effects: increased acceptance angle $\langle \Delta \Omega \rangle$ of the six-pole magnet, and increased dwell time of the atoms in the target volume. As an example we summarize the performance of the source developed at Bonn[7], which incorporates a number of improvements (Fig. 3). The source is entirely pumped by oilfree pumps (two cryopumps, one turbo molecular pump). The copper nozzle of the dissociator, is cooled to 77 K, but the beam temperature is measured to be 120 K, indicating the often observed problem of heat transfer between nozzle surface and atomic beam. The magnet system, consisting of two 16 cm long tapered six-pole magnets, was especially designed for optimum beam transport into a small (3 mm diameter) region some

35 cm from the exit of the six-pole magnet. Contrary to most other sources, provision is made for RF transitions between the two magnets, thus permitting large target polarization in a weak magnetic field, at least for protons (deuterium would require two transition units).

The reported density on the axis of the beam is $n = 6 \times 10^{11} \text{cm}^{-3}$. The integrated target thickness is $t = 4 \times 10^{11} \text{cm}^{-2}$ for $\theta = 90°$ and an estimated

$$t = 2 \times 10^{12} \text{cm}^{-2} \qquad (2.7)$$

for an oblique angle.

Even though the above numbers are based on measurements on operating sources, it must be realized that absolute measurements of atomic beam intensities are rather difficult, and that test measurements are usually reported for a well-groomed source and not necessarily for routine operation. In addition the loss incurred in producing a pure spin state is not included. Also certain compromises may be required to adapt a source to operation on the ring. On the other hand, some gains may result from improved design. A plausible value for target thickness that can be expected for a source based on demonstrated methods for routine 24 hour a day operation is

$$t = 1 \times 10^{12} \text{cm}^{-2}. \qquad (2.8)$$

2.4 Cold Atomic Beams

Here we will consider possible improvements in target thickness from cooling the atomic beam to lower temperatures. Under the simplest assumptions, cooling of the atomic beam as it emerges from the dissociator promises an increased target thickness proportional to $T^{-3/2}$, namely a factor T^{-1} from increased $\Delta\Omega$ (eq. 2.1) and another factor $T^{-1/2}$ from the increased dwell time of atoms in the target region (1/v-factor, eq. 2.4). Thus cooling the beam e.g. from 400 K to 20 K corresponds to a gain by a factor 90 over the conventional room temperature source. Target thicknesses as high as $t = 10^{13} \text{cm}^{-2}$ to 10^{14}cm^{-2} have been predicted on the basis of the above arguments. However, such projections are simplistic, because they neglect a number of fundamental problems:

(1) it is not correct to assume that the forward flux density $(dN/d\Omega)$ from the dissociator remains constant. In first approximation, the flux density decreases as $T^{1/2}$, since gas scattering sets a limit on gas density, and the effusion rate for a given gas density in the nozzle is proportional to v.

(2) problems of heat transfer and recombination in the cold nozzle have to be solved.

(3) the assumption that $\langle \Delta\Omega \rangle$ increases like T^{-1} is not justified if the atoms enter the magnet at $r \neq 0$, as will be increasingly the case as the acceptance angle α_o increases.

(4) it is not clear that a sufficiently achromatic magnet system can be designed to accept the wider range of entrance angles α. Generally, the larger α_0, the larger is on the average the divergence of the beam leaving the magnet, causing increased beam diameter $2r_T$, and correspondingly decreased target thickness (eq. 2.4).

(5) Work with cooled nozzles has shown velocity distributions which are considerably narrower than expected from a Maxwell distribution. Further improvement of this effect may offer a solution to some of the problems mentioned above.

Considerable effort is currently expended on the development of cooled beams. Effective cooling to 40 K has been accomplished by the group at ETH, but no breakthrough in the observed beam density has been demonstrated so far.[8]

An elegant solution to the problem of atomic-beam transport at low temperatures has been suggested by Kleppner.[9] The proposal is illustrated in Fig. 4, which shows atoms emerging from a nozzle at 0.5 K, located just inside a superconducting selenoid. If the field at the nozzle is B, atoms in states 1 and 2 will gain a kinetic energy of $\sim\mu_B B$ in falling from the solenoid field B into a weak field region outside the solenoid. The acceleration has two beneficial effects: forward flux density $(dN/d\Omega)$ of atoms increases because of the added forward velocity component; in addition, the beam becomes quite monochromatic, i.e. the relative spread in velocity $(\Delta v/v)$ is reduced by a factor $1+ (\mu_B/kT) \approx 10$ for $B = 8$ T, $T = 0.5$ K. The small relative velocity spread should make it possible to transport the beam effectively from the nozzle to the target region. Simple estimates, which take into account the $1/v$ reduction in throughput at the dissociator, predict an improvement in target density by a factor ~ 100. I know of no current plans to construct such a source. While the scheme is very promising, even a feasibility test requires substantial cryogenic equipment and manpower.

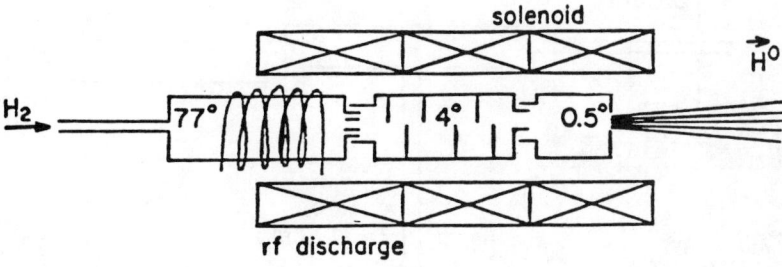

Fig. 4. Schematic of proposed method to produce a cold (~6 K) nearly monoenergetic atomic beam by acceleration of a 0.5 K beam in the fringe field of a solenoid.

3. STORAGE CELLS

3.1 Principles

The principle can be described as dynamic storage of polarized atoms in a bottle, since there is a continuous flow of atoms into and out of the container. Storage bottles for nuclear reaction experiments were first proposed[1] and recently discussed[10] by the group at Wisconsin. While for a free atomic beam high target polarization is assured, this is not the case for a storage bottle where there is opportunity for depolarization in wall collisions. Another intrinsic disadvantage compared to a free atomic beam is that the vessel has a potential of interfering with the circulating beam, and that the reaction products have to penetrate the wall of the vessel. Thus a storage cell is of interest only because of the limited target thickness available from a free atomic beam.

We assume that N atoms per second are injected into a storage cell (Fig. 5) and that these atoms flow out of the cell through the three openings. The density of atoms in the cell, n, will build up to the point where the net flow out per unit time equals N. The flow out, in turn, is determined by the pumping conductance C of the opening. The number of atoms per unit volume in the cell, can thus be expressed as

$$n = N/C. \qquad (3.1)$$

If the openings are tubes of diameter d_i and length ℓ_i, the conductance in cm^3/sec is (Ref. 11)

$$C(cm^3/sec) = 3.81 \times 10^3 (T/M)^{1/2} \sum_i d_i^3 (\ell_i + 1.33\, d_i) \qquad (3.2)$$

where the sum extends over all openings. Here, T is the absolute temperature of the gas, and M the molecular weight. Eq. (3.2) requires corrections if the diameter of the vessel is not large compared to the diameter of the tubes, but these can be neglected for present purposes.

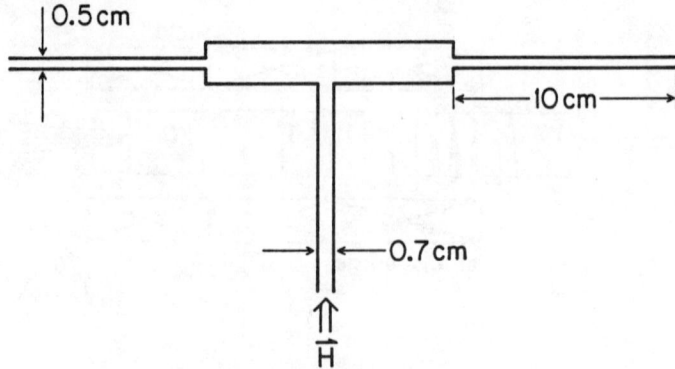

Fig. 5. Schematic of storage cell.

Other quantities of interest are the average dwell time τ of atoms in the cell, and the average number ν of wall collisions the atoms suffer before escape. Obviously, if V is the volume of the cell, and C the pumping conductance, τ will be of the order

$$\tau = V/C \qquad (3.3)$$

If the average speed of the atom is v, the number of wall collisions during time τ is

$$\nu = v \cdot \tau / \lambda \qquad (3.4)$$

where λ characterizes the distance the atom travels between successive wall collisions. We will use average values for each of the quantities on the right-hand side of Eq (3.4), but it should be kept in mind that this is only an approximation. If we use $v = \bar{v} = [8kT/\pi m]^{1/2}$, we obtain

$$v = 1.45 \times 10^4 \ (T/M)^{1/2} \ \text{cm/sec}, \qquad (3.5)$$

and

$$\nu = 3.8 \ V / \sum_i d_i^3 / (\ell_i + 1.33 \ d_i). \qquad (3.6)$$

It is useful to note that ν is a geometric quantity independent of the mass or temperature of the stored atoms.

3.2 Storage Cell at Room Temperature

The simplest arrangement is a storage cell at room temperature, fed by an atomic-beam source. In order to be definite, we use the parameters of the atomic beam shown in Fig. 3. The reported intensity of hydrogen atoms in states 1 and 2 is $N = 3 \times 10^{16}$ atoms/sec within a 0.7 cm diameter.

We assume a cell of the configuration shown in Fig. 5, with a 0.7 cm opening of 10 cm length for entry of the atomic beam and 0.5 cm openings of 10 cm length for the circulating beam. Should the circulating beam initially (e.g. before cooling) require a larger aperture, the cell could be split length-wise so that it can be opened (clam-shell cell). For the above dimensions, the pumping conductance for hydrogen atoms is

$$C = 3.5 \times 10^3 \ \text{cm}^3/\text{sec} \qquad (3.7)$$

which leads to

$$n = 0.9 \times 10^{13} \ \text{cm}^{-3}. \qquad (3.8)$$

For comparison, the maximum density of atoms in the free atomic beam is 6×10^{11} cm^{-3}, or a factor 15 less. To convert n to a target thickness requires an assumption about the acceptable maximum length of the effective target. If we assume a target of 10 cm length, the target thickness for the cell becomes roughly $t = 10^{14} cm^2$. This is an improvement by a factor of about 50 compared to the same beam without storage cell even if an oblique angle θ is assumed between atomic beam and circulating beam.

A cell of this type has a volume $V \simeq 50\ cm^3$, thus

$$\tau = V/C \simeq 15\ msec. \qquad (3.9)$$

This dwell time is short enough that it is possible to reverse the polarization of the sample several times a second.

The most important and also the most difficult questions to consider are recombination and depolarization of atoms in wall collisions. From Eq. (3.6) we find for the average number of wall collisions

$$\nu \simeq 7 \times 10^3 \qquad (3.10)$$

where we assumed $\lambda = 1$ cm. In connection with hydrogen masers, Kleppner et al.[12] have developed wall coatings which permit some 10^4 collisions before depolarization or recombination. Thus there is hope that for the cell considered here wall problems can be handled. However, it must be made clear that the estimated target thickness for a room temperature storage cell of

$$t = 10^{14}\ cm^{-2} \qquad (3.11)$$

involves some open questions and thus this estimate is less conservative than for the case of the free beam.

The above discussion does not mean to suggest that $n = 10^{13} cm^{-3}$ represents a limit any more than do the values for a free atomic beam. Improvements in atomic beam sources will benefit both types of target. There is one important difference, however. An atomic-beam of low velocity is advantageous for a free atomic-beam target, since the density increases as $1/v$. However, for use with a storage cell we are primarily interested in the number of atoms, irrespective of their velocity, and in the diameter and divergence of the atomic beam, since the latter two quantities determine the size of the opening and thus the pumping conductance.

Use of storage cells for nuclear reaction experiments may require thin windows for exit of charged reaction products from the cell. The use of thin Teflon windows suggests itself, since wall coatings of the same material are known to be advantageous. Such windows were used successfully in a first experiment with a dynamically stored target of hydrogen.[1]

3.3 Cold Storage Cells

Since the gas conductance is proportional to $T^{1/2}$, the target density is expected to increase like $T^{-1/2}$ if the temperature of the cell is lowered. Use of a cold cell does not necessarily require that the injected beam be cold, since the atoms will be cooled to cell temperature in the first few collisions, and the heat load from the injected hot atoms is small.

The design of storage cells at various low temperatures requires suitable wall properties to avoid depolarization and recombination. While one might assume that the sticking probability and thus the recombination rate of a given surface increase with decreasing temperature, this is not in fact the case. Measurements[13] on Al surfaces coated with Al_2O_3 show a deep minimum in the recombination rate between 100 K and 140 K, where the measured recombination probability is $<3 \times 10^{-5}$. At ultra-low temperatures (<0.3K) wall coatings of superfluid helium offer a unique solution.

We first consider cells at temperatures easily reached with liquid He. On the basis of conductance alone one predicts for 5 K a gain in density by a factor eight, or $\sim 7 \times 10^{13}$ atoms/cm^3 for the geometry of Fig. 5. However, one needs to consider the effect of atomic hydrogen interaction with a frozen H_2 wall and also the fact that the vapor pressure of H_2 presents a background of unpolarized atoms in the cell.

We first compare the vapor pressure of H_2 (Fig. 6) to the pressure of atomic hydrogen gas of density 10^{14}cm^{-3}. It is seen from the figure that at 4.5 K the equilibrium density of H_2 gas will be roughly 10% of the atomic hydrogen density. The fraction of H_2 background atoms rises sharply with temperature so that already at 4.8 K more than 50% of the atoms in the target arise from H_2. Thus, for the atomic beam intensity assumed here, H_2 background sets an upper limit to the temperature of about 4.5 K.

Consideration of the interaction of atomic hydrogen with frozen H_2 surface places further restrictions on the wall temperature of a storage vessel. The average sticking time t_a of an H atom on frozen H_2 has been studied by Crampton et al.[14] They find a temperature dependence of $e^{39.8/T}$ where 39.8 K is the measured binding energy of H on H_2. Their result for t_a at 4.2 K is 40 nsec. Extrapolation to 1K gives $t_a \simeq 1$ week. Long absorbtion times result in recombination and/or depolarization of the sample gas, so that from this point of view wall temperatures well above 4.2 K are desirable. It is not clear whether a temperature exists where depolarization on the wall and contamination from background gas are both tolerable.

The problem may be substantially reduced by the use of frozen neon walls. As can be seen in Fig. 6, the vapor pressure for Ne at 10 K is similar to that of H_2 at 5 K. The adsorbtion of H on solid Ne has been studied recently by Anderson et al.[15] over the temperature range 6 to 11 K. They observed a binding energy of 34 K for H on Ne, somewhat lower than for H on H_2. On the assumption that depolarization from wall collisions is proportional to t_a, a significant improvement over H_2 walls can be expected by holding the storage vessel at, say, 7.8 K. At this temperature the density of neon atoms in the cell is $4 \times 10^{11} cm^{-3}$.

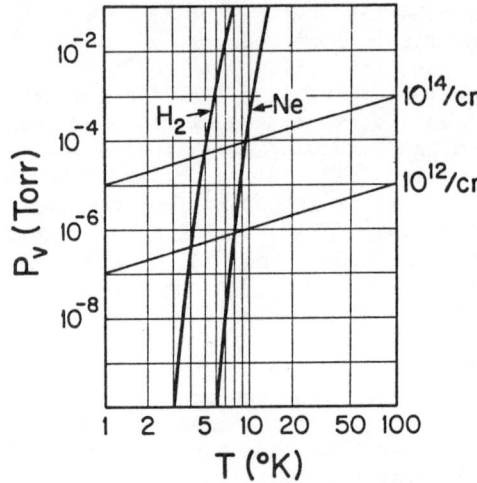

Fig. 6. Vapor pressure of H_2 and Ne vs. temperature (from Ref. 11). For comparison, the pressure corresponding to atomic hydrogen gas densities of $10^{14} cm^{-3}$ and $10^{12} cm^{-3}$ is also shown.

An atomic maser using frozen neon walls has been successfully operated, and storage times as long as 0.1 sec have been observed.[16] Following the results of Ref. 17, the previously reported difficulties[15] in forming a uniform neon coating have been solved by condensing the neon on the walls at temperatures above the triple point. This implies that the coatings would have to be reestablished in a vacuum separate from the circulating high energy beam. It is thus not yet clear whether walls coated with either solid H_2 or solid Ne offer a viable low temperature storage cell.

It would seem more interesting to study in some detail the technical feasibility of a very low-temperature storage cell (e.g. 0.3 K) with a wall coating of superfluid He. A target density of some $3 \times 10^{14} cm^{-3}$ and a target thickness of $10^{15} cm^{-2}$ might be realistic. Much higher densities of polarized hydrogen atoms have been produced in cells at low temperatures, but under conditions which are incompatible with the present requirements of a week magnetic field and access to the cell by an external beam. In fact the high densities were achieved with relatively modest influx of polarized atoms, but very long holding times, so that the conditions are not at all comparable to the wide open cell required for an internal polarized target. Design and tests of a low temperature cell would require a serious commitment, since technical problems and costs are not trivial.

4. CONCLUSIONS

We conclude that accessible technology will provide an atomic beam of hydrogen or deuterium which offers a target thickness of about 10^{12} atoms/cm^2 to the circulating beam of a storage ring. The target polarization for protons is expected to be $P_y \simeq 0.9$. Only a very weak magnetic guide field will be needed to define the spatial orientation of the polarization direction. The production of large and reversible polarization in a weak external field requires two separate spin-separation magnets, with RF transitions between and after the two magnets. The polarization is readily reversed in less than a msec by switching on and off RF oscillators which induce transitions between hyperfine states. For deuterons, large vector polarization and large alignment (P_y, $P_{yy} \simeq 0.9$) can be achieved by using serveral RF transition units.

Work now in progress attempts to increase the target thickness provided by an atomic beam by an order of magnitude through cooling the atomic beam to temperatures of 20-40 K. Larger gains may result through use of a very cold beam (0.5 K) which is accelerated in the fringe field of a superconducting solenoid to increase the brightness of the atomic beam and to decrease the relative velocity spread of the beam and thus the chromatic aberrations of the beam transport system.

We showed that a substantial gain in target thickness results from dynamic storage of the polarized atoms in a vessel through which the circulating beam passes. The number density of polarized atoms in a cell with 10 cm long access tubes of 5 mm diameter for entry and exit of the circulating beam is about 10^{13}cm^{-3}. If the experiment can tolerate a gas target of 10 cm length, the target thickness is about 10^{14}cm^{-2}. It has been proposed to increase the density further by cooling the storage cell, but it is not clear whether problems associated with wall depolarization and vapor pressure of wall coatings can be solved. The dwell time of atoms in a room-temperature cell is of the order 15msec, so that the target polariztion can readily be reversed several times a second. Should it turn out that the target polarization does not survive the roughly 10^4 wall collisions, the number of collisions can be reduced at the expense of target density.

With start of construction of the Indiana cooler ring, experimentalists need to answer the question whether the target densities available from present technology is sufficient to proceed with construction of a polarized gas target, or whether further development of atomic-beam devices or of other methods must precede target construction. On the surface, the densities mentioned above seen marginal. If we assume that the circulating beam provides 10^{16} incident proton/sec upon a target of density 10^{14}cm^{-2}, the luminosity becomes

$$L = 10^{30} \text{sec}^{-1}\text{cm}^{-2}, \qquad (4.1)$$

or roughly 10^8 events/day·mb. While this luminosity is low compared to experiments using conventional solid polarized targets, it would provide an adequate data rate for the observation of cross sections of the order 1 mb/sr. In this connection it should be kept in mind that the cooler ring with a polarized gas target will provide relatively free access, so that large-area detectors covering a solid angle of the order 1 sr can be considered. It would thus appear that the advantages of a polarized gas target over a solid target, which were outlined in the introduction, might well justify the design and construction of a target already at this time.

ACKNOWLEDGMENTS

This work was supported in part by the U.S. Department of Energy. I should like to thank Mr. Tom Wise for discussions of the cold storage cell.

REFERENCES

1. M.D. Barker, G. Caskey, C.A. Gossett, W. Haeberli, D.G. Mavis, P.A. Quin, S. Riedhauser, J. Sowinski and J. Ulbricht, Fifth Int. Symp. on Polarization Phenomena in Nucl. Phys. 1980, (G.G. Ohlsen et al., Eds.), AIP Conf. Proc. No. 69, American Institute of Physics (New York) 1981, p. 931.
2. D.G. Mavis, J.S. Dunham, J.W. Hugg and H.F. Glavish, High Energy Physics With Polarized Beams (M.L. Marshak, Ed.), AIP Conf. Proc. 35, American Institute of Physics (New York) 1976, p. 517.
3. W. Haeberli, Ann. Rev. Nucl. Sci. $\underline{17}$, 373 (1967).
4. Polarized Proton Ion Sources (A.D. Krisch and A.T.M. Lin, Eds.) AIP Conf. Proc. No. 80, American Institute of Physics (New York) 1982; Polarized Proton Ion Sources (G. Roy, Ed.) AIP Conf. Proc. No. 117, American Institute of Physics, (New York) 1984.
5. Polarization Phenomena in Nuclear Physics, Sante Fe 1980, (G.G. Ohlsen et al., Eds.), AIP Conf. Proc. No. 69, American Institute of Physics (New York) 1981.
6. H.F. Glavish, Polarization Phenomena in Nuclear Reactions (W. Grüebler and V. König, Eds.), Birkhäuser (Basel) 1976, p. 844.
7. H.G. Mathews, Ph.D. thesis, Friedrich-Wilhelms-Universität, Bonn, 1979. Magnet Configuration "T2 $_2$."
8. W. Grüebler, Workshop on Polarized Targets in Storage Rings, May 1984, Argonne National Laboratory ANL-84-50, p. 223.
9. D. Kleppner, AIP Conf. Proc. No. 80, American Institute of Physics (New York) 1982, p. 138.
10. T. Wise and W. Haeberli, Workshop on Polarized Targets in Storage Rings, May 1984, Argonne National Laboratory ANL-84-50, p. 249.
11. A. Roth, Vacuum Technology (North Holland, Amsterdam, 1976).

12. D. Kleppner, H.C. Berg, S.B. Crampton and N.F. Ramsey, Phys. Rev. 138, A981 (1965).
13. M.P. Maley, T.J. Bowles, J.C. Browne, T.H. Burritt, J.A. Helffrich, D.A. Knapp, R.G.H. Robertson, M.L. Stelts, and J.F. Wilkerson, Bull. Amer. Phys. Soc. 29, 665 (1984).
14. S.B. Crampton, J.J. Krupczak and S.P. Souza, Phys. Rev. B 25, 4383 (1982).
15. K.E. Anderson, S.B. Crampton, K.M Jones, G. Nunes, Jr. and S.P. Souza, Low Temperature Physics Conference, Karlsruhe, Germany, Aug. 1984.
16. K.M. Jones, Williams College, Williamstown, Massachusetts, private communication.
17. J. Krim, J.G. Dash and J. Suzanne, Phys. Rev. Lett. 52, 640 (1982).

PROSPECTS FOR A DEUTERIUM INTERNAL TARGET, TENSOR POLARIZED BY OPTICAL PUMPING - SPIN EXCHANGE

M. C. Green
Argonne National Laboratory, Argonne, IL 60439

ABSTRACT

The prospects for a tensor polarized deuterium target ($\sim 10^{15}$ atoms/cm^2) appropriate for nuclear physics studies in medium and high energy particle storage rings are discussed. Using the technique of electron spin exchange with an optically pumped sodium (or potassium) vapor, we hope to polarize deuterium at a rate $\sim 10^{17}$ atoms/sec. Predictions for the deuterium polarization for a particular target cell design will be presented leading to the identification of the required optical pumping power and cell wall depolarization probability to attain optimum performance. The technical obstacles to be surmounted in such a target design will also be discussed.

INTRODUCTION

With the aim to carry out measurements of electron-deuteron elastic scattering at high momentum transfer with a tensor polarized deuterium target, we are currently attempting to develop a polarized deuterium gas (internal) target appropriate for use in an electron storage ring.[1-3] We hope to achieve a target thickness in the range of 10^{15} atoms/cm^3 with the deuterium tensor polarization, $t_{20} = 0.3$. Such an internal target should also be suitable for studies with the cooled hadronic beams planned for the IUCF cooling ring.

The deuterium gas density necessary for our target thickness goal is most easily attained by a containment cell placed in the circulating particle beam. Of course, such a cell must have entrance and exit apertures for the beam, through which the gas in the cell leaks out. We estimate that the rate at which gas leaks from such a cell with reasonably sized apertures is $\sim 10^{17}$ atoms/sec. In order to maintain the gas density in the cell, the source of polarized atoms must supply atoms at this rate. "Standard" polarized sources generally produce $\sim 10^{16}$ polarized atoms/sec.[4] Thus, we were lead to consider other techniques for polarizing atoms at the required rate.

As will be discussed in more detail in the following sections, the technique of optical pumping - spin exchange shows some promise of polarizing $\sim 10^{17}$ atoms/sec. While this technique has been used by atomic physicists to polarize a number of different atomic species,[5] the necessary polarization rates and a number of other requirements particular to our application have yet to be demonstrated.

PHYSICAL COMPONENTS AND PROCESSES

In this section the basic components of a system to produce a polarized deuterium target by optical pumping-spin exchange with a Na vapor are discussed; and the atomic and nuclear processes necessary for the production of nuclear polarization are reviewed.

The deuterium polarization process is best idealized as a somewhat ordered sequence of physical interactions at the atomic and nuclear level schematically represented in Fig. 1. The optically pumped Na vapor (electron polarized) transfers angular momentum to the deuterium atom ensemble (previously rf-dissociated) via atomic electron spin exchange. In the absence of a strong magnetic field, the hyperfine interaction transfers a portion of the deuterium electron polarization to the deuterium nuclei. Meanwhile, the Na vapor polarization is replenished and a second spin exchange collision re-polarizes the deuterium electron with hyperfine interactions polarizing the nucleus. After at least two spin exchange collisions, the deuterium nucleus ensemble has some degree of tensor (and vector) polarization. Repeated spin exchange/hyperfine interactions increase the tensor polarization of the ensemble to some limit determined by the degree of Na polarization. For the systems we will consider, spin-exchange collisions occur about once every 10^{-4} sec as compared to the hyperfine mixing time of 10^{-9} sec. Thus our assumption of complete hyperfine mixing between electron-spin exchanges is well justified.

Figure 2 shows the basic components of a spin-exchange polarization system with a dissociator bottle feeding atomic deuterium into a spin exchange cell used for development tests. A liquid Na pot is heated to maintain the desired Na density in the slightly hotter spin-exchange cell (to prevent Na condensation on the cell walls). A set of Helmholtz coils (not shown) provide a weak magnetic "guide" field (5 gauss) aligned along the polarization axis. Polarized deuterium (and some Na) flow out of the end spout. The spout and cell dimensions (discussed later) are chosen so as to contain the deuterium and Na for a time sufficient to allow for the desired number of spin-exchange collisions. This general design is based on the presumed limitation that the deuterium dissociation must take place <u>outside</u> of the polarizing region. We have also assumed that the deuterium emitted through the spout cannot be recycled back into the spin exchange cell because it would most likely be recombined into D_2 nor can it be dissociated again because of Na contamination.

The polarization cell in Fig. 2 can be configured into an internal target cell with entrance and exit apertures for a circulating particle beam (instead of an end spout) with little change in performance characteristics. In the current design, the Na density is ~0.3% of the deuterium density which could be a source of background for some nuclear reaction studies. The reaction products of a nuclear interaction between the beam and target must necessarily pass through the polarization cell walls in such a design. Cell walls can be made thin; however, heat loss from the cell walls must be replenished, and the heat conductance of walls must be taken into consideration.

This technique of tensor polarizing deuterium is initially quite attractive because of its mechanical simplicity (no sextupole magnets or high-speed vacuum pumps) relative to conventional atomic beam sources. However, there are some forseeable technical obstacles in the development of such a target including high-power optical pumping

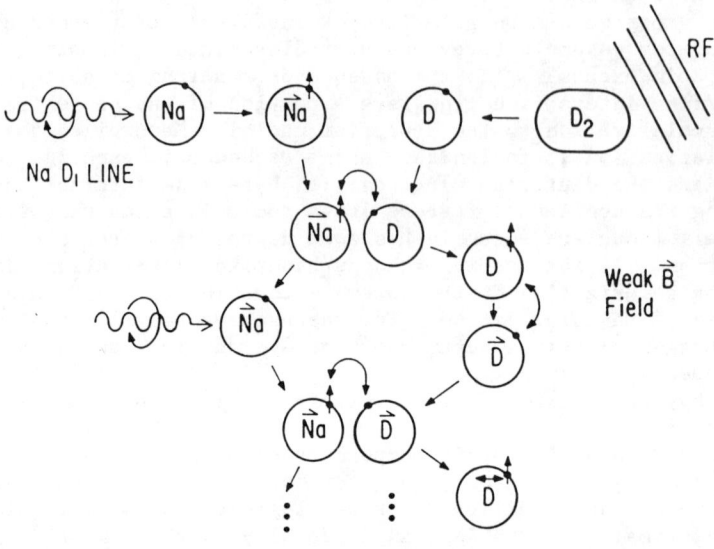

Figure 1. The deuterium tensor polarization process. Optically pumped Na and RF dissociated D_2 undergo repeated electron spin exchange/hyperfine interactions.

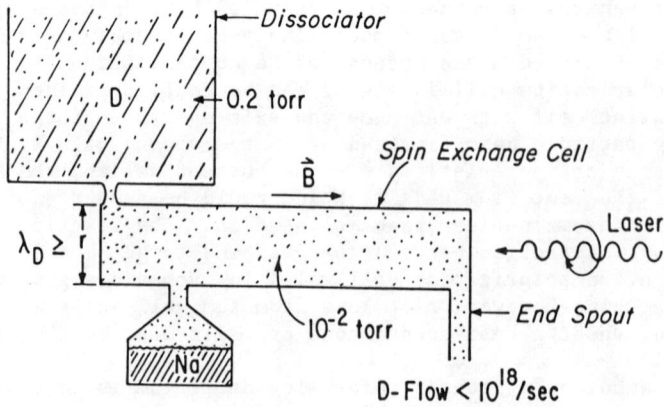

Figure 2. The basic components of a spin exchange polarization cell to produce tensor polarized deuterium.

of a relatively dense alkali vapor, high-efficiency dissociation, and cell wall coatings which sufficiently minimize both D recombination and depolarization.

The Na vapor in the spin exchange cell is spin polarized by a "standard" optical pumping technique. The cell is illuminated by circularly polarized Na D_1 light (5896 Å) (from a dye laser) which excites the $S_{1/2}$-$P_{1/2}$ electron transition in the Na atoms. In effect, the circularly polarized light pumps angular momentum into the Na atom vapor by repeated excitation and spontaneous emission which eventually spin polarizes the outer unpaired electron of the Na atoms as well as their nuclei via the hyperfine interaction. Although optical pumping of Na and other alkali vapors has been studied for many years,[5-8] the efficient pumping of a relatively thick vapor with a high angular momentum transfer rate due to electron spin exchange (without a buffer gas) required for our application, has been mostly unexplored.[9-10]

The initial preparation of the deuterium is to break apart or dissociate the natural diatomic configuration of deuterium. Since all atoms which come out of the dissociator also pass through the spin-exchange cell and eventually contribute to the target thickness, a high dissociation fraction is a prerequisite to obtaining a high target polarization. We are currently pursuing the "standard" technique[11] of radio-frequency (rf) dissociation in an appropriately tuned cavity. While this technique is well known, the dissociators of standard atomic beam sources are operated so as to give a high flux of atoms with the undissociated gas being pumped away after magnetic separation. However, it has been demonstrated that dissociation fractions of 95% are obtainable using rf dissociation.[12]

TARGET PERFORMANCE

In modeling the performance of our cell, we have used the Pauli master equation[13] to describe the coupled rates of interaction between the deuterium and Na spin populations.[14] This approach assumes that there is no quantum mechanical phase coherence between consecutive spin exchange/hyperfine interactions. (For systems of gases having completely random interactions, this assumption seems well justified.) Also, the deuterium flow through the cell has been modelled as simply a depolarization process wherein the spin exchange cell can be viewed as a closed system (see Fig. 3). The average deuterium tensor polarization inside the cell is then given by the equilibrium spin population of this system.

The result of this modeling is shown in Fig. 4 which, for various Na vapor electron polarizations P_{Na}, shows the anticipated tensor polarization of the deuterium t_{20}, versus X_{eff}, the average number of effective Na-D spin-exchange collisions a deuterium atom undergoes before escaping from the cell. Letting X be the average number of Na-D spin-exchange collisions a D atom makes before escape, then

$$X_{eff} = X/(1 + D_p W), \qquad (1)$$

where D_p is the probability that a D atom is completely depolarized by

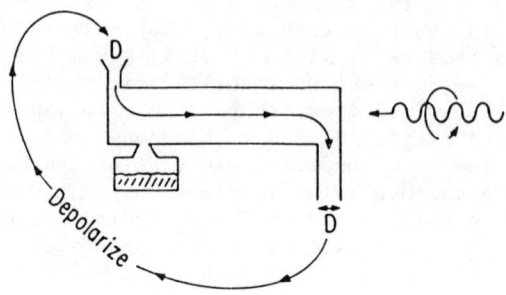

Figure 3. Spin exchange cell viewed as a closed system with D flow as a depolarization process.

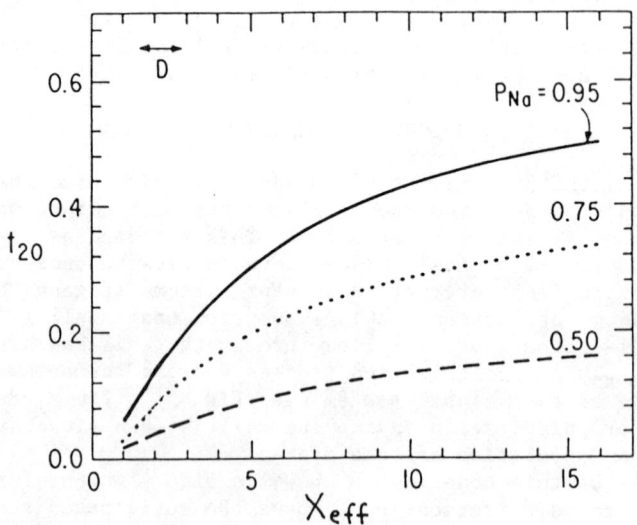

Figure 4. The deuterium tensor polarization, t_{20} versus X_{eff}, the effective number of Na-D spin exchange collisions/D atom, for various Na polarizations.

a wall collision and W is the number of wall collisions a D atom makes before escaping from the cell. Thus, X_{eff} can be viewed as X corrected for wall depolarization processes. From Fig. 4, we see that even with near perfect Na polarization an X_{eff} = 5 yields a tensor polarization t_{20} = 0.3.

The curves in Fig. 4 also give the tensor polarization of the deuterium emitted out of the cell spout to the extent that polarization gradients can be ignored. We expect these gradients to be negligible as long as the cell is operated under conditions of molecular flow;[15] i.e., such that the mean free path for atomic collisions is greater than the diameter of the cell, and the motion of the D atoms is dominated by collisions with the cell walls.

Assuming molecular flow conditions, X (in Eq. 1) can be expressed in terms of cell geometry, the Na-D electron spin exchange cross section (σ_{ex}), and the Na density (n_{Na}) at the chosen operating temperature (T in degrees K); i.e.,

$$X = 4(1 + 2/23)^{1/2} V_c n_{Na} \sigma_{ex} W/A_c , \qquad (2)$$

where A_c and V_c is the cell wall area and volume, respectively. W (given only by cell geometry) is the number wall collisions an atom makes before exiting the cell. A conservative estimate of σ_{ex} is 5 × 10^{-15} cm^2 which comes from measurements of similar atomic reactions.[16] [Notice σ_{ex} (=50 Å^2) is an order of magnitude larger than the Na-D molecular collision cross section.]

We have chosen a set of cell dimensions and Na density (temperature) attempting to minimize depolarizing wall collisions while containing the D atoms for a sufficient number of spin exchange collisions at a moderate Na density.[14] These dimensions are the cell length (2.06 cm) and diameter (0.75 cm), the spout length (1.00 cm) and diameter (0.2 cm). These dimensions imply that an atom makes ~1000 wall collisions before exiting the cell. For a temperature of 232° C (T = 505° K) resulting in a Na density n_{Na} = 3.2 × 10^{12} atoms/cm^3, Eq. 2 yields X = 10. If the cell walls have a depolarization probability for deuterium of $D_p = 10^{-3}$, then X_{eff} = 5 according to Eq. 1.

Thus, independent of the deuterium flow and density, the tensor polarization of the target is limited to 0.3 for a near perfect Na vapor polarization (and $D_p = 10^{-3}$). Of course, the extent of the Na polarization in an operational cell is determined by the rate at which depolarizing spin-exchange collisions (and other losses) can be counteracted by the laser optical pumping process.

Using "low power" optical pumping rate equations in a model framework analogous to those used for the Na-D spin-exchange system, we have estimated that a target thickness of 2.3 × 10^{15} D atoms/cm^2 and a flow of 4 × 10^{17} atoms/sec are attainable. For this deuterium flow and thickness with complete frequency coverage of the Doppler broadened optical transition, a Na wall depolarization probability of 10^{-3}, and the previously discussed design parameters; our estimates of t_{20} versus the incident laser power are shown in Fig 5. We see that the limiting tensor polarization for this cell design is achieved for an incident laser power of ~2 W.

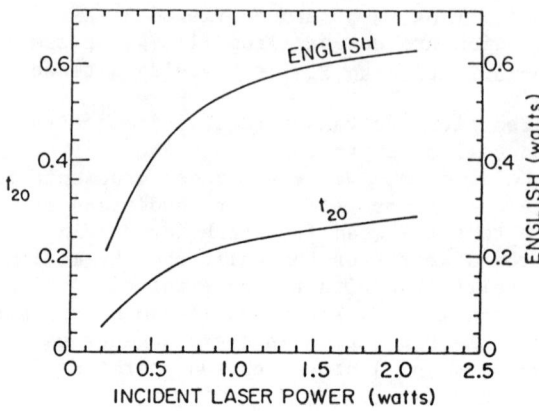

Fig. 5. The expected target tensor polarization (scale on the left) and absorbed English (scale on the right) for the proposed cell dimensions, operating temperature, and other design parameters (giving a target thickness $\sim 10^{15}$ atoms/cm^2) versus incident laser power (see text).

Also plotted in Fig. 5 is the parameter we call "English" which we define as the rate at which angular momentum is transferred through an <u>intermediary</u>, the Na atoms in the cell, to the D and Na flow and the <u>cell walls</u> (analogous to the manner in which the term is used in the game of billiards or pool). This rate is directly related to the power absorbed by the Na vapor in an operating cell, which makes the natural choice of units for "English," power. We see that an incident power of 2 W corresponds to an English of 650 mW for the design parameters discussed earlier. If the Na atoms lose their polarizaion on every wall bounce (instead of 1 out of 1000 bounces), we have computed that the limiting polarization is achieved for an incident laser power of 3 W with an English of 1 W.

As long as molecular flow conditions are maintained (corresponding to deuterium thicknesses $< 8 \times 10^{15}$ atoms/cm^2), the deuterium flow, target thickness, and required incident laser power scale linearly together in our model.

TECHNICAL OBSTACLES TO A WORKING TARGET

In order to produce a tensor polarized deuterium target with t_{20} = 0.3 and a thickness 2.3×10^{15} atoms/cm^2, we require a laser with sufficient power and frequency coverage to optically pump a Na vapor at 232° C to 95% while putting 650 mW of English through this vapor. Also, the cell wall must have a wall depolarization probability $\leq 10^{-3}$ and a recombination probability $\leq 10^{-4}$ at 232° C with the deuterium entering the cell being >90% dissociated.

As for the laser requirements, we estimate that a 4.5 watts/cm^2 (2 W total power) with a 4 GHz bandwidth would do the job with one pass through the spin exchange cell. The current record for power from a dye laser at this wavelength is 40 W with no mention of bandwidth.[15] Commercially available dye laser/argon-ion laser systems are able to provide as much as 5 W in a 10-GHz or <1 GHz bandwidth. We are currently acquiring a 10-GHz system which we hope to adapt to our needs. Our estimates of laser power do not include stimulated emission effects which we believe can only serve to increase the vapor polarization.

The recent work of Weber et al.[10] have shown that a 2×10^{13} atoms/cm^2 thick Na vapor can be optically pumped to a high polarization with 400 mW of laser power. We estimate that if the Na atoms in their cell were completely depolarized on each bounce, their vapor should have absorbed ~130 mW of English. Even though their vapor was twice as thick as in our design, we see that our English requirements (650 mW) due to Na-D spin exchange have yet to be demonstrated. It should be noted that the Na density of 5×10^{12} atoms/cm^3 is near the region for the Weber cell geometry where some[18] predict that "radiation trapping" starts to limit the vapor polarization.

Nearly independent of the necessary optical pumping technology is the requirement of low depolarization and recombination probabilities for the cell wall for deuterium. The relatively high operating temperature of 235° C is most likely a handicap in attaining both of these goals. Barker et al.[19] have reported depolarization probabilities in the 10^{-3} range, while Kleppner et al.[20] have reported recombination probabilities in the 10^{-4} to 10^{-5} range. Both these measurements were done at room temperature with teflon coated cells. Bouchait and Brossel[21] have observed depolarization probabilities for optically pumped rubidium in the 10^{-4} range for some wax based wall coatings at 100° C. As evident from our modeling, a higher wall depolarization of the optically pumped vapor can "in principle" be overcome by increased laser power whereas wall depolarization of the deuterium cannot.

Our choice of Na as the optical pumping vapor was motivated by its low Z to prevent excessive bremsstrahlung losses ($\propto Z^2$) in a stored electron beam.[3] In a hadron ring, a high Z "background gas" may in fact be desirable for efficient cooling of high beam currents;[22] in which case, one could operate with an optically pumped rubidium (or potassium) vapor. For a cell design similar to the Na system, a rubidium (potassium) based spin-exchange system should attain a $t_{20} = 0.3$ at a much lower operating temperature of 125° C (150° C). However, both rubidium and potassium systems require optical pumping wavelengths in the near infrared where recently developed laser dyes have produced less than 1 W of power, thus reducing the target thickness (effectively polarized) to ~10^{15} atoms/cm^2.

ACKNOWLEDGMENTS

I wish to thank my collaborators who have contributed much toward the polarized target project: D. Geesaman, L. Goodman, R. Kowalczyk, J. Napolitano, J. Steward, E. Ungricht, L. Young and B. Zeidman; and in particular, R. Holt for guidance and for originating the spin-exchange target concept and M. Peshkin for laying the theoretical groundwork necessary for the development of the model results presented here.

This research was supported by the U. S. Department of Energy under Contract W-31-109-ENG-38.

REFERENCES

1. R. J. Holt, "Polarized Gas Targets in Electron Rings," CEBAF 1984 Workshop, Newport News, VA, June 1984.
2. R. J. Holt, "Prospect of Polarized Targets in Electron Rings," Conference on the Intersections between Particle and Nuclear Physics, May 1984, Steamboat Springs, CO.
3. R. J. Holt, Proc. of the Workshop on Polarized Targets in Storage Rings ANL-84-50, 103 (1984), Argonne National Lab., Argonne, IL 60439.
4. W. Haeberli, these proceedings.
5. W. Happer, Rev. Mod. Phys. 44, 169 (1972).
6. J. Brossel, A. Kastler, and J. Winter, J. Phys. Radium 13, 668 (1952).
7. W. Franzen and A. G. Emslie, Phys. Rev. 108, 1435 (1957).
8. W. Raith, Z. Physik 163, 467 (1961).
9. Y. Mori et al., Nucl. Instr. and Meth. 220, 264 (1984).
10. E. W. Weber and H. Vogt, Phys. Lett. 103A, 327 (1984). This work shows that a Na vapor of the necessary density and thickness can be optically pumped to high polarization; however their system requires significantly less "English" (discussed in text) than our operating spin exchange cell.
11. J. Slevin and W. Stirling, Rev. Sci. Instr. 52, 1780, (1981).
12. M. P. Maley et al., Bull. Am. Phys. Soc. 29, 665 (1984).
13. W. Peier and A. Thellung Physica 46, 577 (1970); W. Peier Physica 57, 565 (1972); and, L. Van Hove, Physica 21, 517 (1955).
14. M. C. Green, Proc. of the Workshop on Polarized Targets in Storage Rings ANL-84-50, 307 (1984), Argonne National Lab., Argonne, IL 60439.
15. S. Dushman and J. M. Lafferty, Scientific Foundations of Vacuum Technique, John Wiley & Sons, Inc., New York, 1969.
16. R. Franken, R. Sands, J. Hobart, Phys. Rev. Lett. 1, 53 (1964); and L. W. Anderson and A. T. Ramsey, Phys. Rev. 132, 712 (1964). See also theoretical predictions: E. M. Purcell and G. B. Field, Astrophys. J. 124, 542 (1956); and A. Dalgnaro and M. R. H. Rudge, Proc. Roy. Soc. London A286, 516 (1956).
17. Baving et al., Appl. Phys. B29, 19 (1982).
18. L. W. Anderson, Proc. of the Workshop on Polarized Targets in Storage Rings ANL-84-50, 359 (1984), Argonne Natinal Laboratory, Argonne, IL 60439
19. Barker et al., Proc. of Int'l. Pol. Conf., Sante Fe - AIP #69, 931, (1980).
20. D. Kleppner, H. C. Berg, S. B. Crmapton, and N. F. Ramsey, Phys. Rev. 138, A981 (1965).
21. M. A. Bouchait and J. Brossel, Phys. Rev. 147, 41 (1966).
22. R. E. Pollock, Proc. of the Workshop on Polarized Targets in Storage Rings ANL-84-50, 1 (1984), Argonne National Laboratory., Argonne, IL 60439.

SESSION F

ATOMIC AND PARTICLE PHYSICS WITH THE COOLER

PHYSICS WITH ENERGETIC H^0 BEAMS

I. Katayama

RCNP, Osaka University, Ibaraki, Osaka 567, Japan

ABSTRACT

Two atomic experiments of atom-atom collisions using extracted H^0 beams from the electron cooler ring, and of electron capture collisions (Thomas's double scattering) with internal targets in the ring are discussed. The electron cooling system works as a production source of the H^0 beams in the former experiment and as a monitoring device for the emittance of the beam in the latter.

1. INTRODUCTION

Electron cooler rings which are being constructed or considered at Indiana, Uppsala, Jülich and Tokyo will offer a new experimental method not only for nuclear physics but also for atomic physics. It has been discussed that the utilization of an ultra thin target in the ring allows us to perform an atomic collision experiment under the condition in which multiple collisions in the target are negligible. A study of delta ray electrons originating from the collisions of high energy protons with gas targets is still interesting for energy loss theories[1]. The ring has other specific features with which several atomic experiments may become available. In this paper two experiments in atomic physics which are most effectively performed at a cooler ring are discussed. Firstly, high energy neutralized hydrogen atoms H^0 from the electron cooler section allows us to perform atom-atom collision experiments at high incident energies. At Novosibirsk and CERN, it has been demonstrated that the H^0 beams provide a quite powerful diagonastic tool for the electron cooling and the theory has been developed[2]. The intensity of the H^0 beams is around 1×10^4 particles/sec which is sufficiently large for the diagonastic purpose but not large enough for some atomic experiments which use lasers. It might, however, be possible to use the beam for atom-atom collision experiments that have not been tried so far. Among them the electron loss or stripping collision (ionization of H^0 beams) is especially discussed here because of the simplicity of the experiment. Though the electron loss is one of inelastic collisions of the projectile H^0 atom, it offers a rather unique and powerful method to get the momentum transfer integral of the elastic form factor and incoherent scattering function of the target atoms. Secondly, the beam re-circulation technique in the ring allows us to measure Thomas peak in the electron capture reaction by proton beams, which is expected at $0.027°$ (laboratory) in the angular distribution of H^0 atoms at proton energies of 10 to 20 MeV. The peak was observed recently for the first time at Kansas University at proton energy of 7.4 MeV[3]. Although the measurement at higher energies is quite interesting from the theoretical point of view, it would be difficult without the beam re-circulation ring.

Throughout this paper, the author strongly owes to two papers by Shakeshaft and Spruch[4] and by Gillespie[5], which give us very good insight to these atomic collision processes.

2. ATOM-ATOM COLLISIONS USING H^0 BEAMS AT HIGH ENERGY

2.1 Purity of the atomic state of H^0 beams

As is shown later, collision cross section of a H^0 beam with the target atom has a large dependence on the atomic state of the H^0 beam. To get clear results from the atomic collisions with H^0 beams, it is of especial importance to prepare H^0 beams in the ground state. The formation of H^0 atoms in the center of mass system of protons and electrons is due to radiative capture processes at thermal energies. Using the theoretical cross section given by Stobbe,[6] or Bethe-Salpether[7], the rate of the H^0 production has been calculated by Bell and Bell[2]. They concluded that the production rate is sensitive to the velocity distribution of electrons. This means that the measurement of H^0 yield can offer good diagonastic information on the electron cooling. Neuman et al. have discussed that the laser ray irradiation can enhance the production rate of the H^0 by an order of magnitude[8]. Let us see the population of the H^0 atomic states which are formed in the cooler section. Denoting production probability of H^0 atoms in the state of principal quantum number n by (n), it has been shown that the following relation holds i.e., $(1):(2):(3):(4):(5 \text{ to } \infty) = 34:18:11:8:29$[2,8]. The ratios are slightly dependent on the detail of the electron velocity distribution. Lifetimes of the excited states in H^0 atom have been calculated by Hiskes and Tarter for the states of up to n=25 and their results are shown in Fig. 1 for the absence of external fields[9]. If the H^0 beams propagate for some distance, say 20 m, the transit time for the 200 MeV H^0 beam is of the order of 1×10^{-7} sec in the frame moving with H^0 beams (particle

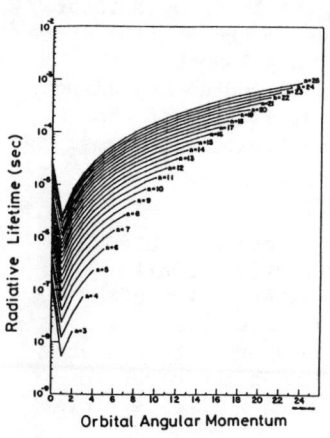

Fig. 1. Radiative lifetimes for the hydrogen atom in the absence of external fields after J.R. Hiskes and C.B. Tarter[9].

frame). This period is long enough for the n = 2(except 2s),3,4 states to decay to the ground state. Higher excited states with n ≥ 6 are hardly expected to decay to the ground state during the period. The 2s state is well known to have a long life time of 130 msec[7]. The possible way of making the 2s state decay to the ground state is to use Stark effect[7]. In the external electric field of 475 V/cm, the life time of the 2s state drastically decreases to 10^{-9} sec, which is an effect of the 2p state mixing into the 2s state. An interesting technique to generate an electric field on the H⁻ atom has been demonstrated by Bryant et al. at LosAlamos[10]. Particles which are travelling with a velocity v in a static magnetic field B (gauss) will see an electric field of 300 βγB V/cm in the particle frame due to the relativistic effect, where β and γ are v/c and $(1-\beta^2)^{-1/2}$, respectively, with c being the light velocity[11]. The electric field strength is shown as a function of H⁰ energy in Fig. 2. Also given in the figure is the flight length to quench the 2s state in the electric field of 475 V/cm in the particle frame. Another interesting application of Stark effect is to use field ionization of H⁰ atoms in higher excited states to remove state-impurities from the beam. If the field-induced changes in atomic structure are ignored, the field at whcih the state of effective quantum numer n* becomes unbound is given as[12,13],

$$F_{ce} = 5 \times 10^9 / (16 Z n^{*4}) \text{ V/cm,} \qquad (1)$$

where Z depends on the orientation of the wave function to the field direction, varying 1 for the 'reddest component' to 1/n* for the 'bluest component'. Given the field of 2×10^6 V/cm corresponding to B = 10kG for 200 MeV H⁰, n* becomes 3.5 to 5.4 depending on Z, which is low enough to expect the complete ionization or decay of the excited states except 2s one in a moderate flight length of H⁰ beams. A more elaborate calculation of the field ionization based on quantum mechanics gives similar results[14]. Another way to increase the state purity of H⁰ beams is to irradiate the beams by a laser ray in a collinear geometry, so that electrons in the excited states may be pushed into the continuum state. The wave length of the laser required to ionize the n=2 state is 300 to 400 nm depending on H⁰ energy. The problem would be in the availability of the enormous power application (0.12 MW) to the continuous H⁰ beams.

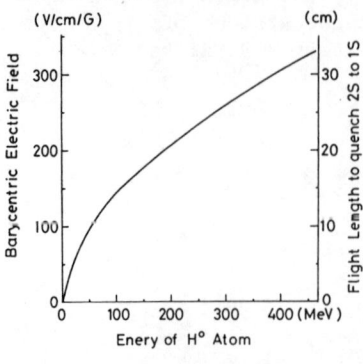

Fig. 2. The barycentric dc electric field which is relativistically generated in an external magnetic field. The same curve gives the flight length to quench the 2s state by Stark effect under the electric of field of 475 v/cm in the particle frame.

2.2 Electron loss collisions using H^0 beams

Quantum mechanical treatment of atom-atom collisions was started for the H^0+H^0 case by Bates and Griffing in 1953[15]. Recently the work has been further developed by Gillespie and Inokuti who discussed collisions of atoms with atomic numbers of less than 18 in the framework of Born approximation[16]. Collisions of hydrogen atom H^0 with neutral target atoms have been treated separately by Gillespie,[5] in which the ionization of H^0 atom (electron loss) is considered. The cross section of electron loss collision at high energy is given in his theory as,

$$\sigma_\ell = 8\pi a_0^2 (1/v_1)^2 [I_{ion,el} + I_{ion,in} + (\gamma_{ion,el} + \gamma_{ion,in})(1/v_1)^2] \quad (2),$$

where v_1 is the projectile velocity in atomic unit, a_0 Bohr radius, I and γ the collision strengths which are described below. The collision strength I is given as

$$I_{ion, el(in)} = \int_0^\infty S_{ion}(K) \, S_{el(in)}(K) \, d(a_0 K)/(a_0 K)^3 \quad (3),$$

where K is momentum transfer and $S_{ion}(K)$ is the ionization transition strength which depends only on the state of the hydrogen atom and is given in Fig. 3 as a function of $a_0 K$. The functions $S_{el}(K)$, $S_{in}(K)$ are the squared form factor for elastic scattering and the incoherent scattering functions for the target atom, respectively. The numerical tabulation of $S_{el}(K)$ and $S_{in}(K)$, based on the non-relativistic Hartree-Fock wave function of electrons, is available for the whole elements[17]. The result does not change so much even when the relativistic Hartree-Fock wave function is employed[16,17]. The approximate evaluations of the second order terms γ in eq.(2) have been given in ref. 5. Notice that the integral given in eq. (3) can be measured by simply observing H^0 yield in the H^0 beam direction. The separation of protons from H^0 beams is quite easy. Proton inelastic collisions provide us also similar expression like eq. (2) (but I has a different form from eq. (3)) at beam direction[18]. But the separation of the beam from the inelastically scattered proton would be impossible.

Investigation of the electron loss collision may have two directions. One is to see the reaction mechanism; the velocity dependence of the measured cross section can tell about the validity of the approximation made in the derivation of eq.(2). The other is to test the atomic model for the target atom. Although the integral of eq.(3) does not show high sensitivity to the atomic model as mentioned above, the calculation should be checked experimentally. In fig. 4 are shown the collision strength (some of them are obtained by extrapolation of the low energy data to high energy limit) together with the theory by Gillespie as a function of the target atomic number[19]. Shell structure of the target atom is evident. Gillespie has shown that electron loss of the H^0 atom in the 2s state, for which $S_{ion}(K)$ is also plotted in fig. 3 for comparison, is more sensitive to the target shell structure than the 1s state[20]. The large difference in $S_{ion}(K)$ between the 1s and 2s state as shown in the figure results in that the electron loss cross section of the 2s state of the H^0 beam in He gas target, for example, becomes twice as large as that of the 1s state[20]. It is, therefore, quite of importance to use the H^0 beams in the ground state for the experiment. It is interesting to note that the high Rydberg states in the H^0 atoms formed in the inelastic collisions may be analysed using the field ionization described in the previous subsection. The collision strength to the Rydberg states has also been given in ref. 5.

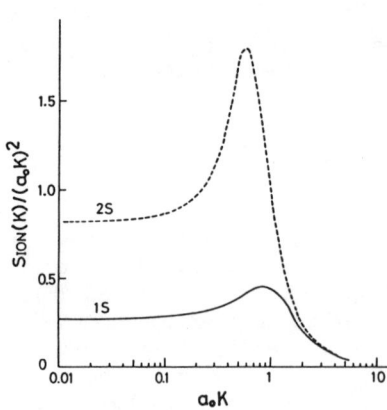

Fig. 3. $S_{ion}(K)/(a_0K)^2$ vs. a_0K for the 1s and 2s states in the hydrogen atom after G.H. Gillespie[5,20].

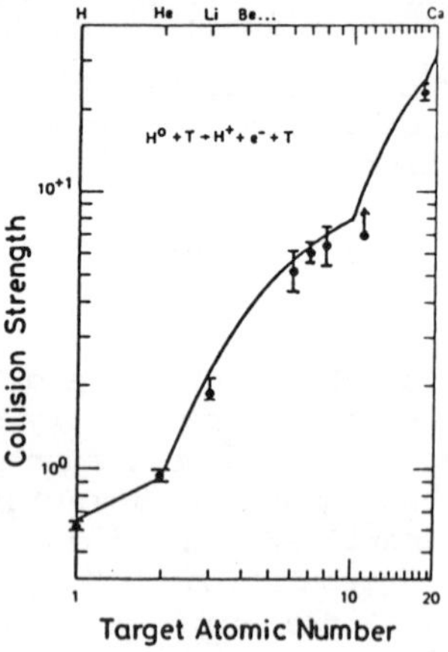

Fig. 4. Collision strength vs. target atomic number for H^0+target → p+e+target. The calculation shown by the solid line and evaluation of experimental data is by G.H. Gillespie[19].

2.3 Experimental arrangements

Three geometries for the collision experiments with H^0 beams are schematically depicted in fig. 5. The first one is for solid targets. Measuring the dependence of H^0 beam intensity on the target thickness, one can derive the electron loss cross section. The target thickness necessary to get a exponential attenuation curve of the H_0 yield can be estimated from the collision strength to range from 10 to 1000 μg/cm^2. If several tens targets can be mounted on a target exchanger as used in Jülich[21], the efficiency of the experiment will increase. Protons due to the electron loss of H^0 beams are easily bent away using a magnet. The protons and H^0 atoms are detected separately and the sum of these intensities gives the H^0 beam intensity. The energies of electrons which are emitted in the electron loss collision in the target or in the field ionization of excited states in the inelastically scattered H^0 beams are 100 to 200 keV in the range of the H^0 beam energies of 200 to 500 MeV. Since the contribution of the delta electrons from the target is negligibly small at this electron energies, the electron detection is another way of measuring the collision cross sections.

The second geometry is similar to the first one except in that the target is a gas target. The third one also employs a gas target filling the whole vacuum vessel. The advantage of the second over the third is in the absence of window materials, which allows us to have a pure H^0 beam on the target. A technical difficulty is in the complexity of evacuation systems required. In the third geometry, a foil of window material is indispensable in order to prevent the gas from flowing into the cooler ring. Some fraction of the H^0 beams are lost by the electron loss collision in the foil. The magnitude, however, can be decreased by using low Z material for the foil[22]. Formvar or VYNS or Zapon films seem to be the candidates for the window[23]. Among them, VYNS has the largest mechanical strength, but it contains small amount of Cl. Formvar and Zapon only consist of C, H, O and N. It is not difficult to prepare a 10 μg/cm^2 thick Formvar foil of about 10 cm^2 in area. Assuming the window material is pure N_2 and thickness is 10 μg/cm^2, one can estimate how the state population of H^0 beams is affected by passage through the foil. The result is shown in fig. 6. The collision strengths for the calculation have been taken from ref. 5.

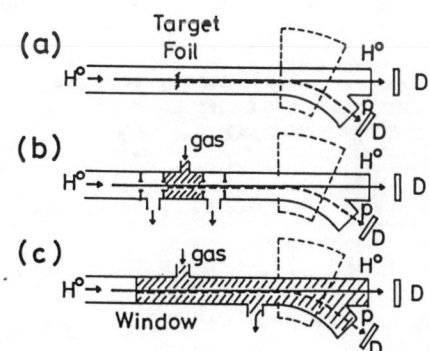

Fig. 5. Three measurement schemes to measure the electron loss cross sections of H^0 beams. Magnets to sweep off protons are shown by dotted lines.

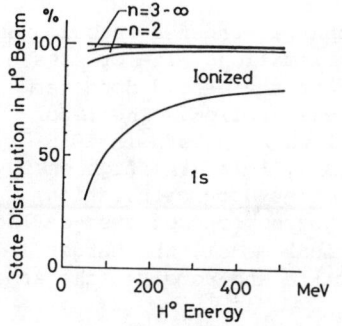

Fig. 6. Ionization or excitation probabilities of the H^0 beam in passing through 10 μg/cm^2 N_2 gas as a function of the H^0 energy.

One apparatus of the third type is being tested at RCNP, Osaka. The method has been developed in Jülich[24]. In the experiment the H^0 atom is produced in the target by proton beams from the cycltoron. Our aim is to measure the total cross sections of electron capture ($p+e \rightarrow H^0$) and electron loss ($H^0 \rightarrow p+e$) in gas targets. The schematic arrangement of the apparatus is shown in fig. 7. The H^0 production is observed at the beam direction as a function of gass pressure. Above a certain pressure, H^0 yields at the entrance of the magnet become constant as a consequence of charge equilibrium. The gas in the chamber of the magnet works as a gas stripper for H^0. The effective region of this stripper can be experimentally determined by putting a thin Havar foil in H^0 beams. A very preliminary result of $P + Ar \rightarrow H^0 + Ar^{1+}$ at 80 MeV is shown in fig. 8. The exponential decay in the high pressure region gives electron loss cross section and from an extrapolation of the curve to zero pressure the ratio of electron capture to loss cross sections is obtained. The experiments is still in progress and in the near future we shall have the cross sections for Ar, Kr, Xe targets at proton energies of 40 to 80 MeV.

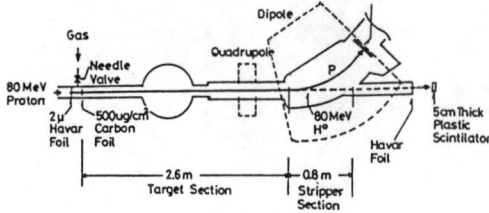

Fig. 7. Schematic arrangements to measure the electron capture ($p+e \rightarrow H^0$) and the electron loss ($H^0 \rightarrow p+e$) cross sections of gas targets at E_p = 40 to 80 MeV (RCNP, Osaka).

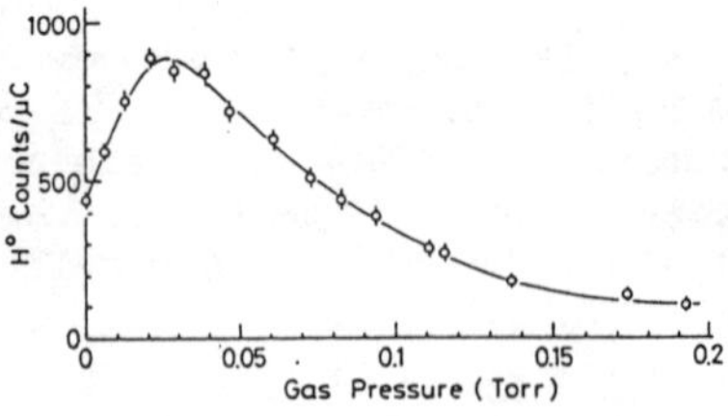

Fig. 8. Preliminary results for the Ar gas at E_p = 80 MeV (RCNP, Osaka).

3. MEASUREMENT OF THOMAS PEAK IN ELECTRON CAPTURE

3.1 Present Status of Electron Capture

Theoretically, electron capture or charge transfer reaction at high energy has been treated by Born approximation. The process at asymptotically high enery has a feature that the second Born term gives considerable contirubtion to the cross section. The importance of the second Born term was first shown by Drisko in 1955[25]. He found that the cross section behaves as v_1^{-11} at the high energy limit. It has been known that the first order Born calculation gives v_1^{-12} dependence [26,27]. In 1927, Thomas proposed a so-called double scattering model[28]. In this classical model, the target electron is first scattered by the projectile and acquires the same

velocity with that of the projectile. Subsequently, being again
scattered by the target nucleus into the projectile direction, the
electron is finally captured in the atomic orbit of the projectile.
It should be noted here that the projectile receives a recoil effect
in the first collision and is scatterred in an angle $\theta_T = (m/M)\sin 60°$,
where m and M are the mass of the electron and projectile,
respectively. The process is schematically shown in fig. 9. The
Thomas model gives a v_1^{-11} dependence for the total cross section.
Shakeshaft and Spruch showed that the coefficient of the v_1^{-11} term
by the second order calculation can be explained correctly by
considering the uncertainty law in Thomas model[4]. The second Born
calculation gives an another effect which has no corresponding
classical process. This is clearly demonstrated in the calcualtion
of the differential cross section by McGuire, Simony, Weaver and
Macek[29]. Since Drisko's work, it has been accepted that higher
order processes in electron capture have to be taken into account
at high energy. There have been many theories on electron capture.
The interrelations between the theories and the limit of the
validity of each theory have been discussed[30,31,32,33].

  Contrasted with the progress of modern theories, experimental
progress is rather slow. Urgent requests of theoreticians are to
measure the angular distribution of the cross section and their
dependences on the projectile velocity. These are more sensitive
to the details of the theories than the angle integrated cross
sections[34,35]. It should be mentioned, however, that the velocity
dependence of the total cross section is not always explained well
by the theories[21]. Recently a big stride has been made by Horsdal-
Pederson, Cocke and Stockli who measured the Thomas peak for the
first time[3]. They observed a clear peak in the angular distribution
at the laboratory angle $\theta_T = 0.027°$ for the $p+He \to H^0+He^+$ reaction at
7.4 MeV. Although the peak position was exactly as predicted by
the theories, the cross section could not be fully explained. It
is still extremely valuable to extend this kind of measurement to
higher energies.

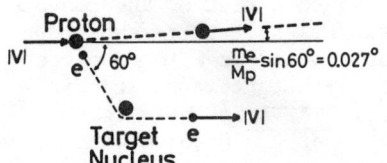

Fig. 9. Schematic scheme of a double scattering model of the electron capture by L.H. Thomas[28].

3.2 Experimental Method to Measure Thomas Peak at High Energy
  Following the feasibility consideration for the measurement of
Thomas peak, a combination of projectile and target and projectile
energy would be protons, He target and $E_p = 10$ to 20 MeV. Beyond
$E_p = 100$ MeV, cross sections becomes too small and relativistic
effect becomes not negligible[36]. Choice of heavier elements for
the target is not appropriate for observing the Thomas peak,

because the relative intensity of the Thomas peak to the first
order Born peak which sohws the maximum at 0° decreases with the
the target atomic number. It should be noted that the H^0 is also
produced by the radiative electron capture (REC) in which target
electron is captured into the projectile atomic orbit by emitting a
photon[4]. The process REC becomes dominant above E_p = 20 MeV for
the He target. The angular distribution of the H^0 of the REC,
however, shows sharply forward peaking and does not overlap with
the Thomas peak. In table 1 are given the total cross sections
calculated from Drisko's equation given below together with the
estimated experimental yields for the p + He case. The H^0 yields
due to the REC is also included:

$$\sigma_{Drisko} = \sigma_{OBK}(0.293 + 2^{-11}5\pi v_1/(Z_p+Z_t)) \quad (4).$$

$$\sigma_{OBK} = 5^{-1} 2^{11} Z_p^5 Z_t^5 v_1^{-12} \pi a_0^2 \quad (5).$$

$$\sigma_{REC} = 3^{-1} 2^7 Z_t Z_p^5 \alpha^3 v_1^{-5} \pi a_0^2 \quad (6).$$

Here, σ_{OBK} is the first order Born calculation by Oppenheimer,
Brinkman and Kramers[26,27]. σ_{REC} is taken from ref. 4 and α is the
fine structure constant. One can not increase the target thickness
above the values given in the table. Since the multiple scattering
in a thick target will smear out the Thomas peak. The technical
difficulty may be in obtaining a beam of extremely good emittance
with divergence angle of less than 0.05 mr. The possible simplest
geometry in which double slits are used at a reasonable distance
results in the reduction of beam intensity on target to the order
of 1×10^{-4} (in the case of the tandem accelerator) and 1×10^{-6}
(in the cyclotron case) of the accelerated beam intensity.   .y.
Numbers given in the last column of the table show how this kind of
experiment is difficult to perform in cyclotron laboratories. It
may be interesting to consider the feasibility of the experiment at
beam re-circulation rings. In the cooler mode, equilibrium
emittance of the beam in the ring is still far larger than that
required. And the beam luminosity is less than that of the conventional one-pass (throw the beams penetrating the target) experiment.
It is, therefore, not clear that the cooling mode can improve the
feasibility of the experiment. The most important is to increase
the brightness of the beam on the target. The brightness is
defined as a ratio of the beam intensity over the beam emittance.
In the Kansas data of angular spreading of the beam penetrating the
target, we notice that the multiple scattering effect is only
observed at the tail of the spectrum and the FWHM of the beam is
not affected by the multiple scattering[37]. The width of a Gaussian
curve which fits the tail component was 0.16 mr which is compatible
with the prediction of 0.10 mr from a multiple scattering formula[38].
The intensity of the tail part was a few percent of the total
intensity. This implies that the most part of the beam could be
used again to irradiate the target. By recirculating the beam, one
can increase the brightness by a factor given as

Table 1

P + He → H^0 + He$^+$ cross sections and estimation of beam time to measure the differential cross section

E_p (MeV)	v_1 (au)	σ_{1st}[1] (b)	σ_{2nd}[2] (b)	σ_{REC}[3] (b)	t[4] (μg/cm²)	N_{1st}[5] (count)	N_{2nd} (count)	N_{REC}[5] (count)	Beam time[6] (pA*hr)
10	19.9	1.13×10^{-2}	5.50×10^{-4}	9.36×10^{-4}	0.44	2.1×10^4	10^3	1.7×10^3	10×1.2×10^2
20	27.9	1.93×10^{-4}	1.38×10^{-5}	1.73×10^{-4}	1.7	1.4×10^4	10^3	1.3×10^4	10×1.2×10^3
30	33.9	1.86×10^{-5}	1.62×10^{-6}	6.52×10^{-5}	3.8	1.1×10^4	10^3	3.9×10^4	10×4.8×10^3
40	38.9	3.63×10^{-6}	3.62×10^{-7}	3.28×10^{-5}	6.8	1.0×10^4	10^3	9.0×10^4	10×1.2×10^4
50	43.2	1.04×10^{-6}	1.15×10^{-7}	1.94×10^{-5}	11	9.0×10^3	10^3	1.7×10^5	10×2.5×10^4
100	58.9	2.52×10^{-8}	3.79×10^{-9}	4.12×10^{-6}	40	6.6×10^3	10^3	1.1×10^6	10×2.0×10^5
200	77.8	8.81×10^{-10}	1.75×10^{-19}	1.03×10^{-6}	146	5.0×10^3	10^3	5.8×10^6	10×1.5×10^6

1). Calculated from the first term (v_1^{-12} dependence) in eq. (4).
2). Calculated from the second term (v_1^{-11} dependence) in eq. (4).
3). Calculated from the equation (6).
4). Multiple scatterings give rise spreading of $\frac{1}{10}$ × Thomas-angle (=0.0027°) in divergence angle of the beam at this target thickness. Estimated from ref. 38.
5). Total count corresponding to σ_{1st} or σ_{REC} when 1000 total count is expected from σ_{2nd}.
6). Estimated machine time to get the total count given to left (target thickness is given at the middle).

$$f = (A+C)/C - (A+C)^2/(NC^2) + (C/\overline{A+C})^{N-1}/N \qquad (7),$$

where A is the acceptance of the storage ring, C the increase in beam emittance due to multiple scattering and N the number of turn in the ring. The acceptance of the ring is determined by a beam collimator in the ring, by which the emittance of the beam on the target is kept constant in such a way that the growing part of the beam in angle by the multiple scattering in the target is stopped by the collimator. It is also important that the injection of the beam in the ring should not induce any betatron oscillation. This will be possible by a longitudinal stacking which, however, results in a deterioration of the beam energy resolution. It is, therefore, important that the ring should be operated so as to give zero dispersion at the target position. Note that the position of the Thomas peak is not influenced by a change in the beam energy. Making use of the data by Kansas and injection method being considered at Indiana, we can estimate the factor of eq.(7) to be around 10. How does the electron cooling work? In this case, the the H^0 beams from the radiative capture process in the electron cooler section can be used to monitor the emittance of the beam. Taking advantage of the increase in the brightness of the beam by beam recirculation as given above, we can expect to perform the measurement at 10 to 20 MeV of proton energies. The experiment at much higher energies, however, still remains not simple without introduction of a breakthrough technique in the injection method.

It is to be mentioned that even at the energy of Kansas experiment, if the beam spreading can be reduced further by a factor of several by the beam recirculation there will be no need to convolute the theoretical curve with the experimental angular resolution of the beam. This allows us to test the theory in much clear way.

4. SUMMARY

Electron cooler rings which are so ardently promoted by the interests in nuclear physics can be also a new experimental facility for atomic physics. Two experiments of atomic collision processes using the cooler ring facility are disussed The H^0 beams from the electron cooling section allow us to play atom-atom collisions at high energy. The "state purification" of the H^0 beams is disussed. The atomic information from the electron loss collisions is similar to that from the proton-atom collision, but the experimental technique for the electron loss collision is much simpler than for proton collision. High luminosity mode of the ring will enable us to observe Thomas peak in the differential cross section of electron capture collision. The experiemtnal data are quite valuable for the stringent test of the current theories of electron capture. Although not discussed here, convoy electrons which are produced by a similar mechanism to the electron capture are also interesting from the theoretical point of view[39].

ACKNOWLEDGEMENTS

The author would like to acknowledge disscusions and comments by Drs. K. Takayanagi, G.H. Gillespie, J.H. McGuire, O. Schult and S. Morinobu. The help for the experiment at Osaka by M. Fujiwara and S. Morinobu is acknowledged.

REFERENCES

1. J.H. Miller, L.H. Toburen and S.T. Manson, Phys. Rev. A27,1337 (1983).
2. M. Bell and J.S. Bell, Part. Accel. 12,49 (1982).
3. E. Horsdal-Pedersen, C.L. Cocke and M. Stockli, Phys. Rev. Lett. 50,1910 (1983).
4. R. Shakeshaft and L. Spruch, Rev. Mod. Phys. 51,369 (1979).
5. G.H. Gillespie, Phys. Rev. A18,1967 (1978).
6. M. Stobbe, Ann. Phys. (Germany) 7,661 (1930).
7. H. Bethe and E. Salpeter, Quantum mechanics of one- and two-electron systems, Handbuch der Physik (Springer, Berlin, 1957), Vol.35, p.88.
8. R. Neumann, H. Poth, A. Winnacker and A. Wolf, Z. Phys. A313, 253 (1983).
9. J.R. Hiskes and C.B. Tarter, UCRL-7088 (1964).
10. H.C. Bryant et al., Phys. Rev. Lett. 38,228 (1977).
11. C. Møller, Theory of Relativity (Oxford Univ. Press, 1952).
12. J.E. Bayfield, Phys. Rep. 51,317 (1979).
13. R.R. Freeman, Atomic Physics 7 (edtd. D. Kleppner and F.M. Pipkin, Plenum Press 1981) p.209.
14. T. Yamabe, A. Tachibana and H.J. Silverstone, Phys. Rev. A16, 877 (1977).
15. D.R. Bates and G. Griffing, Proc. Phys. Soc. A66,961 (1953).
16. G.H. Gillespie and M. Inokuti, Phys. Rev. A22,2430 (1980).
17. J.H. Hubbel, Wm.J. Veigele, E.A. Briggs, R.T. Brown, D.T. Cromer and R.J. Howerton, J. Phys. Chem. Ref. Data 4,471 (1975); with erratum in 6,615 (1977).
18. M. Inokuti, Rev. Mod. Phys. 43,297 (1971).
19. G.H. Gillespie, Private communication
20. G.H. Gillespie, Phys. Rev. A22,454 (1980).
21. I. Katayama, G.P.A. Berg, W. Hürlimann, S.A. Martin, J. Meissberger, W. Oelert, A. Retz, M. Rogge, J.G.M. Römer, G. Gaul and H. Hasai, to be published.
22. G.H. Gillespie, Nucl. Inst. Meth. B2,231 (1984).
23. L. Yaffe, Ann. Rev. Nucl. Sci. 12,153 (1962).
24. I. Katayama, G.P.A. Berg, W. Hürlimann, S.A. Martin, J. Meissburger, W. Oelert, M. Rogge, J.G.M. Römer, J.L. Tain, L. Zemlo and G. Gaul, J. Phys. B17,L23 (1984).
25. R.M. Drisko, Ph. D. Thesis Carnegie Inst. of Tech. (1955).
26. J.R. Oppenheimer, Phys. Rev. 31,349 (1928).
27. H.C. Brinkman and H.A. Kramers, Proc. Acad. Sci., Amsterdam 33, 973 (1930).
28. L.H. Thomas, Proc. R. Soc. 114,561 (1927).
29. J.H. McGuire, P.R. Simony, O.L. Weaver and J. Macek, Phys. Rev. A26,1109 (1982).

30. J. Briggs, J. Macek and K. Taulbjerg, Comments on At. Mol. Phys. $\underline{12}$,1 (1982).
31. J.H. McGuire, IEEE Trans, Nucl. Sci. $\underline{NS-30}$,1100 (1983).
32. J. Macek, Proceedings on 13th ICPEC (1983, Berlin) p.316.
33. J. Macek and S. Alston, Phys. Rev. $\underline{A26}$,250 (1982).
34. R. Shakeshaft and L. Spruch, J. Phys. $\underline{B11}$,L457 (1978).
35. J. Briggs, P.T. Greenland and L. Kocbach, J. Phys. $\underline{B15}$,3085 (1982).
36. B.L. Moiseiwitsch and S.G. Stockman, J. Phys. $\underline{B13}$,2975 (1980).
37. J.H. McGuire, M. Stockli, C.L. Cocke, E. Horsdal-Pedersen and N.C. Sil, Phys. Rev. $\underline{A30}$,89 (1984).
38. Particle Properties Data Booklet 1980, ed. CERN p.81.
39. I.A. Sellin, Atomic Physics $\underline{7}$ (edtd. D. Kleppner and F.M. Pipkin, Plenum Press, 1981) p.455.

NUCLEAR AND ATOMIC PHYSICS EXPERIMENTS WITH COOLED HEAVY ION BEAMS
OF UP TO 500 MeV/u ENERGY

Paul Kienle
Gesellschaft für Schwerionenforschung
D-6100 Darmstadt, Federal Republic of Germany

ABSTRACT

Here I shall try to sketch a research program which we envisage for the time being to be carried out at the ESR storage and cooler ring for heavy ions, planned to be constructed at GSI Darmstadt[1]. These ideas are still in infancy, in fact we plan a little brain-storming in two weeks from now, during which we shall discuss first proposals in detail.

I. HEAVY ION SYNCHRONTRON SIS 18 AND THE EXPERIMENTAL STORAGE RING ESR

As an introduction, Fig. 1 shows a layout plan of the heavy ion synchrotron SIS 18[2] to be built at GSI for acceleration of heavy ions up to an energy of 2 GeV/u connected with the experimental storage ring ESR, which can be used for accumulation, storage and phase space density increase of heavy ions up to uranium with energies between 834 MeV/u (Ne^{10+}) and 556 MeV/u (U^{92+}). The synchrotron with a circumference of 206.4 m and a maximum bending power $B\rho=18$ Tm accelerates heavy ions up to uranium with a cycling rate of 3 Hz (up to 1.2 T) and 1 Hz to the highest energies. The ESR has exactly half the circumference (for preservation of the bunch structure) of SIS 18 and $B\rho= 10$ Tm.

A heavy ion beam accelerated in the UNILAC up to 20 MeV/u, and stripped to an adequate high charge state for the desired energy

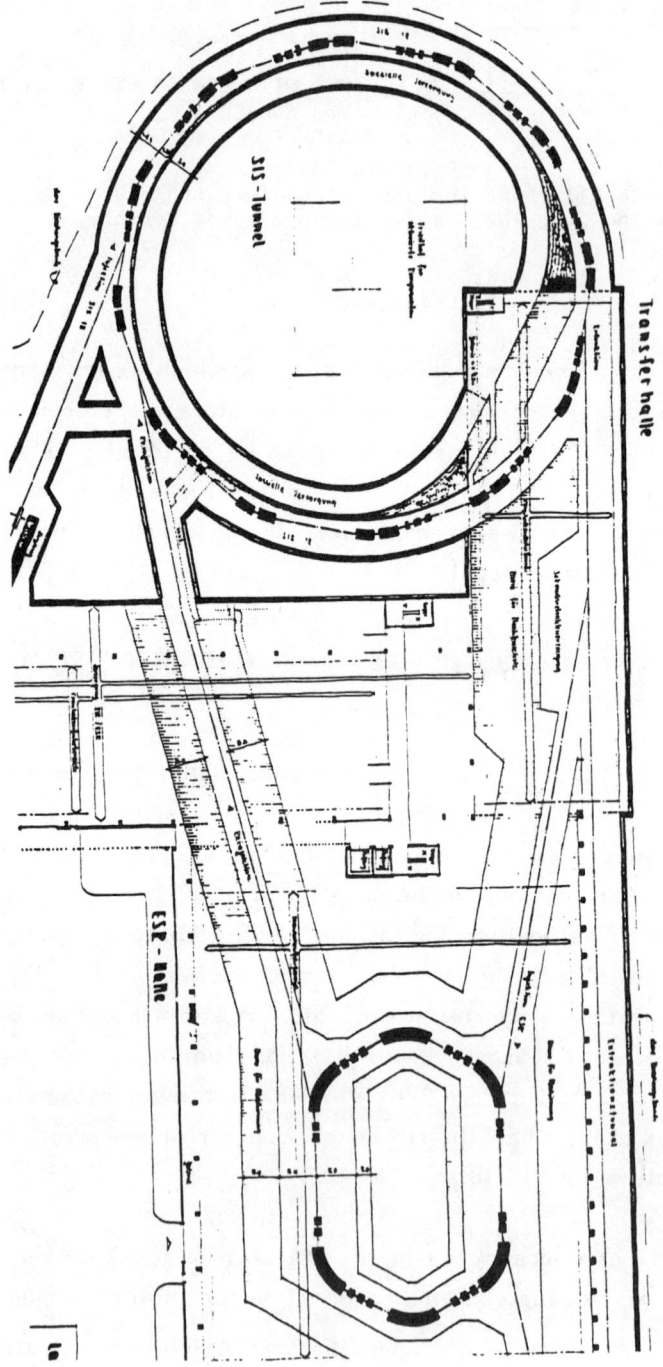

Figure 1. Layout plan of the heavy ion synchrotron SIS 18 and the experimental storage ring ESR.

and intensity, is injected into SIS 18 during 10 to 30 turns and accelerated to maximum energies, depending on the charge states of the ions as shown in Fig. 2.

For uranium ions with a charge state of $q = 78^+$, after stripping behind the UNILAC with a foil target, 1 GeV/u is achieved as maximum energy. Two systems will be installed for beam extraction. A slow extraction over a period of 100 ms and a fast extraction during a single turn for a transfer of the bunch structure into the ESR. Between SIS 18 and ESR the beam may be stripped once more to the highest desired charge state. The ESR with a bending power of $B\rho = 10$ Tm allows to store ions up to U^{92+} with the following maximum energies: Ne^{10+} (834 MeV/u), Ar^{18+} (709 MeV/u), Kr^{36+} (650 MeV/u, Xe^{54+} (609 MeV/u) and U^{92+} (556 MeV/u). The uranium ions can be fully stripped at this energy with an efficiency of 60 % in a Cu-target of several g/cm^2 thickness. The stripping yield increases strongly with decreasing charge, thus one expects a yield of 70 % for Pb^{82+}-ions (574 MeV/u) and already 100 % for Xe^{54}-ions (609 MeV/u). Alternatively one can install a reaction target for Coulomb break up, fragmentation or fission of the beam. The favourable kinematic focussing of the products around the beam direction and velocity allows effective accumulation of radioactive beams with the ESR, which can be operated at $\Delta p/p = \pm 2$ % and a radial acceptance of about 140 x πmm mrad. Fig. 3 shows the layout plan of the ESR, with two 9.5 m long straight experimental sections, in one of which an electron cooling device will be installed.

The other 4 straight sections will be used for the installation of rf cavities, slow and fast extraction elements. The rf cavities are used for acceleration, deceleration and especially also bunching of the beam together with the electron cooling for reduction of the longitudinal phase space. With the fast extraction system one can transfer a highly ionized and cooled beam back to SIS 18 for further acceleration or specially also deceleration. The optics of the ring allows three modes of operation, one specially suited for accumu-

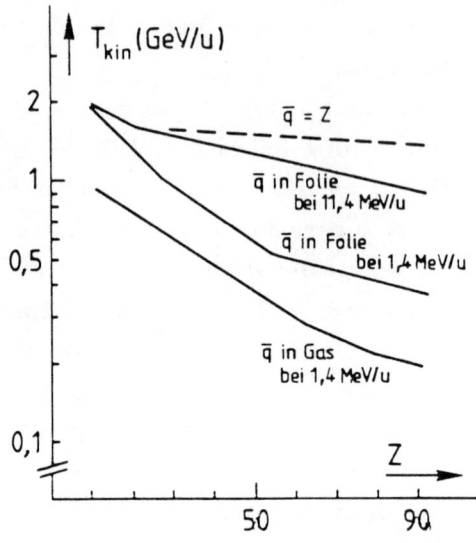

Figure 2. Maximum achieveable energies at SIS 18 as a function of nuclear charge. The energies are given for a gas- or a foil-stripper at an energy of 1.4 Mev/u, resulting in relatively low degrees of ionization. If a second stripper at 11.4 MeV/U is added or if completely ionized particles from the experimental storage ring ESR are reinjected into the synchrotron higher energies can be achieved.

lation, one with no dispersion in the straight sections, which allows multi-charge operation. For a cooled uranium beam, charge states from 89^+ to 92^+ may circulate without any further losses. In a third mode one can inject ions of slightly different momenta, which then may be brought to merge with a well defined angle of about 100 mrad. This may be used to study collisions of <u>two</u> highly ionized beams at fixed target equivalent energies of up to 7.2 MeV/u.

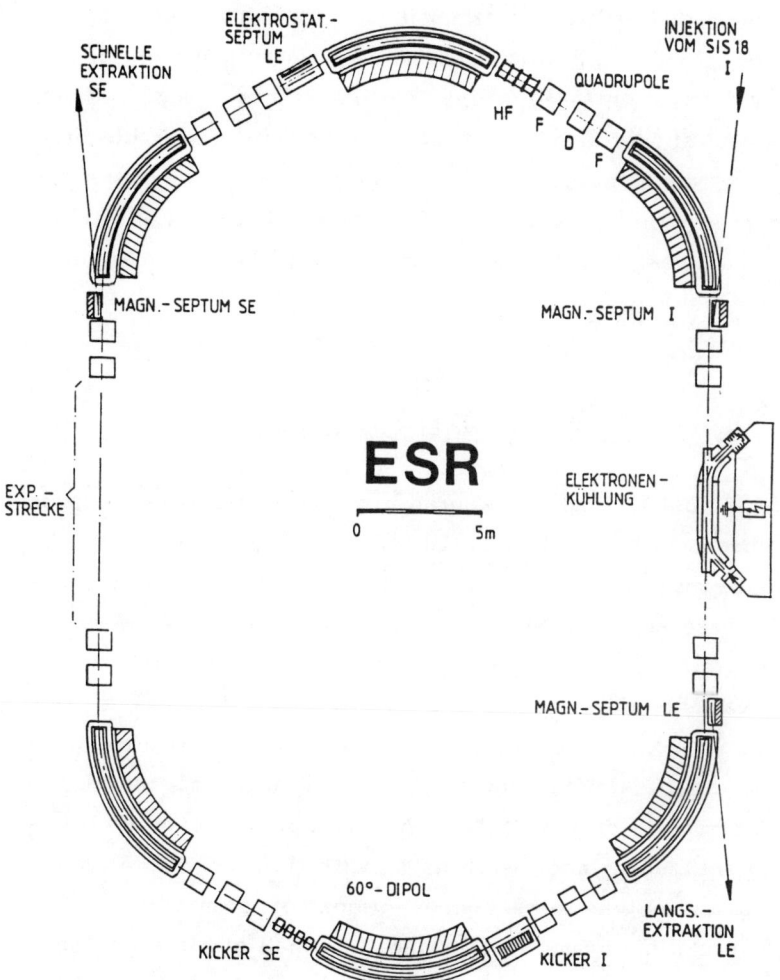

Figure 3. Layout plan of the experimental storage ring ESR with a circumference of 103.2 meters. In two straight sections each with a length of 9.5 m experiments and the facility for cooling can be mounted.

The most important facilities of the ring are various complementary cooling devices which can be operated simultaneously. We consider stochastic cooling for precooling secondary beams and electron cooling to reach radial emittances as small as 0.1 πmm mrad and momentum spreads of 10^{-5}. Both stochastic and electron cooling should be very effective on heavy ions, cooling times below 1 s are expected, and should also in the latter case lead to tolerable beam losses due to recombination, especially for a multi charge operation. In the following we will discuss now a research program with its emphasis in nuclear physics, atomic physics, plasmaphysics and in some applied fields.

II. NUCLEAR PHYSICS EXPERIMENTS

Many unique nuclear physics experiments become practical, because the combined accelerator facility UNILAC + SIS 18 ESR, provides heavy ion beams of nearly any nucleus including radioactive ones, with high phase space density (i.e. radial emittances as low as 0.1 πmm mrad and $\Delta/p \sim 10^{-5}$ with the highest possible particle number), variable time structure from DC to subnanosecond wide pulse trains with high duty factors, variable energies from the Coulomb barrier to 1-2 GeV/u, not only on external targets but also in the circulator mode of the ESR as circulating high current beams interacting with light and electrons or with thin targets. Although we shall focus on the more unconventional experimental program with the circulating beams in the ring, we shall also mention some external target experiments, which need beam tailoring using both rings.

For nuclear reaction studies, at the highest energies, which can be only carried out using external targets, the high beam intensity, low emittance and subnanosecond pulse structure will be specially useful for detection of rare events like subthreshold production of K^+ and $\bar{p}$, which provide interesting signals for cooperative production effects in hot compressed nuclear matter. But also for experiments using combined magnetic and time of flight methods for particle identification the low emittance and subnanosecond pulse structure of the heavy ion beams may be of great importance. We have not thought of many nucleus-nucleus collision studies using the internal target mode, however, for the investigation of target fragmentation with small energy transfer, very thin targets may be required to analyse the low energy heavy fragment. Under these conditions high luminosities of up to 10^{30} cm^{-2}s^{-1} can be reached using 10 ng/cm^2 thin targets of heavy nuclei and 10^{10} stored particles, thus rare production channels can be identified. Furthermore reactions with accumulated very rare radioactive nuclei and thin targets may still be carried out with good luminosities (10^{28} cm^{-2}s^{-1} with 10^4 particles, $A_t=1$ and d= 10 ng/cm^2).

For nuclear structure experiments, the ESR offers a series of possibilities, especially with circulating and cooled radioactive beams. With heavy ion beam energies around 300 - 400 MeV/u one can favourably study all standard quasielastic reactions in inverse kinematics by bombarding atomic beam targets of H, d, t, ^{3}He, ^{4}He, ^{6,7}Li etc., and detecting the light recoil ion at angles ϑ_R around 90°, which corresponds to forward angles in the cm system. For large mass ratios of projectile and target nuclei and Q-values smaller than $E/A_P \cos^2\vartheta_R$ the energy of the recoil is given by $E_R = A_R \cdot E/A_P \cos^2\vartheta_R + 2Q$. For fixed angle ϑ_R, the change of the recoil energy is only determined by the Q-value. Thus by measuring E_R and especially ϑ_R very accurately it seems possible to carry out reaction spectroscopy with a Q-value resolution in the order of 50 keV or less. For this a combination of position sensitive ΔE and high-resolution E-detectors should be sufficient. Thus inverse kinematics allows reaction spectroscopy with medium resolution determined mainly by the emittance of the beam. All improvements of the beam emittance due to better cooling allows more accurate Q-value measurements. This new method is uniquely suited for reaction spectroscopy on exotic nuclei, accumulated as medium energy cooled beams in the accumulator cooler ring. As target one can also use atomic beams with polarized p, d etc. thus gain the full power of this interesting reaction spectroscopy method also for radioactive nuclei.

Accumulated medium energy radioactive beams may be slowed down after cooling and implanted into semiconductor-detectors for high resolution and high efficiency β-γ-spectroscopy on exotic nuclei. If in addition the beam of radioactive nuclei could be polarized - several schemes are in discussion - nmr experiments with the resonance being detected by the destruction of the anisotropy of the nuclear radiation can be performed for determination of magnetic and in favourable cases also quadrupole moments of nuclei far off stability.

In a proposal by Otten and his collaborators [4] collinear laser spectroscopy on Rydberg-states, which may be prepared by laser induced radiative capture, dielectronic recombination or charge transfer, is suggested to investigate Lambshift and the isotope shift also on radioactive nuclei. Using similar techniques, it should be possible to induce M1-transitions between hyperfine levels of few electronsystems of heavy nuclei, the splitting of which becomes in the energy range of optical transitions.

A unique field of nuclear structure physics may come from in beam α- and γ-spectroscopy using neutron deficient or neutron rich ion beams, like ^{56}Ni or ^{8}He, ^{50}Ca etc. to produce nuclei far off stability. Such experiments could be performed externally but also using internal targets for increased luminosity in case of rare beams.

Weak decays of completely ionized radioactive nuclei may be investigated for the first time. There is a particular astrophysics interest in the study of a yet undiscovered decay mode, the β⁻ decay to bound states of highly ionized nuclei, which leads to relatively fast decays of normally stable atoms in a highly ionized plasma. Examples of such nuclei are ^{187}Re or ^{163}Dy, the life times of which may be reduced by many orders of magnitude in a high ionization state [5]. The measurements of endpoints of β-spectra in context with an antineutrino mass determination may be also taken up on completely ionized nuclei to avoid all perturbing atomic effects and at the same time making use of the special kinematic conditions offered by a moving beam of radioactive nuclei with precisely known velocities. More favourable candidates than ^{3}T may be chosen.

III. ATOMIC PHYSICS EXPERIMENTS

The ESR offers probably the prime attraction to precision atomic spectroscopy, the study of atomic processes with highly ionized atoms and some fundamental experiments for QED. This program has an extremely wide scope so that we can sketch it only in the present context.

First let me mention a few interesting processes which can favourably be studied with stored, cooled and highly ionized atoms. The only recombination process for completely stripped ions interacting with free electrons is radiative electron capture. It is of utmost importance for the ion losses in an electron cooling device. This process can be studied for the first time systematically as function of the relative velocity, atomic number and atomic state. There is an interesting proposal by Winnacker et al.[6], to induce radiative electron capture with intense laser light directed along the electron-ion beams in the cooling device, to populate selectively Rydberg states in highly ionized atoms. With a consecutive laser transition precision spectroscopy seems possible. Polarized light may be used to produce polarized atomic as well as nuclear states. Laser induced radiative capture may pave also a road to laser cooling of highly ionized beams. The analysis of the spectral distribution of radiative capture radiation may be used as a diagnostic for the temperature of the cooling electrons [3]. Due to its partial monochromaticity it may become a powerful source of electromagnetic radiation of variable energy up to several 100 keV. It would be circularly polarized in case of the use of polarized cooling electrons.

Another very interesting process is resonant dielectronic recombination [7]. The captured electron excites simultaneously a bound electron in the ion to an excited state. This process may be also used to excite selected states in few electron systems.

Interesting suggestions for investigation of the quasi molecular X-radiation were put forward [8]. In collisions of bare nuclei with atoms or few electron systems at energies with adiabatic collision conditions (close to the Coulomb barrier), the vacancy can be filled with an electron in the initial and final phase of the transition leading to photon emission of the same energy. Thus the transition amplitudes of the ingoing and outgoing channel interfere. From an analysis of the resulting modulation of the MO-X-spectrum as function of the impact-parameter one hopes to measure the binding energies as function of the internuclear distance of quasi molecules up to Z=184. It is also suggested to use this method in nucleus-nucleus collisions to measure the lifetime of short lived nuclear molecules or intermediate systems formed in collisions with nuclear contact; which may lead to longer time delays between the MO-X-transitions in the initial and final channel. These experiments may be performed either with the merging beam technique or by slowing down completely stripped and cooled ions from medium energies to the adiabatic velocity regime.

The merging beam technique using completely stripped uranium-ions may be most favourably applied to the study of spontaneous positron production in the overcritical Coulomb field of longlived nuclear molecules with Z up to 184 formed with cross section of about 100 mb in collisions close to the Coulomb barrier, according to recent experimental indications [9]. The study of this process with bare nuclei would increase the pair creation probability by a factor of 40 and would also allow an unique signature for the formation of the charged vacuum. In this process the electron of a pair created in the high field will be bound in the 1s-state of the Z=184 atom (the positron is emitted and will be detected) and will appear again with high probability in one of the outgoing U-ions as U^{91+} charge state. Thus we suggest positron spectroscopy in coincidence with U^{91+} ions as unique signature for the formation of the charged vacuum in $U^{92+}+U^{92+}$ collisions.

Very interesting suggestions have been made by Deslattes and collaborators [10] to perform high precision X-ray measurements on one, two and three electron systems as function of Z up to uranium. The wavelengths of the X-rays emitted from stored, cooled and slowed down heavy ion beams in the ESR are measured with special focussing X-ray diffraction devices. By measuring the energies as function of the velocity of the circulating beam, the undesired Dopplershifts may be determined and eliminated. Systematic measurements as function of Z and for different isotopes are necessary to extract the various contributions to QED corrections like selfenergy and vacuum polarization but also the undesired nuclear finite size effects.

There are also series of experiments suggested to test fundamental symmetries in electromagnetic interaction [11]. It looks like that T-invariance may be favourable studied with the radiative capture of polarized electrons in completely stripped ions with high precision. Another proposal is concerned with detection of parity violating effects in forbidden transitions (2s-1s) of high Z hydrogen- or heliumlike ions. Again the production of polarized atomic states from which the X-ray emission anisotropy should be measured, would be required.

A very interesting question was asked recently by Schiffer[12] in context with our proposal: Could there be an ordered condensed state in beams of heavy ions? Indeed, one expects as consequence of the large long range Coulomb interaction between the ions at low beam "temperatures" a short range order to be produced leading possibly to a "liquid" or "solid" frozen Coulomb lattice in the beam. The relevant order parameter is the ratio of the Coulomb energy between a pair of particles and the random kinetic energy of the beam particles. This order parameter $\Gamma = [Z^2 e^2/a]/kT$, where a is the distance between two particles, may reach values of up to 100 if a ~ 1 micron and kT ~ 0.5 eV, values which may be reached using electron cooling. This is close to order parameters needed for a phase transition[13]. Very recently Pestrikov[14] claims to have observed ordering effects in a "Coulomb string" of proton beams cooled by electrons.

IV. APPLICATIONS

Many interesting applications of the medium energy heavy ion beams have been suggested[1]. Some of those, which use the specific capability of the ESR to store and bunch high currents of medium energy heavy ions will be mentioned.

A bunched well focussed beam from the ESR stopped in a cylindrical volume of matter (0.3 mm dia, ~ 3 mm long) produces power densities of up to 0.03 TW/mg during 20 ns which may lead to temperatures of 20-50 eV and pressures of 20-100 Mbar in the material depending on the equation of state. Thus it should be possible to study the equation of state at high temperatures (> 10 eV) and dynamic effects like the energy transport by conductivity and thermal radiation and hydrodynamic flow[15].

Intense heavy ion beams are also well suited for pumping sources of lasers as has been shown recently [16]. A rough estimate [17] shows that using a Xe-beam of 1 GeV/u with 10^{11} ions/pulse, bunched to a pulse duration of 5 ns, the pumping of laser transitions in the region of some ten Å should be possible.

It was also suggested [18] to use an uraium beam of 500 MeV/u stored in the ESR traversing a thin gas target for the production of highly charged recoil ions with averaged kinetic energy of a few eV. With a beam of 10^{11} circulating uranium ions at 500 MeV/u one hopes to extract from an argon target with a thickness of 10^{14} atoms/cm^2 a beam of 10^{12} Ar^{18+} ions per sec.

The cooled beams from the ESR are also very well suited for injection into a high energy superconducting colliderring. Due to the small momentum spread, stacking in the longitudinal phase space needs only relatively small aperture of the magnets, which may lead to costsaving magnet designs.

I should like to thank all the many discussion partners which helped us to work out our plans in a very short time.

REFERENCES

1. Die Ausbauplaene der GSI, March 1984
2. SIS-Ein Beschleuniger fuer schwere Ionen hoher Energie, GSI-Bericht 82-2
3. B. Franzke et al., Zwischenbericht zur Planung des Experimentier-Speicherrings (ESR) der GSI, GSI-SIS-INT/84-5 August 1984

4. E. Otten, Workshop on the Physics with Heavy Ion Cooler Ring, May 29/30, 1984, Heidelberg.
5. K. Takahasi and K. Yokoi, Nucl. Phys. A404, 578(1983)
6. A. Winnacker (see ref. 4)
7. McLaughlin and Hahn, Phys. Letters 88A, 394(1982)
8. I. Tserruya (see ref.4)
9. J. Schweppe et al., Phys. Rev. Letters, 51, 2261 (1983)
   M. Clemente et al., Phys. Letters, 137B, 41(1984)
10. R.D. Deslattes (see ref. 4)
11. D. Bosch (see ref. 4)
12. J.P. Schiffer, private note, April 1984
13. J.P. Hansen, Phys. Rev. A8, 3096(1973), ibid A8, 3110 (1973), ibid. A11, 1025 (1975)
14. Karlsruhe cooler workshop, October 1984
15. Moeglichkeiten zur Untersuchung dichter, heisser Materie mit einem Schwerionensynchrotron, K. Beckert et al., GSI-Report 83-10- (1983)
16. A. Ulrich et al., Appl. Phys. Lett. 42, 782 (1983)
17. A. Ulrich (see ref. 4)
18. H. Schmidt-Boecking (see ref. 4)

LASER APPLICATION IN ELECTRON COOLING

H. Poth[*]
Institut für Kernphysik,
Kernforschungszentrum Karlsruhe,
Fed. Rep. Germany

ABSTRACT

Electron cooling has received continuously growing attention in the last years. Many projects to store light and heavy ions operating between a few and some hundred MeV/N are planned. In all these rings electron cooling is an essential ingredient to improve and maintain the beam quality. Apart from this task electron cooling keeps the promise of being very useful for the study of atomic physics aspects in a domain hitherto difficult to reach.

This paper describes a collinear beam arrangement for studying non-destructively the properties of an electron beam, both in the absence and in the presence of the cooled ion beam.

It allows for a detailed study of the electron cooling device prior to its installation in the storage ring. Applied during the cooling phase it can reveal further details of the cooling process. The described technique can be used to form exotic states of atoms and to study their properties. A variant of the outlined method can be applied to produce a beam of antihydrogen atoms allowing for a test of fundamental symmetries and matter-antimatter interaction.

---

[*] Visitor at CERN, Geneva, Switzerland.

1. INTRODUCTION

Application of electron cooling is compulsory for high resolution and high-luminosity experiments with ion beams. The requirements of high cooling rates imposes stringent conditions on the electron beam. The achievement of efficient cooling demands a precise knowledge of the electron beam properties. In this paper we shall discuss a method of studying the quality of a cooling electron beam in a non-destructive way.

Although electron cooling of proton beams has already been studied in pioneering experiments [1], there remain a number of open questions concerning the cooling of ion beams for future cooler projects. We shall outline the usefulness of a collinear laser-electron-ion beam arrangement for the study of particular aspects of the cooling process.

The electron cooling of ion beams opens up very interesting possibilities for the investigation of electron-ion interactions at very low relative energies. It will be shown here that the application of a laser beam in an arrangement similar to that used for electron cooling diagnostics could pave the way for a new generation of atomic physics experiments.

Electron capture by protons is well known. It accompanies the electron cooling process. Hydrogen atoms are formed which escape from the ring. If in an antiproton storage ring a positron beam is overlapped with the antiproton beam, the recombination leads to the formation of an antihydrogen atomic beam. The recombination can be enhanced by exposing the beams to a suitable light field. In this way an antihydrogen beam of sufficient intensity for many experiments might be achieved, as will be described.

A glossary of the notation used in the formulas can be found at the end of this article.

2. ELECTRON BEAM DIAGNOSTICS

Knowledge of the electron beam properties is crucial for the successful application of electron cooling. The essential parameters are the electron temperature, the density distribution, and the velocity profile of the electrons across the beam. There exists a variety of possibilities for determining these parameters, but in many cases they are destructive. Alternatively, they can be deduced indirectly from the cooling process itself. The following describes a method for the measurement of some important electron beam properties in a non-destructive way. Moreover, it provides the possibility to study the electron beam prior to its installation in the cooler ring.

The technique was originally proposed by W. Kells [2]. A laser beam is shot so that it meets the electron beam head on. In the electron cooler the laser light of frequency v_L is Thompson-scattered. The scattered light is shifted in frequency to $v_s = v_L(1 + \beta \cos \theta_L)/(1 - \beta \cos \theta_s)$ owing to the fact that the electrons are moving with velocity βc. It is expedient to study the back-scattered photons ($\cos \theta_s = +1$). The intensity and frequency spread of the back-scattered light contain the relevant information about the electron beam. Since the laser meets only a fraction of the electron beam, the electron velocity and density can be spatially mapped.

The number of back-scattered photons is given by

$$N_s = (d\sigma/d\Omega) \, N_L n_e(r) \, \Delta\Omega L \ .$$

It is an advantage to use a pulsed laser for this type of experiment and to measure the back-scattered photons within the appropriate time window only. This helps to suppress the background and to improve the peak-to-noise ratio considerably. Since the Thompson cross-section, the electron density, and the observation solid angle are fixed, the count rate depends essentially on the number of incoming photons, as can be seen from the above formula. High-power pulsed lasers have in general, however, a low repetition rate. In order to increase the average photon intensity, a special technique can be applied [3], which is to pass the laser pulse several times through the electron beam (Fig. 1). This is achieved by having the electron beam inside a cavity.

As can be seen from the above formulas, the number of back-scattered photons depends on the electron density, and their frequency spread on the electron velocity for a given solid angle. The

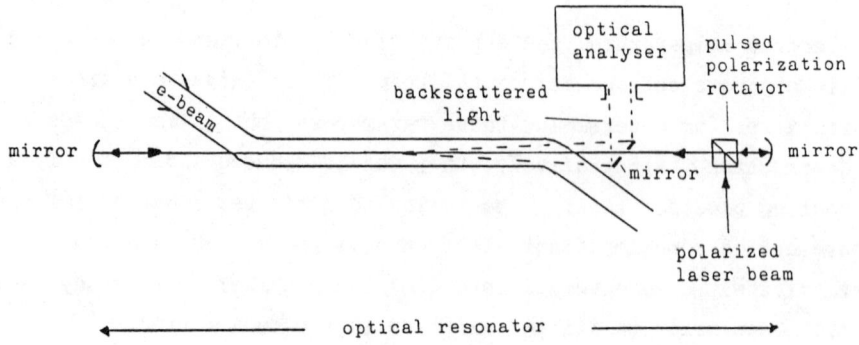

Fig. 1: Schematic collinear electron-laser beam arrangement for electron beam diagnostics.

frequency of the back-scattered photons changes with the angle θ_s. Hence the solid angle determines the frequency resolving power. The photons can be frequency-analysed in an optical system and counted by a photomultiplier. For the measurement of the electron beam density distribution, only a rough frequency analysis is necessary to suppress the background. In order to obtain spatial information, the laser beam is moved across the electron beam. It is meaningful to use a dye laser for this measurement, since for a number of reasons it is preferable to have the possibility of tuning the laser frequency as desired, without having to touch the electron beam.

Planned electron coolers typically aim at electron densities of 10^8 cm^{-3}, with cooling sections of about 2 m length. Commercially available Eximer-pumped high-power dye lasers yield pulse energies of 50 mJ at a pulse length of 20 ns and a repetition rate of 250 Hz. Operating at 500 nm, this gives photon rates of 3×10^{19} s^{-1}. Using the cavity arrangement, one may be able to increase this number by about an order of magnitude. With a suitable analyzing system (Fabry-Perot), one could cover an average solid angle $\Delta\Omega L = 2 \times 10^{-1}$ cm (2×10^{-2} cm) for a frequency resolving power of 5×10^{-4} (5×10^{-5}). This would yield an analyzable photon rate of 1400 (140). An electron velocity equivalent to $\beta = 0.5$ was assumed.

3. ELECTRON COOLING DIAGNOSTICS

In electron cooling of positive ions, cooling electrons are occasionally captured by ions. In the case of photons, hydrogen atoms are formed; these leave the storage ring and can be detected. This neutral beam is an excellent tool for controlling the cooling process. Whilst the rate of the hydrogen formation is related to the electron temperature, the properties of the neutral beam reflect those of the circulating proton beam.

The cross-section for spontaneous electron capture by a naked ion into an atomic level with main quantum number n is given by [4]:

$$\sigma_n = 1.96 \, \pi^2 \, \alpha \, \lambda_e^2 E_1^2 / [nE_e(E_1 + n^2 E_e)] \; .$$

The spontaneous capture rate is then $r_n^{spon} = n_e v_e \sigma_n$. Note that

$$v_e^2 \sigma_n = 10 \text{ mb } z^2 c^2 / n \quad \text{for} \quad E_e \ll E_1 \; .$$

Since the electrons of the cooling beam have a certain velocity profile $f(\vec{v}_e)$, the cross-section has to be averaged over this distribution to obtain the capture rate as observed in electron cooling:

$$r_n^{spon} = n_e \int f(\vec{v}_e) v_e \sigma_n d^3\vec{v}_e \; ,$$

which can be reduced to

$$r_n^{spon} = 10 \text{ mb } z^2 \, n_e \, c^2 n^{-1} \int [f(\vec{v}_e)/v_e] \, d^3\vec{v}_e \quad \text{for} \quad E_e \ll E_1 \; .$$

This integral is easily evaluated [5] for a Maxwellian and a flattened electron velocity distribution, respectively, resulting in

$$r_n^{spon} = 10 \text{ mb } z^2 \, \frac{\sqrt{\pi} \, c n_e}{n} \sqrt{\frac{m_e c^2}{2 T_{e_\perp}}} \times \begin{Bmatrix} 1 \text{ flattened} \\ 2/\pi \text{ Maxwellian} \end{Bmatrix} \text{distribution} \; .$$

As seen from this equation, the spontaneous capture rate is determined by the over-all transverse electron beam temperature $T_{e_\perp}$.

For simplicity we have restricted ourselves to captures into a specific atomic level. The above relation has to be modified (but only slightly) if the integral capture rate is considered [6].

Note also that in radiative electron capture, monoenergetic photons are emitted which may be useful for various applications [7].

In order to obtain more detailed information on local beam temperatures, we consider an arrangement similar to the one in the Thompson scattering experiment, but now using the laser to stimulate radiative electron capture. In this set-up the laser beam hits the ion beam head on as it passes through the cooling section.

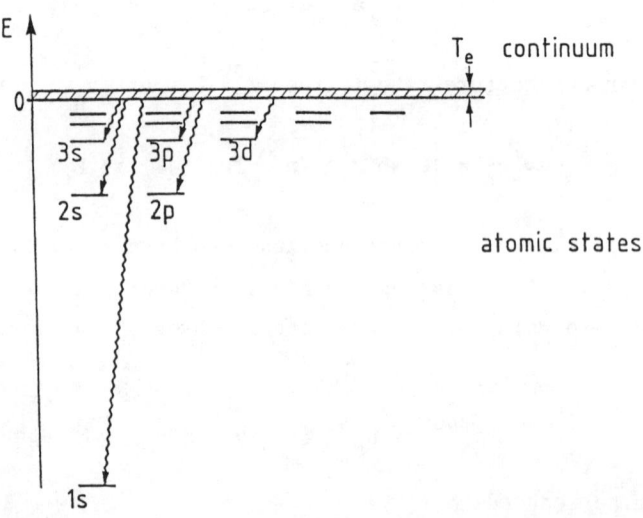

Fig. 2: Level scheme of a fully stripped ion in a cooling electron beam as seen from the ion rest-frame.

In contrast with the spontaneous capture, where all electrons of the continuum (Fig. 2) take part, now only those electrons are captured which fulfil $m(v_e^2/2) = h\nu - E_1/n^2$. When evaluating the induced capture rate [5], we find

$$r_n^{ind} = \pi^2 \lambda_e^2 \left(\frac{m_e c^2}{h\nu}\right)^2 \frac{\hbar}{T_{e_\perp}} \phi \sqrt{\frac{T_{e_\perp}}{T_{e_\|}}} \exp\left(-\frac{v_e^2}{\Delta_\perp^2}\right) r_n^{spon}.$$

This formula shows that through the determination of the induced capture rate the ratio of longitudinal to transverse electron temperature can be determined at the point where the other two beams overlap with the electron beam. Moreover, by changing the laser frequency and hence v_e, one can scan across the electron velocity distribution. If the ion beam is not cold, the ion velocity distribution also enters into the formula and can be studied. Considering a maximal useful photon flux ϕ (it is limited by the occurrence of re-ionization transitions to some 10^{25} photons $cm^{-2} s^{-1}$), we get $10-10^3$ more induced captures than spontaneous ones into the same atomic level, depending on the ratio of $T_{e_\perp}$ to $T_{e_\|}$. This has a considerable impact on other applications, as will be discussed below.

4. ATOMIC PHYSICS EXPERIMENTS

4.1 Formation of antihydrogen

Whenever a stored antiproton beam meets a collinear positron beam of equal velocity, positron capture takes place analogously to the process described in the previous section. Then antihydrogen atoms are formed and a beam of antihydrogen emerges from the positron-antiproton overlap region. It is as well collimated and monoenergetic as the antiproton beam itself.

The study of antihydrogen and its interaction with ordinary matter is, beyond any doubt, of fundamental interest.

The formation of antihydrogen through spontaneous positron capture by antiprotons was first suggested by the Novosibirsk group [8] and then elaborated in more detail by H. Herr et al. [9]. The spontaneous $\bar{H}$-formation rate is $R_{\bar{H}} = r^{spon} \eta N_{\bar{p}} \gamma^{-2}$. It is essentially limited by the quality and intensity of low-energy positron beams. Several schemes for slow positron production have been studied in detail in Ref. 9. They estimate that suitable positron beams of $1-200$ cm^{-3} density could be achieved. This has to be compared with cooling electron beam densities of typically 10^8 cm^{-3}, producing 10^{-5} hydrogen atoms per proton and second in a cooling section which occupies η = 5% of the proton storage ring. Although the presently available positron beams have rather low densities when considered for the production of antihydrogen, the rapid progress made in the development of thermalized positron sources [10] is encouraging. Moreover, an arrangement similar to that described in the previous section could be used to considerably enhance the positron capture by antiprotons. We have estimated [5] that with a laser beam giving a photon flux ϕ of 3×10^{25} cm^{-2} s^{-1}, the capture rate can be enhanced by roughly a factor of 100 during the pulse duration. Then, even with a positron beam of 1 cm^{-3}, about 1 $\bar{H}$ per second can be produced if 5×10^{11} $\bar{p}$ are stored and cooled. Furthermore there is the advantage that the laser pulse imposes its time structure on the $\bar{H}$ beam, and hence the formation time of the $\bar{H}$ atom is known.

The most obvious experiment one would like to do with antihydrogen is the measurement of the 2s-2p Lamb shift. For this purpose the 2s-level population has to be enhanced. The partial spontaneous capture rates are [5] $r_{1s}:r_{2s}:r_{2p}$ = 100:15:40. Induced capture to the n = 2 level would populate the angular momentum states according to their statistical weight. The decay of the 2p state to the 1s state is immediate, whilst the positron would remain in the 2s level provided there is no Stark quenching. (An electric field of 0.5 kV cm^{-1} would be sufficient to quench the 2s to the ground state in a flight path length of 10 cm if the antihydrogen atom travels with β = 0.3.) This could already be provoked by the motional electric field originating from the transverse magnetic field $B_\perp$ in the positron beam guiding field. This may, however, be compensated by an electric field (E = 300 $\beta\gamma B_\perp$) which would not affect the $\bar{p}$ beam and would only disturb the positron beam in a controllable way. The measurement of the Lamb shift, for instance, could then be performed in the straight section between the cooler and the next quadrupole by inducing the 2s → 2p transition, or quenching and detecting the 2p → 1s transition photons. A Lamb shift experiment with a fast hydrogen beam produced through spontaneous capture of cooling electrons was performed in Novosibirsk [11].

It is clear that many technical problems have to be solved before one arrives at a useful rate for $\bar{H}$ production. But in principle there is no obstacle, and many of the underlying problems can be studied in electron cooling of a proton beam using the triple-beam arrangement.

4.2 Formation of Rydberg atoms

In electron cooling of naked ions, spontaneous capture takes place with a rate given in Section 3:

$$r^{spon} = \Sigma \, r_n^{spon} = n_e \alpha_r^{spon} \, z^2 \, .$$

The recombination coefficient α_r has been determined experimentally to be about 2×10^{-12} cm^3 s^{-1} for typical electron cooling arrangements, in good agreement with theory. As is apparent from the capture cross-section, the rate scales with n like n^{-1}-n^{-3}. This means that most of the electrons are captured into low n-states. Let us consider, for instance, Ca^{20+}. The ground-state capture rate is $r_1^{spon} = 2.4 \times 10^{-2}$ s^{-1} per ion and 10^8 cm^{-3} electrons. In the n = 42-state, for example, it is 42 times lower. By irradiation with laser light of appropriate frequency, however, captures into this level can be induced, leading to a much higher level population. With the formula of Section 3 for induced recombination, an enhancement factor of $g = r_{42}^{ind}/r_{42}^{spon} = 420$ can be estimated, i.e. a 10 times higher population than in the ground state ($\phi = 10^{25}$ cm^{-2} s^{-1}, $T_{e_\perp} = 0.2$, $\sqrt{T_{e_\perp}/T_{e_\parallel}} = 30$) during the laser pulse duration. The total capture rates, however, barely increase, since the gain factor g has to be multiplied by the duty factor of the laser, which in the case considered here is about 5×10^{-6}. This means that

$$r_{42}^{ind} = g \, \frac{r_1}{42} \, 5 \times 10^{-6} = 1.2 \times 10^{-7} \text{ s}^{-1} \text{ per ion and } 10^8 \text{ cm}^{-3} \text{ electrons} .$$

Assuming that 10^{10} ions can be stored in a ring with an electron cooling section occupying 5% of the ring circumference, we end up with a formation rate for Rydberg atoms of nearly 100 s^{-1}.

This rate may also be further increased if one gives up the concept of a tuneable dye laser and uses a fixed-frequency high-power laser to deliver a higher average number of photons. The energy-matching can then be achieved by tuning the ion and the electron velocity correspondingly.

The lifetime of Rydberg atoms is rather long and may in some cases exceed the revolution time of the ions in the ring. This is advantageous for those states that are not field-ionized in the bending magnets of the storage ring. Since the naked ion beam can be separated from the hydrogen-like beam, the formation rate of the Rydberg atoms can be measured at a suitable position in the ring as a function of the laser frequency and within a time window corresponding to the laser pulse duration.

In a triple beam arrangement, having the cooling section within a laser resonator as described here, also photons travelling parallel to the two particles (i.e. reflected from the second mirror) can be used for induced capture. They are red-shifted in energy, and address even higher levels. In this way one may be able to form Rydberg states which can be magnetically stripped. Again they are separated from the primary beam and can be identified in a suitable detector set-up.

4.3 Production and study of metastable helium-like ions

The 2^3S_1 states of light helium-like ions are metastable. The lifetime for B^{3+} in this state is for instance 150 ms. This is long compared to the revolution time of an ion in a typical ring (µs). Production and storage of such ionic states provides a unique way to study these systems under clean conditions. In the following, only a brief outline of the basic idea to use a triple-beam set-up in this context is given.

Consider a hydrogen-like stored ion beam cooled by electrons. Through laser-stimulated electron capture, the ion beam can be down-charged by one unit. With a suitable ring lattice the two charge states of the ion beam can be stored. The n = 2 states will be populated according to their statistical weight (25% singlet, 75%

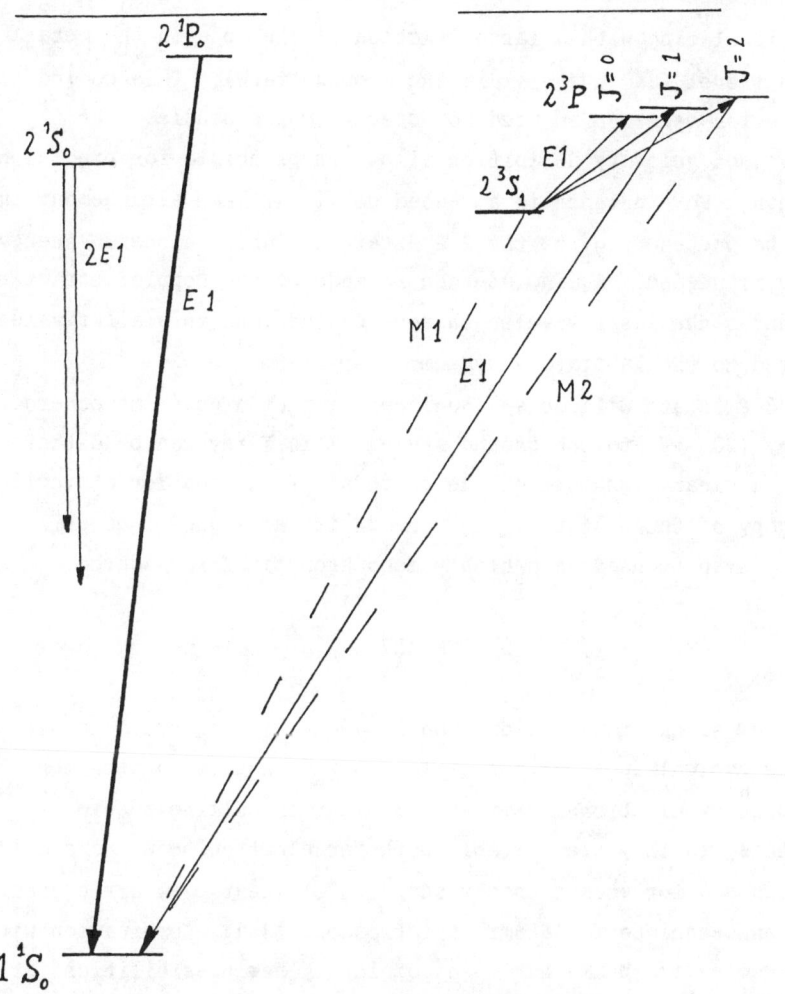

Fig. 3: Lowest states of a light helium-like ion. The strength of the arrows indicates the strength of the transitions.

triplet). While the 2^3P and the singlet states will decay fast, ions in the 2^3S_1 will survive thousands of revolutions because of the slow M1-transition (Fig. 3). The $^3S_1$ state is not quenched in the magnetic field because of its large separation from the P states. Hence, several revolutions after the laser pulse, a helium-like ion beam is circulating with a large fraction of the ions in the metastable $^3S_1$ level (and the rest in the ground state). This cooled metastable ion beam can be used for spectroscopic studies. The extremely good velocity definition allows in principle for precision experiments. For instance in a second collinear beam arrangement the ions can be pumped, e.g. to the 2^3P_1 state for which a photon energy of 4.4 eV is needed. Again, use can be made of the Doppler effect which reduces the laser wavelength seen by the ions to smaller values as compared to the laboratory system.

The 2^3P_1 state will decay under emission of a quasi-monoenergetic soft X-ray (203 eV) to the ground state. This X-ray can be detected and gives a clear signature of the process. It allows for a precise spectroscopy of these states. In Ref. 12 it was argued that this method may even be used as a highly monochromatic X-ray source.

5. SUMMARY

In view of the many cooler ring projects, it was tried to outline a few typical applications of lasers in conjunction with ion beams cooled by electrons. Emphasis was put on collinear beam arrangements, as they are suitable both for electron beam and cooling diagnostics and for atomic spectroscopy. Collinear beam arrangement is a proven technique for atomic spectroscopy [13]. Combination with electron-cooled ion beams may open completely new possibilities for atomic physics experiments.

ACKNOWLEDGEMENTS

Thanks for stimulating discussions are due to C. Habfast, H. Herr, D. Möhl, R. Neumann, H. Pilkuhn, A. Wolf and A. Winnacker.

GLOSSARY

- c = speed of light
- E_e = electron energy in the ion rest frame
- E_1 = electron binding energy in a hydrogen-like ion ground state
- L = length of the cooling section
- m_e = electron mass
- n = main quantum number in a hydrogen-like ion state
- n_e = electron density
- N = number of stored ions
- N_L = number of photons per laser pulse
- N_s = number of back-scattered photons
- r = distance from the electron beam axis
- r_e = classical electron radius
- $T_{e_\parallel}$ = longitudinal electron temperature
- $T_{e_\perp}$ = transverse electron temperature
- v_e = electron velocity in the ion rest frame
- Z = charge of the naked ion
- $\hbar$ = Planck's constant
- βc = velocity of the electrons (ions)
- γ = $(1 - \beta^2)^{-1/2}$
- $\Delta_\perp$ = transverse electron velocity spread
- $\Delta\Omega$ = solid angle for the observation of back-scattered photons
- η = fraction of storage ring occupied by electron cooling
- θ_L = angle between laser and electron beam (= 0 in the case considered here).
- θ_s = angle between scattered photon and electron beam
- ν_L = laser light frequency
- ν_s = frequency of the scattered photons
- p = photon flux of the laser beam
- λ_e = Compton wavelength of the electron
- $d\sigma/d\Omega$ = backward differential cross-section for Thompson scattering $[r_e^2 (1+\beta)(1-\beta)^{-1}]$

REFERENCES

[1] G.I. Budker et al., Part. Accel. 7 (1976) 197.
  M. Bell et al., Nucl. Instrum. Methods 190 (1981) 237.
  R. Forster et al., IEEE Trans. Nucl. Sci. NS28 (1981) 2386.

[2] W. Kells, Fermilab Technical Memorandum, TM 771 (1978).

[3] A. Wolf, KFK-Primärbericht 11.01.02 P17 B, Kernforschungszentrum Karlsruhe (1982).

[4] H. Bethe and E. Salpeter, Quantum mechanics of one- and two-electron systems, in Handbuch der Physik (Springer Verlag, Berlin, 1957), Vol. 35, p. 88.

[5] R. Neumann et al., Z. Phys. A313 (1983) 253;
  R. Neumann, Laser-induced electron-ion recombination, Proc. Workshop on Electron Cooling and Related Applications (ECOOL84), Karlsruhe, 1984, ed. H. Poth (to appear as KFK-Bericht 3846, Karlsruhe, 1985).
  A. Winnacker, Laser-induced electron capture, in Proc. Workshop on the Physics with Heavy-Ion Cooler Rings, Heidelberg, 1984 (ed. D. Habs et al.).

[6] M. Bell and J. Bell, Part. Accel. 12 (1982) 49.

[7] H. Poth and A. Wolf, Phys. Lett. 97A (1983) 135.

[8] G.I. Budker and A.N. Skrinsky, Sov. Phys.-Usp. 21 (1978) 277.

[9] H. Herr et al., Proc. Workshop on Physics with Low-Energy Cooled Antiprotons, Erice, 1982, eds. U. Gastaldi and R. Klapisch (Plenum Press, New York, 1984), p. 659.

[10] A.P. Mills, Jr, Appl. Phys. 23 (1980) 189, and Brookhaven National Bulletin Vol. 38, No. 39, Oct. 1984.

[11] W.W. Parkhomchuk, private communication.

[12] H. Pilkuhn and H. Poth, An ion beam lamp for monochromatic X-rays, Proc. Workshop on Electron Cooling and Related Applications (ECOOL 84), Karlsruhe, 1984, ed. H. Poth (to appear as KFK-Bericht 3846, Karlsruhe, 1985).

[13] A.C. Müller et al., Nucl. Phys. A403 (1983) 234.

NEW POSSIBILITIES WITH RECOILLESS KINEMATICS
USING HIGH QUALITY PROTON BEAMS

Kurt Kilian
CERN, EP Division, Geneva, Switzerland

ABSTRACT

Recoilless or "magic" kinematics which has been used successfully in hypernuclear physics has peculiar features in nucleon-nucleon interactions. It should provide a powerful tool to study low energy pion scattering, low energy pion nucleus interactions, η-nucleus interactions or the recoilless excitation of Δ or N^* resonances in nuclei.

INTRODUCTION

Under certain kinematical conditions one can produce in an inelastic two body reaction

(1) $$1 + 2 \to 3 + 4$$

a new particle (say 4) at rest in the laboratory system. This happens at a "magic" beam momentun p_1^{mag} when particle 3 goes forward to 0° and overtakes exactly the momentum of the beam particle 1. The magic momentum p_1^{mag} depends on the masses m_i (i=1,..4) of the reaction partners

(2) $$p_1^{mag} = \sqrt{\frac{[(m_1^2 + m_2^2) - (m_2-m_4)^2]^2 - 4m_1^2 m_3^2}{4(m_2 - m_4)^2}}$$

Only the difference of the target mass m_2 and the mass of the "recoilless" particle m_4 appears in eq. 2. This means that we can embed the elementary reaction eq. (1) in the nuclear environment of a residual nucleus R without changing the "magic" condition

(3) $$1 + \binom{R}{2} \to 3 + \binom{R}{4}^*$$

The elementary reaction (equ. 1) which is at high energy, just creates in a direct single step process a nucleon hole by removing particle 2 from the target. The residual nucleus R is left in a simple hole configuration. Into this well defined system R now the particle 4 is introduced in a smooth, cold way with small relative energy ε and small momentum transfer q. In the limit of q=0 the orbital quantum numbers of the removed particle 2 and the introduced particle 4 will be the same and (R+4)* states will be populated which have the same orbital configuration as the target.

By changing the beam momentum or scattering angle the relative energy ε and momentum transfer q between particle 4 and the residual nucleus R can be tuned, and various (R+4)* particle hole configuration can be selectively populated.

In a missing mass experiment (measuring beam particle 1 and outgoing particle 3) one can reconstruct the excitation spectra of (R+4)* and study the interaction of particle 4 with the nuclear system R.

It was suggested quite some time ago to study the Λ-nucleus interaction (hypernuclei) using the elementary interaction $K^- + n \to \pi^- + \Lambda$ at its magic K^- momentum around 550 Mev/c embedded in nuclei [1]. Such experiments were performed in the last decade. They have proven that magic, recoilless kinematics is really an excellent tool [2].

In this paper I want to draw attention to some very interesting aspects of magic kinematics in connection with nucleon-nucleon interaction. A proton beam allows to create on a nucleon recoillessly one or more pions, heavier mesons (η, ρ, ω..) or just "excitation energy". (See Table 1). Performed on bound nucleons one can introduce the new objects in a cold way into a nucleus. The low energy, low momentum transfer interaction of those objects with a well defined and cool nuclear system can be studied. To my knowledge this has never been tried so far.

Table 1: Kinematic quantities for "magic" conditions in proton-nucleon interactions. When "particle 3" is a two-nucleon system then 10 MeV/c² have been added to take into account the relative NN energy. Recoilless hyperon production is shown for comparison.

Elementary reaction				threshold momentum	"magic" quantities		
					momentum	energy	mass
beam 1 +	target 2 →	scatt 3 +	recoilless 4	p_1^{th} MeV/c	p_1^{mag} MeV/c	T_1^m MeV	$\sqrt{s}$ MeV/c²
p	n	d	"50 MeV"	436,8	443.3	99.5	1927
p	p	d	π^+	788.8	828.65	287.5	2027
p	p	(pp)	π^0	809.8	849.4	327.4	2034
p	n	(pp)	π^-	820.3	862.1	335.9	2039
p	p	d	$\pi^+\pi^0$	1202.7	1363.4	716.7	2206
p	p	(NN)	"300 MeV"	1308	1512	841	2258
p	n	d	η^0	1978.4	3037.9	2241.2	2782
p	p	(pp)	η^0	2014.5	3105.5	2305.8	2802
p	N	(NN)	ω	2699	8463	7576	4212
K^-	n	π^-	Λ^0	–	530.9	231.3	1.578
K^-	n	π^-	Σ^0	–	284.5	76.1	1.482

KINEMATICAL AND EXPERIMENTAL CONSIDERATIONS

The basic kinematical facts can be explained with the reaction $pp \to d\pi^+$. The elliptical polardiagrams of the π^+ momenta at various beam momenta are plotted in Fig. 1. The (low momentum) pion and (high momentum) deuteron distributions add up to the beam momentum.

(4) $\qquad \vec{p}_\pi + \vec{p}_d = \vec{p}_1 \qquad (p_\pi^\perp = -p_d^\perp, \quad p_\pi^L + p_d^L = p_1)$

Fig. 1 shows that, close to threshold, pions are always thrown forward by the center of mass (c.m.) movement. At high beam momenta the velocity (energy) of pions in the c.m. system becomes high enough that they can go backward in the laboratoy system. The minimum (maximum) momentum pions are always at 180° (0°) in c.m. and go backward (forward) in the cm system. The magic beam momentum is reached when 180° backward pions in the c.m. system just compensate the forward boost of the c.m. system and appear at rest in the laboratory system. They are related to deuterons which are going forward with beam momentum at 0°. Deuteron angles larger than zero correspond to finite transverse deuteron momenta. These are balanced by opposite going transverse pion momenta.

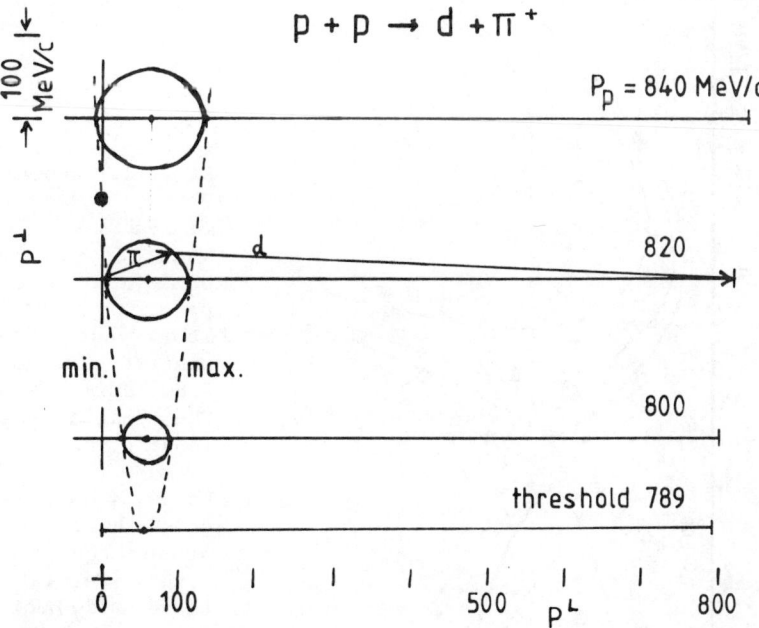

Fig. 1: Polar diagrams for pion and deuteron momentum vectors in the reaction $pp \to d\pi^+$. The dependence of the maximal and minimal pion momentum on the beam momentum is indicated by the dashed curve. Backward pions have small energy for beams between 820 and 840 MeV/c.

If one allows for pion momenta up to a certain maximum (say 16.7 MeV/c which corresponds to a kinetic energy $\tau_\pi = 1$ MeV) then these are only available in a limited range of beam momenta (Fig. 2). Those low energy pions come only from a part $\Delta\Omega$ of the c.m. solid angle around 180° which is purely kinematically defined and plotted in Fig. 2. Multiplying this solid angle $\Delta\Omega$ with the momentum dependent $d\sigma/d\Omega$ (extrapolated according to ref.3) gives us a partial cross section $\Delta\sigma = \Delta\sigma(p_1)$ for low energy ($\tau_\pi < 1$ MeV) pion production (Fig. 2) which goes up to 3 μb at $p_1 \simeq 820$ MeV/c. The total pp cross section here is only ~ 25 mb and dominated by elastic scattering. The low energy pions come with a rather high branching ratio[4]. Kinematically they can be clearly distinguished from elastic events. The pp → ppπ° reaction is the third open channel. Its cross section is much smaller here.

Let us imagine that the pp → dπ+ reaction is embedded in a nucleus. Assume that those ≤ 1 MeV pions interact strongly with the residual nucleus R in the form of say 2 proton emission. Then this signature will appear with an energy dependent cross section like $\Delta\sigma$ in Fig. 2. It might be misinterpreted as a resonance effect in the direct channel while it would be in reality just the proof that low energy π+R interactions are best studied at the magic kinematics.

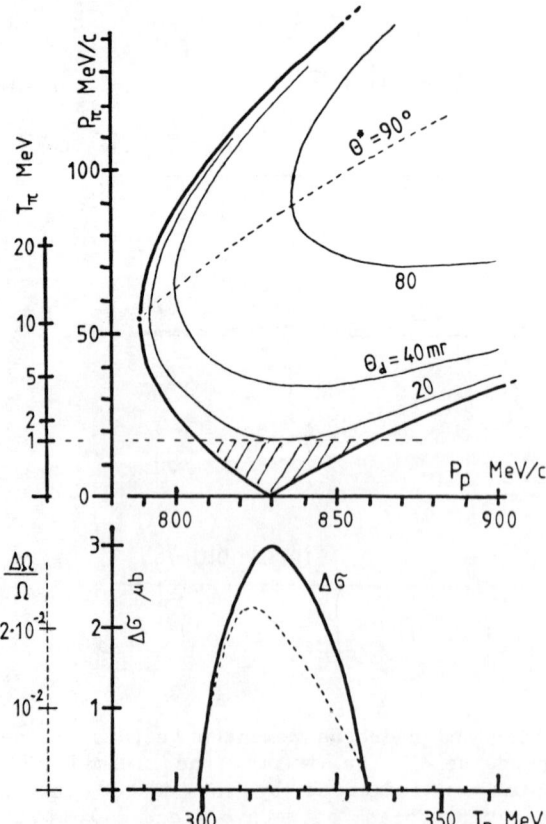

Fig. 2: Upper part: Pion laboratory momentum p_π as a function of proton beam momentum p_p in pp→dπ+. The solid lower curve is the minimal pion momentum (180° emission in cm). The "magic" momentum is at p_p=829 MeV/c.

Lower part: Partial cross section $\Delta\sigma$ for production of pions with <1 MeV lab energy ($\Delta\sigma=(d\sigma/d\Omega)\Delta\Omega$). The relevant $\Delta\Omega$ is shown as dashed curve.

The above kinematical considerations are not changed if we consider as scattered particle 3 not a bound deuteron but 2 independent nucleons N and N'. Their total momentum will replace p_3 $\vec{p}_3 \equiv \vec{p}_N + \vec{p}_{N'}$ and the mass m_3 has to be replaced by $\sqrt{s_{NN'}} \simeq m_N + m_{N'} + \varepsilon_{NN'}$. Recoilless reactions of the type

(5) $\qquad\qquad\qquad$ N + N → (NN') + X

have attractive features in our context

* They allow to get pions or other mesons of all charge states especially π^- (e.g. in p + n → pp + π^-)
* They allow to have the (NN') system with internal odd angular momentum thus negative parity. In contrast to pp → dπ^+ where parity conservation needs $\Delta\ell = 1$ the reaction NN → (NN') + π allows to replace particle 2 (in equ. 2) by a pion with $\Delta\ell = 0$.
* In a recoilless reaction p + N → (pp) + X the two individual protons can be well separated in a magnetic spectrometer from the primary beam since they have both about half the beam momentum while deuterons at magic momentum can only be separated by time of flight from the primary beam.

There are obvious signatures related to recoilless production p + N → NN + X.

* Deuterons or (NN) combinations which go forward with a total momentum close to the beam momentum.
* The recoilless particles X interact with the residual nucleus R. If they are in bound states with R then e.g. quasi-deuteron absorption can occur giving two nearly back to back nucleons.
If X is in a resonance state a pion can be emitted in all directions with a characteristic energy.

POSSIBLE APPLICATIONS OF MAGIC KINEMATICS WITH PROTON BEAMS

A) Scattering experiments with tagged very low energetic $\pi^\pm$

The lab momentum distribution of pions in an elementary reaction pp → dπ^+ or pp → NNπ can be tagged by measuring the scattered d or 2N system. For a secondary pion scattering on another target (or in the same hydrogen target) then the pion direction and momentum is known. In comparison with experiments on normal pion beams one has the advantage that very short distances between pion production and pion scattering are feasible so that pion decays in flight are negligible even at very low momenta. The decay length of a 1 MeV $\pi^\pm$ (16.7 MeV/c) for example is 94 cm. This leads to negligible decays in flight in a double scattering setup with typically 10 cm size but it is prohibitive for a say 10 m long standard pion beam where only a fraction of $2.4 \cdot 10^{-5}$ of the pions would survive.

B) Study of hadronic pion nucleus systems

The pion is at the same time mediator of NN interaction and partner in πN interactions. Implanting a real low energy pion in a nucleus could therefore lead to a complex but also interesting situation. Of course pions are absorbed in a nucleus but the absorption width Γ is much smaller than the pion mass. About Γ≈10 MeV can be estimated from π-R potentials[5]. If pion-nucleus resonant or bound states exist or not can be checked in a missing mass experiment

$$p + \binom{R}{N} \to \frac{d}{(or\ NN)} + \binom{R}{\pi}^* \quad (6)$$

where the mass spectrum of the (R+π)* system is determined from the forward p → d or p → (NN) kinematics. One optimizes the production of (Rπ)* systems by working at the magic momentum. If there is a (R+π)* resonant system than also monochromatic pions should be observable in all directions from the (R+π)* system at rest with a yield which is dependent on the beam momentum like Δσ in fig. 2. The maximum pion yield should occur at a beam momentum which is "magic" for the production of the pion mass and the excitation energy of the (R+π)* system. There might be experimental indications of this situation[6] in experiments where one sees a "flash" of low energy pions at 90° in inclusive p+Cu interactions around 350 MeV. One should extend these experiments and check if deuterons or 2N systems go forward with beam momentum in coincidence with the pion flash.

C) Study of deeplying π^- atomic states

Pionic atoms are a well known example of bound pion nucleus systems. They are normally studied by stopping π^- in a target and observing the X-ray cascade of π^- captured in the Coulomb field of target nuclei (from outside in). The deeplying atomic states in heavy nuclei where hadronic distortions are strongest are not accessible this way since just the hadronic absorption of π^- becomes so much faster than the radiative transitions that the X-ray cascade dies out. The recoilless production of the type

$$p + \binom{R}{n} \to pp + \binom{R}{\pi}^* \quad (7)$$

should be very favourable in populating deeplying atomic states from inside out[5]. The position and width just of the hadronically distorted deeplying states could be determined in a missing mass experiment with the trigger signatures described above. One can get information about pion nucleus low energy interaction in the central high density region of a nucleus.

D) η mesons in nuclei

One can repeat the arguments for recoilless production of η mesons. The total pp → dη cross section in the magic region around 3050 MeV/c is about 100 to 200 μb[6]. η energies below 1 MeV can be obtained for proton beam momenta between 2810 MeV/c < p_1 < 3360 MeV/c in a part of the solid angle $\Delta\Omega/\Omega$ ≤ $7.6 \cdot 10^{-4}$. The η production is symmetric in the c.m. system and should be enhanced in 0° and 180° direction which would enhance the production in the recoilless region.

The η is an isoscalar. In contrast to the isovector pion and it should probe different aspects of meson nucleus interactions. With its higher mass of 550 MeV/c it could allow to use its absorption inside nuclei in some respect like central low energy $\bar{p}$ annihilation.

E) Recoilless excitation of nuclei

In (p,d) or (p, 2N) reactions one can also transfer pure excitation energy without recoil into nuclei.

(8) $p + \binom{R}{N} \rightarrow ^d_{(or\ NN)} + \binom{R}{exc}$

For deeplying hole states (instead of a 140 MeV pion one considers a ≃ 50 MeV hole) the magic momentum would be around 450 MeV/c (see Table 1). At this momentum, selected hole states should stand out very clearly since the remaining nucleus is disturbed minimally[9].

Finally one could introduce without stress on the nuclear structure energies which allow to excite in a resonant way Δ or N* resonances inside nuclei (≈ 300 or ≈ 500 MeV respectively). If there are modifications of width and mass of these resonances by the nuclear medium, they should be measurable. A signature for those resonances might be the rather isotropic emission of energetic pions.

CONCLUSION

Recoilless kinematics in proton experiments seems to open up exciting new possibilities. It may be that the predicted effects have not been seen so far because nobody has searched for them. With the good beam quality in the relevant energy range which should be available at the cooler ring of IUCF and then also in Uppsala and Jülich one could try out some of the ideas described here.

Aknowledgement: I should like to thank T.E.O. Ericson, J. Julien and W. Weise for many stimulating discussions.

REFERENCES AND FOOTNOTES

1) M.I. Podgoretski, Zh.Eksp. & Theor. Fiz. <u>44</u> (1963) 695.
   H. Feshbach and A.K. Kermann, Preludes in theoretical physics (North Holland Amsterdam 1966) p 260.
2) See for example: Proc. of the Int. Conf. on Hypernuclear and Kaon Physics, Heidelberg June 1982.
   B. Povh, Ann. Rev. Nucl. Sci. <u>28</u> (1978) 1.
   K. Kilian in Proc. Meson Nuclear Physics Conf. Houston 1979 (AIP Conf. Proc. No 54) p 666.
3) C.M. Rose, Phys. Rev. <u>154</u> (1967) 1305.
4) These slow pions can make experimental problems. They might stop in the hydrogen target and decay into μ^+. The muons may hit detectors even in backward direction and simulate "large angle pions" where there are none. These decay muons would appear with a yield similar to the $\Delta\sigma$ curve in fig. 2 which then could be misinterpreted as a s-channel resonance at $\sqrt{s}=2020$ MeV. Two pions recoillessly produced, absorbed and decaying would fake a "dibaryon" resonance at ~ 2200 MeV.
5) This can be deduced from π nucleus potentials, given e.g. in M. Ericson and T.E.O. Ericson, Ann. of Phys. <u>36</u> (1966) 323. and W. Weise, private communication.
6) V.A. Krasnov at al., Phys. Lett. <u>108B</u> (1982) 11
   J. Julien et al., Phys. Lett. <u>142B</u> (1984) 340.
7) Formation of pionic atoms from "inside out" has been discussed by G.T. Emery, Phys. Lett. <u>60B</u> (1976) 351 using threshold production. In the reaction $^{13}C(p\pi^-)^{14}O$ used as example by Emery the recoil on the nuclear system is quite high (≥ 550 MeV/c) and the π^- production at threshold is small.
8) E. Pickup at al, Phys. Rev. Lett. <u>8</u>(1962) 329.
   Compilation of cross sections CERN-HERA 84-01 (1984).
9) This recoilless excitation was suggested by T.E.O. Ericson. See also T.E.O. Ericson and K. Kilian, PANIC, Particles and Nuclei X Int. Conf. Heidelberg, July 1984, Book of Abstracts, Vol II contribution N9.

SOME PARTICLE-PHYSICS PROBLEMS FOR COOLER RINGS

E. Hagberg and S. Kullander

The Gustaf Werner Institute
University of Uppsala
Box 531, S-751 21 Uppsala, Sweden

ABSTRACT

A collaboration has been formed proposing a program of particle physics experiments using the unique features of a cooled storage ring beam. The main points of the program are 1. The study of rare π^0 decays, 2. a search for dibaryon states and 3. the measurement of the neutrino mass.

INTRODUCTION

Storage rings with cooled beams are well suited for some of the relevant problems of contemporary elementary particle physics. A collaboration consisting of groups from Poland and Sweden has been formed in order to carry out a series of experiments at the CELSIUS facility in Uppsala[1]. The experiments proposed are on rare neutral pion decays, particularly the positron-electron mode, searches for narrow dibaryon states and as a second or third generation experiment, neutrino mass measurements. The pion decay experiment profits from the thin targets which will be used in CELSIUS. The severe background from gamma-ray conversion in the target will thus be reduced compared with other measurements made so far. Moreover, accurate measurements can be made of the recoils from the reaction proposed as a source for the pions.

Dibaryon states will be searched for in the invariant mass spectrum of two protons resulting from proton interactions with hydrogen. The energy precision of the circulating beam and the small energy loss in the target by the two decay protons will help in obtaining a best possible energy resolution. Moreover, the possibility of tuning the energy of the ring continuosly will be valuable for these measurements. In the following we develop these two problems in some more detail and we also mention briefly how neutrino mass measurements could be made.

DIBARYON STATES

The information on possible dinucleon states, mostly from inclusive experiments, has been summarized in the surveys of Yokosawa[2] and Makarov[3], where extensive references are given. The candidates have masses in the 2100-2500 MeV range. In recent years, exclusive or semiexclusive experiments have been performed using bubble chambers. Candidates for dinucleon states (pp, pn and nn states) have been found at masses around 1936, 1960, 2030 and 2140 MeV[4]. The bubble chamber experiments, give complementary information to the inclusive counter experiments, their advantage being the information on particle correlations, their serious drawback being the limited statistics.

Apart from the conventional approaches to the interpretation of dinucleon states, many six-quark states have been predicted in quark models. Among them, narrow dibaryons with width in the MeV range can, most probably, be understood only as real six-quark exotics. Several predictions for narrow or even stable dibaryon states can be found in the litterature[5,6]. The existence of such states within the energy range of CELSIUS can not be excluded from present experimental data. Conclusive experimental information concerning the existence of six-quark dinucleon states would be of great importance for our understanding of the structure of hadrons and would put some severe constraints on current quark models.

An extension of this experiment could include the possibility to establish the existence of 'demon deuterons'. The demon deuteron is a hypothetical metastable six-quark state consisting of three diquarks[7]. It was first introduced to explain the 'anomalons', nuclear fragments with an enhanced reaction cross section that emerges from high energy heavy-ion collisions. The predicted mass is around 2 GeV/c^2 and the quantum numbers are extraneous to the NN-channel so the main decay mode would be electromagnetic to $np\gamma$. The demon life-time is therefore expected to exceed 10^{-10} s. A search for narrow demon states can be performed within the experimental program for dibaryons and the result would help in clarifying the present confused experimental situation concerning anomalons.

NEUTRAL PION DECAYS

Two neutral pion decay modes are of particular interest[8]. One is the internal conversion or Dalitz decay, $\pi^0 \to e^+e^-\gamma$. The invariant mass distribution of the lepton pairs in this decay gives information on the quark struc-

ture of the form factor governing the decay. The total
branching ratio for this decay is about 1.2 percent.
However, the sensitivity to the form factor is greatest
for invariant masses between, say, $0.4\, m_\pi$ and $0.9\, m_\pi$
which reduces the relative rate to $\sim 10^{-4}$. The other
decay of interest to is the rare $\pi^0 \to e^+e^-$ decay, which is
helicity suppressed in electromagnetic interactions and
therefore sensitive to "exotic" contributions such as
Higgs-like particles or hypothetical pseudoscalar partners to the Z^0 [8,9]. Two experiments have reported inte-
restingly large values, more than a factor two above the
one predicted by an electromagnetic decay mode, but with
insufficient precision. Since the branching ratio is
only of the order of 10^{-7}, this type of experiment is
quite difficult, demanding both high luminosity and low
background.

EXPERIMENTAL ASPECTS

A large yield of π^0's can be obtained from the reaction
$pd \to {}^3He\,\pi^0$. The total cross section is of the order of 50 μb.
At present we envisage to detect the photons, positrons
and electrons from the decay in a detector array with
large solid angle (Figure 1). This same apparatus will

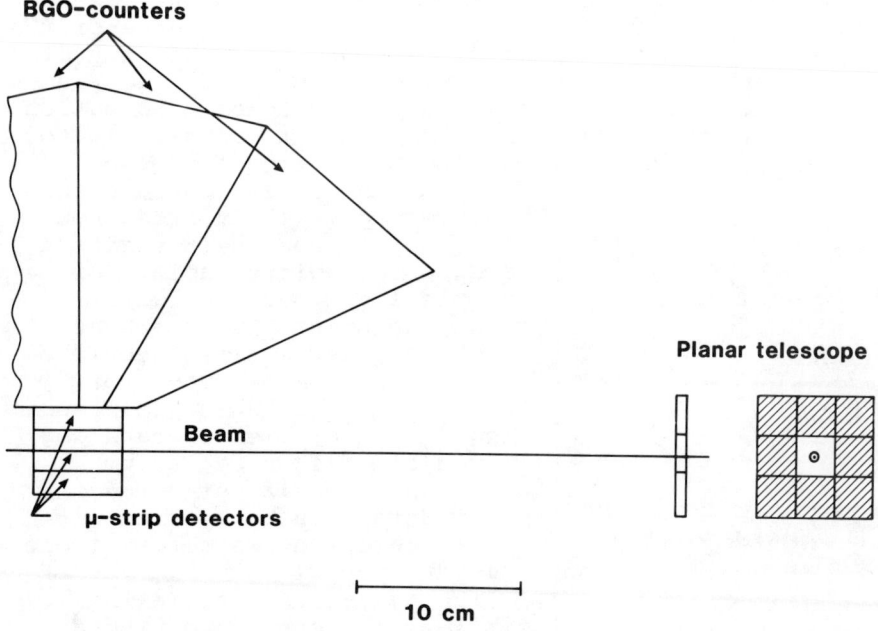

Fig. 1. Schematical layout of the experimental apparatus.

also be used to detect and measure two protons from possible dibaryon decays. The particles will be localized in silicon strip detectors and their energy will be measured using dense, probably BGO, scintallotors arranged in a cylindrical geometry. The set-up will also contain a detector for forward-going recoils. It will be made of silicon strips for the position measurement and ΔE-E detectors for the identification and the energy measurement. The proposed set-up will give a mass resolution of a few MeV/c^2 around the pion mass, necessary for the proper identification of a positron-electron pair.

Due to muliple coulomb scattering in detector material it is necessary to use the precise knowledge of the target position together with the position information from one detector plane. Thin wire targets discussed in a report to this conference [10] are unsuitable in our case because of lack of knowledge of the vertical coordinate. One way of combining the need for high luminosity with precise information about the target position is to use liquid or frozen drops of deuterium. These have to pass the beam very quickly in order not to evaporate and their positions have to be registed using for example laser beams. Figure 2 shows the principle of a scheme which to a large extent could be taken over from one used for fluorescence activated cell sorting[11]. The problem of handling liquid and frozen deuterium drops has been investigated in great detail in connection with fusion research. We refer to one relevant paper[12].

The possibility of making small-size drops traversing a beam with velocities in the range 10 to 100 m/s seems possible from our known information so

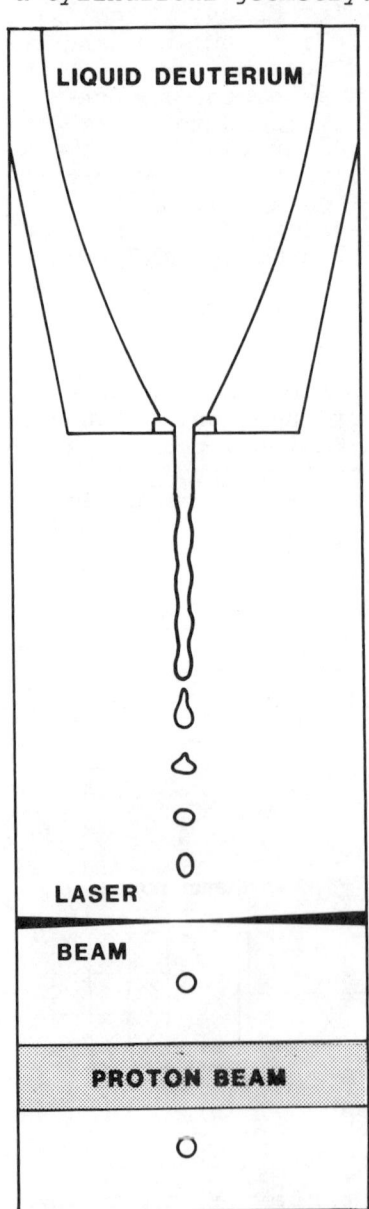

Fig. 2.

far. The heating and charging of drops should be manageable at the aimed luminosities of 10^{32} cm^{-2}s^{-1}. Drops of 20 μm size in a beam containing 10^{10} protons and 2 mm in diameter will give the required luminosity.

NEUTRINO MASS MEASUREMENTS

Storage rings offer some unique possibilities to measure the neutrino mass[13]. Tritons can be stored and their decay products measured thus avoiding the uncertainties from atomic and molecular excitations. If an energy resolution of 10^{-4} for the circulating triton beam can be obtained very accurate mass determinations are possible. Unfortunately the required number of circulating tritons must be at least 10^{12} for such an experiment to be feasible. Future developments of storage rings will show its possibility.

REFERENCES

1. B. Badelek et al., Letter of intent for a research program on elementary particle physics at CELSIUS, 1984-03-30.
2. A. Yokosawa, ANL Report HEP-CP-80-01.
3. M.M. Makarov, Sov. Physics Uspekhi 136, 185 (1982).
4. I.P. Zielinski, CELSIUS Note 84-1 and references therein
5. M. Jändel, CELSIUS Note 83-23 and references therein.
6. K. Beshliu et al., Dubna report D1-83-815.
7. S. Fredriksson, CELSIUS Note 83-24.
8. L. Bergström, CELSIUS Note 83-22 and references therein.
9. L. Bergström, "Note on the anomalous Z^0 decays", to appear in Phys. Lett. 3.
10. H.O. Meyer, Contribution to this conference.
11. L.A. Herzenberg, R.G. Sweet, L.A. Herzenberg, Sci. Am. 234, 108 (1976).
12. P.B. Parks, R.J. Turnbull, C.A. Foster, Nuclear Fusion 17 3 (1977).
13. S. Kullander, CELSIUS Note 83-25.

WORKSHOP SUMMARY

SUMMARY PRESENTED AT THE 1984 IUCF WORKSHOP
ON
NUCLEAR PHYSICS WITH STORED, COOLED BEAMS

Robert A. Eisenstein[*]
University of Illinois
Urbana, Illinois 61801

INTRODUCTION

This year's IUCF conference, the fourth in what has become an anticipated yearly event, was devoted to the discussion of an exciting new technology and its probable impact on our study of nuclear physics. This technology, that of stored, cooled particle beam accelerators, has already been implemented at a few facilities, and interesting physics results have begun to emerge. Plans for the installation of other new machines are progressing at several institutions. In view of the transitional nature of these developments, the conference was divided half-and-half between technical and nuclear physics considerations. By the end, it was clear that the development of these facilities would likely have a dramatic impact on many areas of physics. Although some physics results were presented, the conference was mostly about the future. The intense interest in the subject was indicated by the world-wide list of institutions represented.

COOLER DESIGN AND MOTIVATION

The development of these accelerators has been driven, in the word of Bob Pollock, by a few "basic truths". The fact is that many of today's most interesting nuclear physics experiments require substantial improvement in beam quality. This includes one or more of the following areas: energy resolution, signal-to-noise ratio, beam purity, beam emittance, or intensity. In addition, experiments must be done in a reasonable time, and this requires luminosities $L \approx 10^{-31}$ cm^{-2} s^{-1}. Beam cooling, using techniques described below, can provide great improvements in beam energy resolution and emittance, although the cooling thus requires beam storage technology, and, for experiments done using internal beams, the development of thin internal targets with electron densities less than 5×10^{-16}/cm^2. In spite of formidable technical difficulties, the results can be dramatic: in the case of antiproton beams, such storage provides the time for unwanted short-lived contaminants to decay away, leaving behind an unprecedented $\bar{p}$ beam of 100% purity, very small spot size (1 mm^2) and very high intensity. Thus, such storage rings are capable of providing intermediate energy $\bar{p}$, p, d, t, and heavy ion beams of a quality more akin to a Van de Graaff accelerator than a synchrotron. And, in spite of the expense of the technology, such rings could in fact provide a very cost-effective access to intermediate energy physics for many laboratories.

[*] This work was supported by the U.S. National Science Foundation

At present, cooling techniques are limited to dealing with about 10^{10} particles in circulation at once. For a β of ~0.75 and a ring of 80 m circumference, this amounts to a circulating current of order 4-5 mA. Two major types of cooling techniques are presently in use, one involving a feedback mechanism (stochastic cooling) for sharpening the beam momentum, and the other using a co-rotating electron beam (electron cooling) to absorb the transverse momentum of the heavier ions. For orientation purposes, the methods are briefly described below; for reasons described there it is clear that it would be advantageous to be able to use both methods simultaneously. Until now, it has not been demonstrated that this is possible. Thus, for the future, two of the most interesting machine-related questions are just what the maximum limits are for the circulating particle currents, and whether or not the two cooling methods can be used together.

The principle of stochastic cooling was described at the conference in the contributions of Fred Mills and Dieter Mohl; briefly stated, the method involves a sensor to detect horizontal and transverse betatron oscillations, which feeds a wide-band amplifier that in turn provides a correction, or "kicker" signal at a later point in the ring (see figure). The cooling times for particles in the ring go approximately as the following expression:

$$\tau = (N/W) [2g - g^2(M+U)]^{-1},$$

where N is the number of circulating particles, W is the bandwidth, g the fraction of the beam corrected by the cooling, M the mixing coefficient, and U the noise factor. For longitudinal cooling, the sensor is used in conjunction with an rf cavity in the above scheme. It is seen from the above equation that the cooling times are inversely proportional to the bandwidth, and directly proportional to the number of particles to be cooled. Typical times for transverse and longitudinal cooling of present-day beams are of order 100 seconds; this clearly represents a limit to increasing the beam intensity for such a cooling scheme, unless the bandwidth can be increased commensurately. It is of interest that this method provides a "frictional damping force" equivalent to about 7×10^{-4} eV/turn, while electron cooling technique can provide 50-100 times more.

Electron cooling can be described schematically in thermodynamic terms as a mixing of a "hot gas" of beam particles with a "cold gas" of electrons. The physical arrangement is depicted below, and

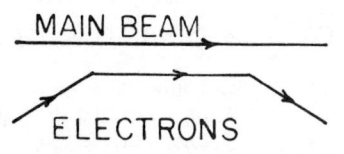

involves a beam of electrons co-rotating over a span of 1 or 2 meters with the particle beam to be cooled. The electron beam current is typically of order 1 A, and its energy is adjusted so that the longitudinal electron speed is the same as the particle beam speed. In this frame, the transverse momentum of the heavy particle is transferred to the electron, resulting in a cooled beam. While this method can be quite rapid, it is limited in usefulness due to the very large power requirements for the electron beam as the primary beam energy increases. This is illustrated in the following table (for proton beams):

Beam Momentum (GeV/c)	Power Required (kW)
0.05	0.7
0.10	2.9
0.30	25.0
0.60	95.0
2.00	690.0

On the other hand, the cooling time τ is proportional to $v^3_{Relative}$, which indicates that electron cooling improves more rapidly as it goes on! Indications are that typical cooling times of order 1 second can be obtained in the low-energy regime. Such cooling proceeds simultaneously in all three planes, which is not always useful, for example in the slow extraction mode being used at LEAR.

In the table below, following D. Mohl,[1] we summarize the advantages and disadvantages of the two cooling methods.

Hardware and Space Requirements

	Electron Cooling	Stochastic Cooling
Hardware	High power e⁻ beam (cooling all planes)	Fast electronics (3 systems, 1 each plane)
Space	One large space	Several smaller spaces
Vacuum Problems	Gas desorption	Many rf feedthroughs
Change of Energy	Retune gun	Retune all lines and filters
Change Work Point	Easily done	Rearranging, or more elements

Dependence on Machine Parameters

	Electron Cooling	Stochastic Cooling
p̄ Intensity	Weak, but untested for $N > 10^3$	Strong ($\tau \sim N$, e.g. 1 sec for 10^7, and 1 day for 10^{13})
p̄ Energy	Strong ($\beta^4 \gamma^5$), e.g. 1 sec for 0.1 GeV/c 1 day for 2.0 GeV/c	Weak
Beam Size and $\Delta p/p$	Works best for cool beams	Works best for hot beams

Physicists at CERN are in the process of constructing LEAR, a low energy antiproton ring featuring a synchrotron design and both electron and stochastic cooling. The table below presents a comparison of some of the impacts of the two methods on the operation of LEAR.

Application to LEAR

	Electron Cooling	Stochastic Cooling
Stretcher Mode	Hard to separate transverse and longitudinal. Hard to preserve beam quality during extraction and transport.	Well adapted (independence of planes) but needs 1-2 min.
Internal Targets	Very desirable. Allows thick targets.	Thin targets with 10^{10} p̄ per 10^4 sec possible.
H⁻ Beams	Desirable but care to avoid stripping	Too slow at more than 10^8 H⁻/pulse
Minicollider	High-energy cooling necessary to have $L > 10^{29}$	Necessary

The status of the various cooler projects throughout the world is presented on the next page. As can be seen from the table, the method is in wide use over a broad spectrum of energies and beam particle species.

A very informative way of contrasting these facilities with the pre-cooler era was presented by Bob Pollock, and is summarized in the figure below. As can be seen from this figure, luminosities (in

Status of Cooler Facilities

Institute	Project Name	Beam Type	Max. Energy (MeV)	Cooling Method	Completion Date
CERN	LEAR	$\bar{p}$	1275	S,E	1983
	AA	$\bar{p}$	2690	S	1981
	ACOL	$\bar{p}$	2690	S	----
FNAL	AS	$\bar{p}$	8000	S	1985
IUCF	COOLER	p	500	E	1986
UPPSALA	CELSIUS	p-Ar	1300 300/A	E	1986
DARMSTADT	ESR	p-U	822 104/A	S,E	----
JULICH	COSY	p	500	E	----
INS-TOKYO	TARN	p	7	S	1984
	TARN II	p-Ne	1300 450/A	S,E	1986

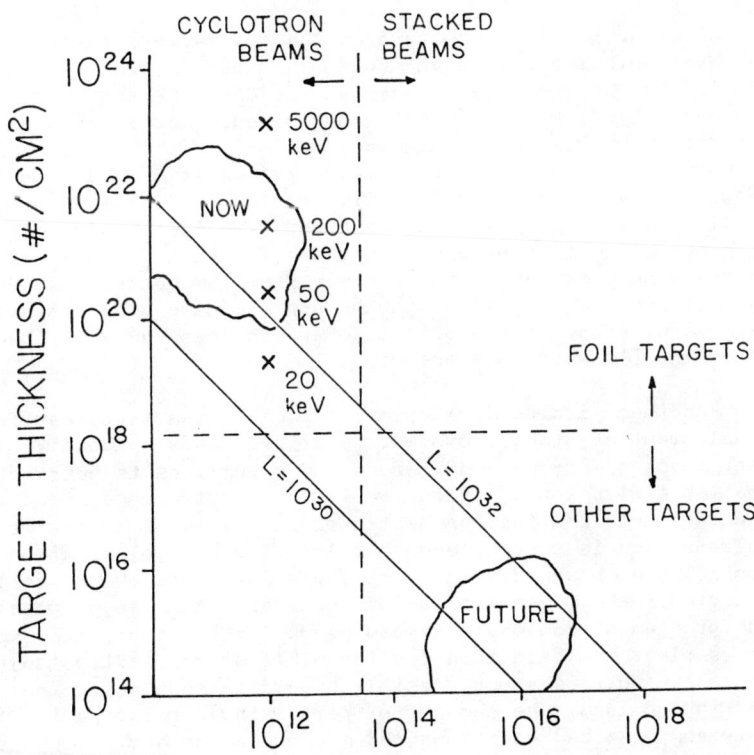

excess of 10^{32}) are available now, which make a certain class of nuclear physics experiments available to us. The problem arises when one wishes to improve significantly the beam energy resolution; to do this requires very thin internal targets and cooled beams. Keeping the luminosity up then requires the kind of currents that are only available from circulating beams. Nuclear physics experiments on a reasonable time frame will require beams of order 10^{16} particles per second, and internal target thicknesses of order 10^{15} particles/cm^2.

A number of discussion were presented at the conference on the status of various internal target designs which could fulfill these requirements. The primary design goal is to obtain high luminosity, and in this regard the most useful target at present is the fiber strand target. The principal problem with this design is to find a way to cool the target, and to keep the degradation of phase space under control. The Monte Carlo calculations of Hans Meyer[2] have indicated that the use of oscillating fibers might help to maintain the beam. Target densities of order 10^{16} can easily be obtained. S. Kullander discussed the possibility of using an "ink jet" technique to form droplets of reasonable density. Other dust and pellet mechanisms were also described.[3]

Approaching the problem from the low target density side, one has available the atomic and molecular beams approach. The application of these techniques at the CERN ISR and S$\bar{p}$pS was discussed by H. Macri. Development of such targets has also been undertaken by workers at Argonne,[4] Celsius,[5] the IUCF,[6] and Wisconsin.[7] In all cases, the problems are two-fold: on the one hand, the target density is still too low ($\sim 10^{12}$/cm^2) for use in routine experiments, and on the other, the maintenance of high machine vacuum is a problem. Rather extensive further development needs to be undertaken; it is clear from the outset that these targets require an advance in the state of the art, and are quite expensive. They are important for $\bar{p}$ machines, however, because they allow the $\bar{p}$'s which are scarce, to be reused.

Another machine development topic of great interest is the development of stored, cooled, <u>polarized</u> beams. Ideally, we would want beams of protons, deuterons and other ions to be readily available with known, large, variable polarizations. The problem has been well studied for synchrotrons, and it is clear that transverse polarization is stable, and that longitudinal polarization can probably be obtained using techniques involving the so-called Siberian Snake. The storage of such beams thus seems possible, but the problem of cooling the beam needs further study and experiment. It is clear that electron cooling will not depolarize the beam, but there are worries about possible hyperfine depolarization. An intriguing idea, the thought of performing a polarization transfer between <u>polarized</u> electrons and a circulating antiproton beam, to produce <u>polarized</u> antiprotons, seems not to be feasible.

We turn now to a brief discussion of the physics applications possible with this exciting new technology.

ATOMIC PHYSICS

The possibilities for using cooled, stored beams for atomic physics experiments was discussed by P. Kienle and I. Katayama. Many of the experiments envisioned feature the use of cooled heavy ion beams, and the plans of the GSI-Darmstadt, the MPI-Heidelberg, and TARN-II (Tokyo) were outlined. Some of the interesting physics possibilities include study of β-decay to Coulomb bound states, the radiative capture of cooling electrons by the fully ionized circulating ion beam, and the possibility of laser-induced radiative capture. The intriguing idea of Greiner and Rafelski, of producing e^+e^- pairs in the very strong electric field obtained in uranium-uranium collisions, was discussed in the context of the storage of two co-rotating U^{238} bunches. The study of atomic collisions using H^o beams was described briefly, as were the new possibilities for studying hydrogenic systems with beams of anti-hydrogen formed at LEAR.[8]

NUCLEAR PHYSICS

We now turn to an examination of the possibilities for nuclear physics studies with cooled, stored particle beams. The two principal new features offered by the cooler facilities involve extremely good energy resolution at high beam energy, and the possibilities for studying spin physics using polarized beams. So, as J. Heisenberg pointed out in his talk, the purpose is certainly not to measure the properties of 50,000 energy levels, but instead to use the high intrinsic energy resolution and resulting good signal to noise ratio to study the behavior of certain chosen levels under new experimental conditions, e.g. high incident beam energy and with varying degrees of initial state polarization.

Of special interest in this regard is the possibility of studying spin degrees of freedom in nuclear reaction physics at these energies. The strength function associated with such elementary modes of excitation as single particle/hole states and giant resonance phenomena could be profitably explored in this way. The technology will make possible such capabilities as polarized atomic beam targets, variable duty factor lifetime measurements, tagged neutron production, fine energy-tune searches for resonance behavior, and a careful examination of nuclear reaction thresholds. As an example of the power of such new instrumentation, spectra taken with the Big Karl spectrograph at Julich were presented.[9] This device is a QQDDQ arrangement and has variable dispersion, which can be continuously changed from ~26 cm/% to -2 cm/%, making possible studies at levels of precision commensurate with experimental goals. The energy resolution $\Delta E/E$ is of order 10^{-4}. A similar device is being constructed at IUCF for use with the cooler there.

As an example of the kind of physics that can be studied with facilities of this type was presented at the Conference by J. Redish, who discussed for probing the NN force at short range by

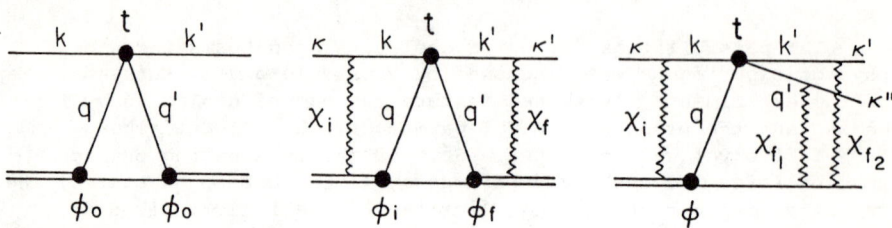

Role of the same amplitude $t(k,k',\theta,E)$ in (p,p), (p,p') and (p,2p) processes.

complementary studies of the reactions (p,p), (p,p') and (p,2p). These processes are based on the same internal diagram as shown in the figure above. Redish and Stricker-Bauer[10] have recently attempted to do a benchmark calculation of these processes using a "standard model" -- non-relativistic nucleons, a potential taken from the NN two-body data, and inclusion of the Δ and π -- to be compared in detail with some of the newer ideas involving Dirac small components, mean-field theories, quarks, and so on. By confronting these notions with the detailed data, one hopes to learn much more about the nature of effective interactions, and the role of the above ideas, especially where spin degrees of freedom are concerned.

One has always in mind the kind of precision that has been available for many years using electron scattering as a probe of nuclear structure, and there is hope[11,12] based on the work done at the HRS at LAMPF, and at Saturne, that hadronic probes at high energies will provide much in the way of clean new information about both nuclear structure and the underlying dynamics. Heisenberg reminded us once again of just how much has been learned in recent years in the electron scattering field, and of the necessity to use all of our experimental information in a complementary way in order to improve our understanding. In recent work[11] it has been possible, through detailed studies of the transition densities measured in electron scattering, to identify the roles played by mechanisms such as core polarization, relativistic corrections, and so on.

RADIATIVE PROCESSES AND BREMSSTRAHLUNG

Harold Fearing summarized the present situation with respect to radiative processes and bremsstrahlung studies. In spite of the long history of this subfield, there are many areas in which new experiments using cooled beam facilities could have a major impact on our understanding.

In all such work, one seeks to take advantage of the fact that the electromagnetic interaction is much better known and also much weaker than the strong force. In addition, there are several global principles of physics that link various aspects of the problem together: Siegert's theorem connecting charge conservation to current flow, Low's soft photon theorem connecting radiative and non-radiative processes, and overall, gauge invariance. In order to keep the nuclear structure complications to a minimum, one ofter deals with very light systems to study to basic amplitudes as directly as possible. Examples of these processes that were discussed are $pp \rightarrow pp\gamma$, $np \rightarrow np\gamma$, and $np \rightarrow d\gamma$.

In the case of pp bremsstrahlung experiments, we hope to shed light on the nature of the off-shell NN force. This subject has a long history; it is indicative of the difficulty of the experiment that the recent very nice work at TRIUMF is not able to distinguish between several of the best recent calculations. New experiments measuring $d\sigma/d\Omega$ and the analyzing power $A_y(\theta)$ in the energy range 100-200 MeV would help a lot in this regard. In the case of $np \rightarrow np\gamma$, in which the photon is emitted from the exchanged pion, any experiment would be pioneering work.

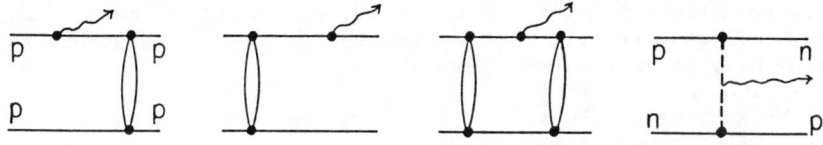

Diagrams important in pp and pn bremsstrahlung.

There is also interest in the study of similar reactions on the A=2-4 systems. The idea is to embed the basic NN process in a slightly larger system where the wave functions are rather well known. The hope is that one can generate a good microscopic description of the process. There is good recent experimental work from the IUCF and TRIUMF on the process $pd \rightarrow {}^3He\ \gamma$, which represents a consistent systematic study. An interesting problem arises in the interpretation of the IUCF study of $d\sigma/d\Omega$ and tensor analyzing power for the process $dd \rightarrow {}^4He\ \gamma$, for in this case the physics involves more than single-particle capture, and there are significant questions about how one writes the relevant currents in this case.

Finally, there is much interesting work to be done on the (p,γ) process in "real" nuclei, in addition to what has already been done at the IUCF and Bates. The process involves high momentum transfers, as does the (p,π) process, and will probably be subject to a similar set of problems in interpretation. In the former case, however, the fact that one side of the reaction involves a γ-ray should help considerably. These studies should help our further understanding of the population of stretched states. More experimentation in the regime 150-200 MeV proton energy, especially with measurements of the analyzing power and $d\sigma/d\Omega$, would be very helpful.

FEW-BODY PROBLEMS

We turn now to a quick review of an area in which the advent of the coolers can be expected to have a large and immediate impact. The study of proton interactions with nucleons, deuterons, tritons, ^{3}He and ^{4}He is especially important in nuclear physics, for it is there that we may test multiple scattering theories in more tractable form. The current and prospective situation in this area was reviewed at the conference by H. Spinka and M. Bleszynski.

As is well known, the NN interaction can be characterized by 5 complex amplitudes. Thus, a complete description of the process requires 10 measurements at each energy. In pp scattering the angular range extends from 0-90°, while for np scattering it is 0-180°. Many angles need to be measured to determine each partial wave amplitude. In spite of what one might think, pp data are sparse in the energy region from 150 to 300 MeV. In the region 100 to 250 MeV, more forward angle data are needed. In the np case, the data are fewer and generally poorer in quality. Thus, there is no way to do a complete model-independent analysis below 450 MeV. Also, there are no $\Delta\sigma_L$ data available for that channel. In order to perform such meaurements, a polarized beam and target and polarimeter are necessary. For the forward angle measurements, gas targets would be ideal, and the possibility of performing np experiments with a tagged neutron beam is extremely interesting.

The subject of inelasticities in NN scattering needs further exploration. It would be of value to study in detail the processes $pp \to d\pi^+$ and $np \to d\pi^0$, which are related through charge independence. The more general question of which partial waves are important in $NN \to NN\pi$ has not been satisfactorily answered.

A related issue concerns the interaction of protons with deuterons, tritons, ^{3}He and ^{4}He. As in the above-mentioned cases, careful measurements of the differential cross section and spin-dependent quantitites would be of help in constraining the freedom available now in our model-building efforts. The appeal of these systems is of course the fact that they represent the simplest application of multiple scattering theory in nuclear physics. It is clear from work done in recent years that both spin and relativity play an important role down to energies as low as 100 MeV, and spin is universally important. In the case of pd scattering, an amplitude consisting only of single and double scattering contributions works well only at low momentum transfers, and the relativity contributions are known to be large. A more complete experimental determination of the scattering matrix for this system would be quite useful. Similarly, even in the case of p-^{4}He scattering, which is a well-studied problem, measurement of the ratio Re f(0°)/Im f(0°) as well as spin-dependent quantities such as $\Delta\sigma_L$ and $\Delta\sigma_T$ would be extremely interesting.

THRESHOLD PHENOMENA

In the course of the conference, Torleif Ericson reminded us of the fact that a storage ring circulating a cooled beam is naturally a threshold machine. With its precise and continuously variable beam energy and large current, detailed studies of reactions near threshold become an exciting possibility. His remarks were illustrated by a brief examination of the case of pion production. The behavior of the cross sections for π^0, π^+, and π^- in the threshold region are shown to the left. The π^0 cross section rises like $(T-T_{Thresh})^{1/2}$ as befits s-wave production, while the π^+ cross section is suppressed by the Coulomb penetrability. On the other hand, the π^- cross section remains finite

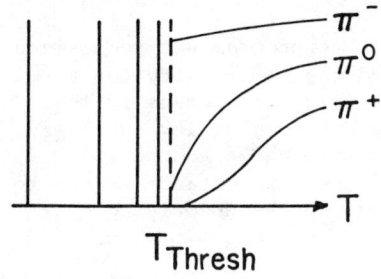

at threshold because of the correspondence principle connection to the sub-threshold Coulomb bound states. In the past this has led to novel suggestions for recoilless π-atom production, even on excited nuclear states, and for searches of the elusive deeply bound pionic atom 1s states. Another application of a threshold reaction would be to search for the process $dd \to \pi^0 + {}^4He$, which is forbidden by isospin conservation.

Another very interesting application for a storage ring would be to search for the recoilless production of π-mesons in nuclei. K. Kilian discussed the possibility of using the reactions $pp \to d\pi^+$ and $pn \to d\pi^0$ to create narrow states in nuclei via the reaction $p+(Rp) \to d+(R\pi^+)^*$; in this expression R is the residual core. As in the case of the prototypical reaction (K^-,π^-), one searches the missing mass spectrum for the states of interest. On free nucleons, the recoilless condition for pions occurs at 826 and 806 MeV, respectively.

ELEMENTARY PARTICLE PHYSICS

S. Kullander presented a brief discussion of some possibilities for studying elementary particle processes at one of these facilities. We mention here three experiments: an examination of rare π decays; a search for dibaryon resonances; and the possibility of measuring the ν mass.

The pion rare decay modes can be studied via the $pd \to {}^3He\ \pi^0$ reaction, where one examines the π^0 decay for signatures of e^+e^- or $e^+e^-\gamma$ decays. Near threshold one can separate π^0 production from charged mesons, and possibly use detection of the ${}^3He$ as an additional experimental constraint.

The dibaryon search could be conducted by examining, as a function of energy, the reactions $pp \to pp$, $pp \to pp\gamma$, and searching for resonance behavior in these different channels.

The neutrino mass measurments could be made by a study of the reaction $^3H \rightarrow {}^3He + e^- + \nu_e$. A novel scheme to search for the ν mass, by observing triton decay in CELSIUS, was discussed. The advantages are a greater precision in the measuremnt, and the fact that molecular effects, which usually are a problem in studies like this, are absent.

Finally, the present status of the LEAR program was summarized by K. Kilian. The $\bar{p}$ beams at LEAR afford the opportunity for a broad-scale study of low-energy QCD. Of special interest is the question of $\bar{p}p$ annihilation, which is very strong at low energies. In terms of quark diagrams, the contrast between $\bar{p}p$ and pp is extremely interesting. It is clearly of great importance to try to connect NN descriptions in terms of mesons to the underlying quark pictures.

CONCLUSIONS

Several things are clear from the proceedings of this conference. Within a rather short time period there will be a lot of very high quality medium energy physics experimental apparatus available. There is the exciting prospect of stored, cooled particle beams of protons, antiprotons, light and heavy ions, possibly with polarization, coupled with internal targets and excellent detectors. The conference featured many speakers who identified a very wide spectrum of excellent physics possibilities to be done. Nonetheless, the situation is such that the technology seems at this point to be ahead of the experimental uses for it. If true, this may seem worrying, but it seems to me that this situation has many precedents, as it is very often the case that the most exciting experiments are not at all envisioned when facilities like these initially become available. Some of the most interesting work done at the IUCF, namely the study of giant Gamow-Teller resonances and the study of Dirac phenomenology, was almost certainly not foreseen.

So, it seems to me that there is ample justification to proceed with the development of these facilities. It is indeed rare that an experimental facility can have an impact on such a large variety of physical problems. As we have seen, this range includes machine development physics of the most interesting sort, as well as the scientific spectrum from atomic physics, to nuclear physics and few-body systems, to quark and meson degrees of freedom, to elementary particle physics, to experiments on the nature of fundamental symmetries. These are indeed most interesting prospects, as the first results have confirmed. We await more with enthusiasm!

REFERENCES

1. D. Möhl, in <u>Physics at LEAR with Low-Energy Cooled Antiprotons</u>, U. Gastaldi and R. Klapisch, eds., Plenum Press, New York, 1984.

2. See H. Meyer, this conference.

3. C. Ekström, this conference.

4. M. Green, this conference.

5. S. Kullander, this conference.

6. W. Lozowski, this conference.

7. W. Haeberli, this conference.

8. K. Kilian, this conference.

9. G. Berg, this conference.

10. J. Redish and K. Stricker-Bauer, to be published.

11. J. Heisenberg, this conference.

12. S. Wallace, Adv. Nuc. Phys. <u>12</u>, 135 (1981), J. W. Negele and E. Vogt, eds., Plenum Press.

CONFERENCE PARTICIPANTS

J. Adams, University of Maryland, College Park, MD 20742, USA
M. Asmar, Indiana University, Bloomington, IN 47405, USA
S. Aziz, Indiana University, Bloomington, IN 47405, USA
A. Bacher, Indiana University Bloomington, IN 47405, USA
S. Banks, University of Melbourne, Parkville, Victoria, Australia
P. Barnes, Carnegie-Mellon University, Pittsburgh, PA 15213, USA
R. Bent, Indiana University, Bloomington, IN 47405, USA
G. Berg, KFA Julich, Julich, West Germany
T. Berggren, Lund Institute of Technology, Lund, Sweden
I. Bergqvist, University of Lund, Lund, Sweden
L. Bland, Indiana University, Bloomington, IN 47405, USA
M. Bleszynski, UCLA, Los Angeles, CA 90024, USA
J.D. Bowman, Los Alamos National Lab, Los Alamos, NM 87545, USA
R. Byrd, Indiana University, Bloomington, IN 47405, USA
Z.-J. Cao, Indiana University, Bloomington, IN 47405, USA
A. Chabert, GANIL, Caen-Cedex, France
N. Chant, University of Maryland, College Park, MD 20742, USA
Q. Chen, Indiana University, Bloomington, IN 47405, USA
V. Cupps, Indiana University, Bloomington, IN 47405, USA
W. Daehnick, University of Pittsburgh, Pittsburgh, PA 15260, USA
R. Eisenstein, University of Illinois, Champaign, IL 61820, USA
C. Eckstrom, NFL-Studsvik, Nykoping, Sweden
T. Ellison, Indiana University, Bloomington, IN 47405, USA
G. Emery, Indiana University, Bloomington, IN 47405, USA
T.E.O. Ericson, CERN, Geneva, Switzerland
M. Fatyga, Indiana University, Bloomington, IN 47405, USA
H. Fearing, TRIUMF, Vancouver, B.C., Canada
C. Foster, Indiana University, Bloomington, IN 47405, USA
D. Friesel, Indiana University, Bloomington, IN 47405, USA
J. Gering, Indiana University, Bloomington, IN 47405, USA
C. Glover, Oak Ridge National Lab, Oak Ridge, TN 37830, USA
C. Goodman, Indiana University, Bloomington, IN 47405, USA
M. Green, Argonne National Lab, Argonne, IL 60439, USA
U. Hacker, KFA Julich, Julich, West Germany
W. Haeberli, University of Wisconsin, Madison, WI 53706, USA
H. Hagedoorn, Eindhoven Univ. of Tech., Eindhoven, The Netherlands
R. Haight, Lawrence Livermore Lab, Livermore, CA
D. Habs, Max-Planck Inst. f. Kernphysik, Heidelberg, West Germany
M. Harakeh, University of Groningen, Groningen, The Netherlands
T. Hasegawa, University of Tokyo, Tokyo, Japan
K. Hatanaka, RIKEN (IPCR), Saitama, Japan
J. Heisenberg, University of New Hampshire, Durham, NH 03824, USA
H. Herr, CERN, Geneva, Switzerland
R. Holt, Argonne National Lab, Argonne, IL 60439, USA
M. Huber, University of Erlangen, Erlangen, West Germany
W-Y.P. Hwang, Indiana University, Bloomington, IN 47405, USA
W. Jacobs, Indiana University, Bloomington, IN 47405, USA
E. Jaeschke, Max-Planck Inst. f. Kernphysik, Heidelberg, West Germany

A. Johansson, University of Uppsala, Uppsala, Sweden
W. Jones, Indiana University, Bloomington, IN 47405, USA
B. Karlsson, University of Uppsala, Uppsala, Sweden
I. Katayama, RCNP Osaka University, Ibaraki, Japan
T. Katayama, University of Tokyo, Tokyo, Japan
P. Kienle, GSI, Darmstadt, West Germany
K. Kilian, CERN, Geneva, Switzerland
E. Korkmaz, Indiana University, Bloomington, IN 47405, USA
M. Kovash, University of Kentucky, Lexington, KY 40506, USA
S. Kullander, University of Uppsala, Uppsala, Sweden
K. Kwiatkowski, Indiana University, Bloomington, IN 47405, USA
B. Lay, University of Melbourne, Parkville, Victoria, Australia
D. Low, Indiana University, Bloomington, IN 47405, USA
W. Lozowski, Indiana University, Bloomington, IN 47405, USA
M. Macfarlane, Indiana University, Bloomington, IN 47405, USA
M. Macri, INFN-Genova/CERN, Geneva, Switzerland
S. Martin, KFA Julich, Julich, West Germany
T. Mayer-Kuckuk, University of Bonn, Bonn West Germany
H. Meyer, Indiana University, Bloomington, IN 47405, USA
D. Miller, Indiana University, Bloomington, IN 47405, USA
F. Mills, Fermi National Accelerator Lab, Batavia, IL 60510
D. Mohl, CERN, Geneva, Switzerland
H. Nann, Indiana University, Bloomington, IN 47405, USA
L. Nilsson, University of Uppsala, Uppsala, Sweden
C. Olmer, Indiana University, Bloomington, IN 47405, USA
K. Omata, University of Tokyo, Tokyo, Japan
A. Opper, Indiana University, Bloomington, IN 47405, USA
W.K. Pitts, Indiana University, Bloomington, IN 47405, USA
P. Poffenberger, TRIUMF, Vancouver, B.C., Canada
R. Pollock, Indiana University, Bloomington, IN 47405, USA
H. Postma, Tech. Hogeschool Delft, Delft, The Netherlands
H. Poth, CERN, Geneva, Switzerland
E. Redish, University of Maryland, College Park, MD 20742, USA
D. Reistad, University of Uppsala, Uppsala, Sweden
P. Riley, National Science Foundation, Washington, DC 20550, USA
W. Rodney, National Science Foundation, Washington, DC 20550, USA
P. Schwandt, Indiana University, Bloomington, IN 47405, USA
B. Serot, Indiana University, Bloomington, IN 47405, USA
R. Siemssen, University of Groningen, Groningen, The Netherlands
P. Singh, Indiana University, Bloomington, IN 47405, USA
A. Sinha, Indiana University, Bloomington, IN 47405, USA
T. Sloan, Indiana University, Bloomington, IN 47405, USA
K. Solberg, Indiana University, Bloomington, IN 47405, USA
J. Sowinski, Indiana University, Bloomington, IN 47405, USA
J. Speth, KFA Julich, Julich, West Germany
B. Spicer, University of Melbourne, Parkville, Victoria, Australia
H. Spinka, Argonne National Lab, Argonne, IL 60439, USA
E. Steffens, MPI Heidelberg/CERN, Geneva, Switzerland
E. Stephenson, Indiana University, Bloomington, IN 47405, USA
J. Stetson, Indiana University, Bloomington, IN 47405, USA

E. Sugarbaker, Ohio State University, Columbus, OH 43212, USA
T. Taddeucci, Indiana University, Bloomington, IN 47405, USA
J. Templon, Indiana University, Bloomington, IN 47405, USA
T. Throwe, Indiana University, Bloomington, IN 47405, USA
G. Tibell, University of Uppsala, Uppsala, Sweden
I. van Heerden, Univ. of the Western Cape, Bellville, South Africa
S. Vigdor, Indiana University, Bloomington, IN 47405, USA
V. Viola, Jr., Indiana University, Bloomington, IN
P. von Rossen, KFA Julich, Julich, West Germany
G. Wagner, University of Tubingen, Tubingen, West Germany
G. Walker, Indiana University, Bloomington, Indiana 47405, USA
T. Ward, Indiana University, Bloomington, IN 47405, USA
L. Westerberg, University of Uppsala, Uppsala, Sweden
H. Willard, National Science Foundation, Washington, DC 20550, USA
S. Wissink, Indiana University, Bloomington, IN 47405, USA

AUTHOR INDEX

Berg, G.	2
Bleszynski, M.	220
Dillig, M.	188
Eisenstein	333
Ekstrom, C.	106
Ericson, T.E.	172
Fearing, H.	138
Green, M.	268
Haeberli, W.	251
Hagberg, E.	327
Heisenberg, J.	47
Herr, H.	68
Holmquist, B.	106
Hopf, H.G.	188
Huber, M.	188
Johansson, A.	14
Katayama, I.	278
Katayama, T.	24
Kienle, P.	291
Kilian, K.	319
Kullander, S.	327
Macri, M.	89
Meyer, H.O.	76
Moehl, D.	180
Pollock, R.E.	34
Poth, H.	305
Redish, E.	112
Reistad, D.	14
Spinka, H.	198
Steffens, E.	241
Sterner, H.	106
Wagner, G.	157

AIP Conference Proceedings

		L.C. Number	ISBN
No. 1	Feedback and Dynamic Control of Plasmas	70-141596	0-88318-100-2
No. 2	Particles and Fields - 1971 (Rochester)	71-184662	0-88318-101-0
No. 3	Thermal Expansion - 1971 (Corning)	72-76970	0-88318-102-9
No. 4	Superconductivity in d- and f-Band Metals (Rochester, 1971)	74-18879	0-88318-103-7
No. 5	Magnetism and Magnetic Materials - 1971 (2 parts) (Chicago)	59-2468	0-88318-104-5
No. 6	Particle Physics (Irvine, 1971)	72-81239	0-88318-105-3
No. 7	Exploring the History of Nuclear Physics	72-81883	0-88318-106-1
No. 8	Experimental Meson Spectroscopy - 1972	72-88226	0-88318-107-X
No. 9	Cyclotrons - 1972 (Vancouver)	72-92798	0-88318-108-8
No. 10	Magnetism and Magnetic Materials - 1972	72-623469	0-88318-109-6
No. 11	Transport Phenomena - 1973 (Brown University Conference)	73-80682	0-88318-110-X
No. 12	Experiments on High Energy Particle Collisions - 1973 (Vanderbilt Conference)	73-81705	0-88318-111-8
No. 13	π-π Scattering - 1973 (Tallahassee Conference)	73-81704	0-88318-112-6
No. 14	Particles and Fields - 1973 (APS/DPF Berkeley)	73-91923	0-88318-113-4
No. 15	High Energy Collisions - 1973 (Stony Brook)	73-92324	0-88318-114-2
No. 16	Causality and Physical Theories (Wayne State University, 1973)	73-93420	0-88318-115-0
No. 17	Thermal Expansion - 1973 (lake of the Ozarks)	73-94415	0-88318-116-9
No. 18	Magnetism and Magnetic Materials - 1973 (2 parts) (Boston)	59-2468	0-88318-117-7
No. 19	Physics and the Energy Problem - 1974 (APS Chicago)	73-94416	0-88318-118-5
No. 20	Tetrahedrally Bonded Amorphous Semiconductors (Yorktown Heights, 1974)	74-80145	0-88318-119-3
No. 21	Experimental Meson Spectroscopy - 1974 (Boston)	74-82628	0-88318-120-7
No. 22	Neutrinos - 1974 (Philadelphia)	74-82413	0-88318-121-5
No. 23	Particles and Fields-1974 (APS/DPF Williamsburg)	74-27575	0-88318-122-3
No. 24	Magnetism and Magnetic Materials - 1974 (20th Annual Conference, San Francisco)	75-2647	0-88318-123-1
No. 25	Efficient Use of Energy (The APS Studies on the Technical Aspects of the More Efficient Use of Energy)	75-18227	0-88318-124-X
No. 26	High-Energy Physics and Nuclear Structure - 1975 (Santa Fe and Los Alamos)	75-26411	0-88318-125-8
No. 27	Topics in Statistical Mechanics and Biophysics: A Memorial to Julius L. Jackson (Wayne State University, 1975)	75-36309	0-88318-126-6
No. 28	Physics and Our World: A Symposium in Honor of Victor F. Weisskopf (M.I.T., 1974)	76-7207	0-88318-127-4

AIP Conference Proceedings

No. 29	Magnetism and Magnetic Materials - 1975 (21st Annual Conference, Philadelphia)	76-10931	0-88318-128-2
No. 30	Particle Searches and Discoveries - 1976 (Vanderbilt Conference)	76-19949	0-88318-129-0
No. 31	Structure and Excitations of Amorphous Solids (Williamsburg, VA., 1976)	76-22279	0-88318-130-4
No. 32	Materials Technology - 1976 (APS New York Meeting)	76-27967	0-88318-131-2
No. 33	Meson-Nuclear Physics - 1976 (Carnegie-Mellon Conference)	76-26811	0-88318-132-0
No. 34	Magnetism and Magnetic Materials - 1976 (Joint MMM-Intermag Conference, Pittsburgh)	76-47106	0-88318-133-9
No. 35	High Energy Physics with Polarized Beams and Targets (Argonne, 1976)	76-50181	0-88318-134-7
No. 36	Momentum Wave Functions - 1976 (Indiana University)	77-82145	0-88318-135-5
No. 37	Weak Interaction Physics - 1977 (Indiana University)	77-83344	0-88318-136-3
No. 38	Workshop on New Directions in Mossbauer Spectroscopy (Argonne, 1977)	77-90635	0-88318-137-1
No. 39	Physics Careers, Employment and Education (Penn State, 1977)	77-94053	0-88318-138-X
No. 40	Electrical Transport and Optical Properties of Inhomogeneous Media (Ohio State University, 1977)	78-54319	0-88318-139-8
No. 41	Nucleon-Nucleon Interactions - 1977 (Vancouver)	78-54249	0-88318-140-1
No. 42	Higher Energy Polarized Proton Beams (Ann Arbor, 1977)	78-55682	0-88318-141-X
No. 43	Particles and Fields - 1977 (APS/DPF, Argonne)	78-55683	0-88318-142-8
No. 44	Future Trends in Superconductive Electronics (Charlottesville, 1978)	77-9240	0-88318-143-6
No. 45	New Results in High Energy Physics - 1978 (Vanderbilt Conference)	78-67196	0-88318-144-4
No. 46	Topics in Nonlinear Dynamics (La Jolla Institute)	78-057870	0-88318-145-2
No. 47	Clustering Aspects of Nuclear Structure and Nuclear Reactions (Winnepeg, 1978)	78-64942	0-88318-146-0
No. 48	Current Trends in the Theory of Fields (Tallahassee, 1978)	78-72948	0-88318-147-9
No. 49	Cosmic Rays and Particle Physics - 1978 (Bartol Conference)	79-50489	0-88318-148-7
No. 50	Laser-Solid Interactions and Laser Processing - 1978 (Boston)	79-51564	0-88318-149-5
No. 51	High Energy Physics with Polarized Beams and Polarized Targets (Argonne, 1978)	79-64565	0-88318-150-9
No. 52	Long-Distance Neutrino Detection - 1978 (C.L. Cowan Memorial Symposium)	79-52078	0-88318-151-7
No. 53	Modulated Structures - 1979 (Kailua Kona, Hawaii)	79-53846	0-88318-152-5

AIP Conference Proceedings

No. 54	Meson-Nuclear Physics - 1979 (Houston)	79-53978	0-88318-153-3
No. 55	Quantum Chromodynamics (La Jolla, 1978)	79-54969	0-88318-154-1
No. 56	Particle Acceleration Mechanisms in Astrophysics (La Jolla, 1979)	79-55844	0-88318-155-X
No. 57	Nonlinear Dynamics and the Beam-Beam Interaction (Brookhaven, 1979)	79-57341	0-88318-156-8
No. 58	Inhomogeneous Superconductors - 1979 (Berkeley Springs, W.V.)	79-57620	0-88318-157-6
No. 59	Particles and Fields - 1979 (APS/DPF Montreal)	80-66631	0-88318-158-4
No. 60	History of the ZGS (Argonne, 1979)	80-67694	0-88318-159-2
No. 61	Aspects of the Kinetics and Dynamics of Surface Reactions (La Jolla Institute, 1979)	80-68004	0-88318-160-6
No. 62	High Energy e^+e^- Interactions (Vanderbilt, 1980)	80-53377	0-88318-161-4
No. 63	Supernovae Spectra (La Jolla, 1980)	80-70019	0-88318-162-2
No. 64	Laboratory EXAFS Facilities - 1980 (Univ. of Washington)	80-70579	0-88318-163-0
No. 65	Optics in Four Dimensions - 1980 (ICO, Ensenada)	80-70771	0-88318-164-9
No. 66	Physics in the Automotive Industry - 1980 (APS/AAPT Topical Conference)	80-70987	0-88318-165-7
No. 67	Experimental Meson Spectroscopy - 1980 (Sixth International Conference, Brookhaven)	80-71123	0-88318-166-5
No. 68	High Energy Physics - 1980 (XX International Conference, Madison)	81-65032	0-88318-167-3
No. 69	Polarization Phenomena in Nuclear Physics - 1980 (Fifth International Symposium, Santa Fe)	81-65107	0-88318-168-1
No. 70	Chemistry and Physics of Coal Utilization - 1980 (APS, Morgantown)	81-65106	0-88318-169-X
No. 71	Group Theory and its Applications in Physics - 1980 (Latin American School of Physics, Mexico City)	81-66132	0-88318-170-3
No. 72	Weak Interactions as a Probe of Unification (Virginia Polytechnic Institute - 1980)	81-67184	0-88318-171-1
No. 73	Tetrahedrally Bonded Amorphous Semiconductors (Carefree, Arizona, 1981)	81-67419	0-88318-172-X
No. 74	Perturbative Quantum Chromodynamics (Tallahassee, 1981)	81-70372	0-88318-173-8
No. 75	Low Energy X-ray Diagnostics - 1981 (Monterey)	81-69841	0-88318-174-6
No. 76	Nonlinear Properties of Internal Waves (La Jolla Institute, 1981)	81-71062	0-88318-175-4
No. 77	Gamma Ray Transients and Related Astrophysical Phenomena (La Jolla Institute, 1981)	81-71543	0-88318-176-2
No. 78	Shock Waves in Condensed Matter - 1981 (Menlo Park)	82-70014	0-88318-177-0
No. 79	Pion Production and Absorption in Nuclei - 1981 (Indiana University Cyclotron Facility)	82-70678	0-88318-178-9
No. 80	Polarized Proton Ion Sources (Ann Arbor, 1981)	82-71025	0-88318-179-7
No. 81	Particles and Fields - 1981: Testing the Standard Model (APS/DPF, Santa Cruz)	82-71156	0-88318-180-0

No. 82	Interpretation of Climate and Photochemical Models, Ozone and Temperature Measurements (La Jolla Institute, 1981)	82-071345	0-88318-181-9
No. 83	The Galactic Center (Cal. Inst. of Tech., 1982)	82-071635	0-88318-182-7
No. 84	Physics in the Steel Industry (APS.AISI, Lehigh University, 1981)	82-072033	0-88318-183-5
No. 85	Proton-Antiproton Collider Physics - 1981 (Madison, Wisconsin)	82-072141	0-88318-184-3
No. 86	Momentum Wave Functions - 1982 (Adelaide, Australia)	82-072375	0-88318-185-1
No. 87	Physics of High Energy Particle Accelerators (Fermilab Summer School, 1981)	82-072421	0-88318-186-X
No. 88	Mathematical Methods in Hydrodynamics and Integrability in Dynamical Systems (La Jolla Institute, 1981)	82-072462	0-88318-187-8
No. 89	Neutron Scattering - 1981 (Argonne National Laboratory)	82-073094	0-88318-188-6
No. 90	Laser Techniques for Extreme Ultraviolet Spectroscopy (Boulder, 1982)	82-073205	0-88318-189-4
No. 91	Laser Acceleration of Particles (Los Alamos, 1982)	82-073361	0-88318-190-8
No. 92	The State of Particle Accelerators and High Energy Physics (Fermilab, 1981)	82-073861	0-88318-191-6
No. 93	Novel Results in Particle Physics (Vanderbilt, 1982)	82-73954	0-88318-192-4
No. 94	X-Ray and Atomic Inner-Shell Physics-1982 (International Conference, U. of Oregon)	82-74075	0-88318-193-2
No. 95	High Energy Spin Physics - 1982 (Brookhaven National Laboratory)	83-70154	0-88318-194-0
No. 96	Science Underground (Los Alamos, 1982)	83-70377	0-88318-195-9
No. 97	The Interaction Between Medium Energy Nucleons in Nuclei - 1982 (Indiana University)	83-70649	0-88318-196-7
No. 98	Particles and Fields - 1982 (APS/DPF University of Maryland)	83-70807	0-88318-197-5
No. 99	Neutrino Mass and Gauge Structure of Weak Interactions (Telemark, 1982)	83-71072	0-88318-198-3
No. 100	Excimer Lasers - 1983 (OSA, Lake Tahoe, Nevada)	83-71437	0-88318-199-1
No. 101	Positron-Electron Pairs in Astrophysics (Goddard Space Flight Center, 1983)	83-71926	0-88318-200-9
No. 102	Intense Medium Energy Sources of Strangeness (UC-Santa Cruz, 1983)	83-72261	0-88318-201-7
No. 103	Quantum Fluids and Solids - 1983 (Sanibel Island, Florida)	83-72440	0-88318-202-5
No. 104	Physics, Technology and the Nuclear Arms Race (APS Baltimore - 1983)	83-72533	0-88318-203-3
No. 105	Physics of High Energy Particle Accelerators (SLAC Summer School, 1982)	83-72986	0-88318-304-8

AIP Conference Proceedings

No. 106	Predictability of Fluid Motions (La Jolla Institute, 1983)	83-73641	0-88318-305-6
No. 107	Physics and Chemistry of Porous Media (Schlumberger-Doll Research, 1983)	83-73640	0-88318-306-4
No. 108	The Time Projection Chamber (TRIUMF, Vancouver, 1983)	83-83445	0-88318-307-2
No. 109	Random Walks and Their Applications in the Physical and Biological Sciences (NBS/La Jolla Institute, 1982)	84-70208	0-88318-308-0
No. 110	Hadron Substructure in Nuclear Physics (Indiana University, 1983)	84-70165	0-88318-309-9
No. 111	Production and Neutralization of Negative Ions and Beams (3rd Int'l Symposium, Brookhaven, 1983)	84-70379	0-88318-310-2
No. 112	Particles and Fields - 1983 (APS/DPF, Blacksburg, VA)	84-70378	0-88318-311-0
No. 113	Experimental Meson Spectroscopy - 1983 (Seventh International Conference, Brookhaven)	84-70910	0-88318-312-9
No. 114	Low Energy Tests of Conservation Laws in Particle Physics (Blacksburg, VA, 1983)	84-71157	0-88318-313-7
No. 115	High Energy Transients in Astrophysics (Santa Cruz, CA, 1983)	84-71205	0-88318-314-5
No. 116	Problems in Unification and Supergravity (La Jolla Institute, 1983)	84-71246	0-88318-315-3
No. 117	Polarized Proton Ion Sources (TRIUMF, Vancouver, 1983)	84-71235	0-88318-316-1
No. 118	Free Electron Generation of Extreme Ultraviolet Coherent Radiation (Brookhaven/OSA, 1983)	84-71539	0-88318-317-X
No. 119	Laser Techniques in the Extreme Ultraviolet (OSA, Boulder, Colorado, 1984)	84-72128	0-88318-318-8
No. 120	Optical Effects in Amorphous Semiconductors (Snowbird, Utah, 1984)	84-72419	0-88318-319-6
No. 121	High Energy e^+e^- Interactions (Vanderbilt, 1984)	84-72632	0-88318-320-X
No. 122	The Physics of VLSI (Xerox, Palo Alto, 1984)	84-72729	0-88318-321-8
No. 123	Intersections Between Particle and Nuclear Physics (Steamboat Springs, 1984)	84-72790	0-88318-322-6
No. 124	Neutron-Nucleus Collisions - A Probe of Nuclear Structure (Burr Oak State Park-1984)	84-73216	0-88318-323-4
No. 125	Capture Gamma-Ray Spectroscopy and Related Topics 1984 (Internat. Symposium, Knoxville)	84-73303	0-88318-324-2
No. 126	Solar Neutrinos and Neutrino Astronomy (Homestake, 1984)	84-63143	0-88318-325-0
NO. 127	Physics of High Energy Particle Accelerators (BNL/SUNY Summer School, 1983)	85-70057	0-88318-326-9
No. 128	Nuclear Physics with Stored, Cooled Beams (McCormick's Creek State Park, Indiana, 1984)	85-71167	0-88318-327-7

RAYMOND H. FOGLER LIBRARY